Reinforced and Prestressed Concrete Design

The Complete Process

Reinforced and Prestressed Concrete Design
The Complete Process

EUGENE J. O'BRIEN and ANDREW S. DIXON

Department of Civil Structural and Environmental Engineering, Trinity College, Dublin

KML Consulting Engineers, Dublin

Chapter 8 written in collaboration with:
ROBERT E. LOOV
Department of Civil Engineering, University of Calgary

Advisory Editor
ANTHONY J. FITZPATRICK
Ove Arup & Partners, London

Longman
Scientific &
Technical

Longman Scientific & Technical
Longman Group UK Limited
Longman House, Burnt Mill, Harlow
Essex CM20 2JE, England
and Associated Companies throughout the world

Copublished in the United States with
John Wiley & Sons, Inc., 605 Third Avenue, New York,
NY 10158

First published 1995

British Library Cataloguing in Publication Data
A catalogue entry for this title is available from the British Library.

ISBN 0-582-21883-7

Library of Congress Cataloging-in-Publication data
O'Brien, Eugene, 1958–
    Reinforced and prestressed concrete design : the complete process
/ Eugene O'Brien and Andrew Dixon.
        p.   cm.
    Includes bibliographical references and index.
    ISBN 0-470-23365-6
    1. Reinforced concrete construction.   2. Prestressed concrete
construction.   3. Structural analysis (Engineering)   I. Dixon,
Andrew, 1968–   .   II. Title.
TA683.027   1994
624.1'8341—dc20   93-40446 CIP

Set by 16 in 10/12pt Monotype Lasercomp Times 569

Produced through Longman Malaysia, VVP

Contents

Preface

In his early years as a designer, the first author was asked to design a reinforced concrete floor slab for a plant room in the attic of a hotel. The room had no door, being accessed through a square opening in the slab from the room below, and the only location available for this opening was in an area where the slab moments were very high. He grappled with this problem for a great deal of time, looking for structural solutions, until he realized that he needed an overview of the problem – why was this opening needed anyway? A number of telephone calls later the alternative emerged. Access to the plant room could be achieved through a doorway from elsewhere in the attic instead of through a hatch from below and the troublesome opening could in fact be omitted.

All of this taught him the philosophy behind this book, namely that every member of the design team must understand the complete design process – the thinking behind all the decisions that relate in any way to his/her contribution. Thus we have, in one volume, covered all aspects of the design process from initial conception of the structural alternatives through the process of analysis and on to the detailed design traditionally taught in concrete design courses. We have sought to strike a balance between the very practical – knowing how to **apply** code clauses – and the very theoretical – doing what are essentially brain teasers. Normally we favour a theoretical approach – the detailed interpretation of code clauses makes very boring reading, particularly when the clauses are based on empirical evidence. However, we feel that there are a whole host of topics relating to the design process, some of them quite practical, with which the graduate needs at least a passing familiarity if he/she is to have an overview of the complete process. For example, he/she needs an understanding of where the spans and section dimensions come from, in order to understand where his/her contribution fits into the whole process. This way, he/she will appreciate the implications of thickening a slab or deepening a beam, will be more likely to suggest rational changes to the structure and the risk of errors in the design office will be greatly reduced.

We have put together what might, at first sight, appear to be rather a strange collection of material – some qualitative design (load paths, etc.), quite a lot of analysis, rules of thumb and methods of sizing up members as well as some conventional reinforced and prestressed concrete design. We did this because we see the traditional separation of analysis and design as artificial and have found our students graduating much confused over the distinction between an applied load and a capacity to resist it. We feel that they need to have all the material in one book in order to understand the interrelationships between conceptual design, analysis and detailed design of concrete structures.

We also feel that many textbooks are lacking on some very essential practical points. For example, an explanation of the calculation of wind load is somewhat unsatisfactory, being largely based on empirical evidence. Nevertheless it is a very necessary evil in the design office and we feel that all students should have some exposure to such basic essentials before graduating. This will give them some familiarity with the concepts before they are faced with real structures and will reduce the risk of a misinterpretation of the code.

Other problems which we had with many books are that they are unrealistic for today's market. Many topics taught in structures curricula, most of them in the analysis family, are never used afterwards by design engineers. To some extent this is necessary, as a concept as fundamental to many areas as virtual work ought to be taught, even if practising engineers will never use it. However, there are many other topics that are not important either in providing foundations for further study or for everyday application in design offices. In the same vein, we feel that presenting unrealistic examples with a view to setting clever problems in the examinations is a mistake. For example, more than a cursory look at methods of calculating deflections by hand makes no sense when very few designers ever avail of them. In contrast, a discussion of the factors which influence the choice of Young's modulus and second moments of area for a computer calculation of deflection is quite essential.

So, to those who would criticize and say that young engineers can learn preliminary sizing or qualitative design after they graduate, we say 'no' – they need to have at least a brief look at these topics to see how everything fits together and they can spend their free postgraduate time calculating deflections by hand for structures with obscure geometry or, indeed, doing *The Times* crossword!

The principal part of the Eurocode for concrete is currently available for use in parallel with British Standard, BS 8110. It was scheduled to replace that document completely in 1995, but there appear to be some delays in this timetable. While many engineers are resisting this change, it would seem wrong to educate our students of today using a standard which will be withdrawn soon after they graduate. Hence we have based this book completely on the Eurocodes, EC1 for loading and EC2 for concrete. We have found that, while a little time is required to become familiar with the new notation, they are very useful and comprehensive documents. We hope that our book will help to ease your way as gently as possible into these new codes of practice.

Acknowledgements

A casual comment made many years ago planted the seed of an idea which later led to the writing of this book. Credit for that comment goes to Nick Ryan. The idea would probably never have become anything more than that had it not been for the inspiration and enthusiastic support of Simon Perry.

Many others have helped in various ways: Denis Wall and Nick Ryan with Eurocodes, Michael McGowan and John Flanagan with figures, Walt Dilger and Graham Bowring with the securing of photographs, Damien Keogh and Tony Fitzpatrick with early versions of the cover design and Peter Mitchell with the index. Photographs are reproduced with the kind permission of CCL Systems Ltd., Decon Canada and Brendan Dempsey of Trinity College Dublin.

The following publishers are acknowledged for permission to reproduce tables and/or figures: E & FN Spon for material from *Bridge Deck Behaviour* by E. C. Hambly and from *Reinforced Concrete Designer's Handbook* by Reynolds and Steedman; Prentice-Hall for material from *Reinforced Concrete: Mechanics and Design* by J. G. McGregor; The Concrete Society for material from *Concrete* by Chane and Clapson; John Wiley & Sons for material from *Prestressed Concrete Structures* by Lin and Burns. The National Standards authority of Ireland is acknowledged for permission to reproduce many figures and tables from Eurocode 1, Drafts, N105, N106A and N118, 1993 and from Eurocode 2, ENV 1992-1-1: 1991. Extracts from BS 8110 are reproduced with the permission of BSI, Linford Wood, Milton Keynes MK14 6LE, from whom complete copies can be obtained. Extracts from *Manual for the Design of Reinforced Concrete Building Structures* are reproduced with the kind permission of the Institution of Structural Engineers from whom complete copies can also be obtained.

Special thanks to the authors of the **STRAP** analysis package which has been used extensively, particularly in Chapters 4 and 5.

This book is dedicated to Sheena and Orfhlaith.

WARNING: The Eurocodes, in general, and EC1 in particular, are still draft standards. Readers may find that revisions have been made after the time of going to press.

Part I

Structural Loading and Qualitative Design

Introduction

In this part of the book, what is arguably the most important aspect of structural design is dealt with, namely qualitative design. Structural failures are rarely a result of a calculation error in which, say, a stress is thought to be 10 per cent less than its actual value. More commonly, failure results from the omission of reinforcement altogether from a critical connection or confusion over the role that each structural member plays in resisting the applied loads.

Chapter 1 presents an overview of the complete design process from conception to finished drawing. In addition, a description is given of the function of various types of structural member and structural system and the ways in which they resist load. The chapter also presents the factors which affect the choice of reinforced or prestressed concrete as an appropriate structural material. In Chapter 2, the ways of combining structural members to form complete structures are described. Also, a qualitative explanation is given of how loads are carried through the various structural members to the ground. Finally, in Chapter 3, the principal sources of loading are described. Poor decisions in the provision for load can result in a structure that is unsafe on the one hand or is too expensive to construct on the other.

It is hoped that, from this part of the book, the reader will develop a qualitative understanding of how a structure resists loads. In addition, an explanation is given of the nature of loads and the means by which they are quantified.

1

Fundamentals of Qualitative Design

1.1 The design process

Design in any field is a logical creative process which requires a wide amalgamation of skills. As a complete process, structural engineering design can be divided into three main stages:

1. Conceptual design
2. Preliminary analysis and design
3. Detailed analysis and design

With the exception of detailed analysis, a comprehensive treatment of which is beyond the scope of this book, the three stages are dealt with here in Parts I, II and III, respectively. The first stage, described in detail below, consists of the drawing up of structural schemes which are safe, buildable, economical and robust. The second stage consists of performing preliminary calculations to determine if the proposed structural schemes are feasible. Rules of thumb are used to determine preliminary sizes for the various members and approximate methods are used to check these sizes and to estimate the quantities of reinforcement required. In the third stage of the complete design process, the adequacy of the preliminary member sizes is verified and the quantities of reinforcement calculated accurately. The process consists firstly of **analysing** the structure. That is to say, the distributions of bending moment, shear force, etc., due to all possible combinations of applied loading, are determined. The various structural members are then **designed**. This is the process by which the capacity of each member to **resist** moment, shear, etc., is compared with the values due to applied loading. If the capacity is inadequate, the quantity of reinforcement and/or the member size is increased. Following completion of these three stages, drawings and specifications are prepared for the construction of the chosen structure.

Conceptual design, which forms the first stage of the design process, involves the identification of design constraints and the synthesis of structural schemes which comply with these constraints. The fundamental constraints are things like the allowable budget, site and size restrictions, provision for safe access, final appearance and utility. These are normally specified by the client's brief or by the body, such as the local planning council, which has given permission for the commencement of the project. More detailed constraints emerge during the multi-disciplinary design process. For example, the architect involved in the development of a large hotel may request a large open space in the central foyer to enable free movement of people in the area. If columns are allowed, as illustrated in Fig. 1.1(a), the span lengths and hence the beam depths will be modest (the required beam depth tends to be proportional to the span length). If, on the other hand, no columns are to be used, the span length and hence the beam depth will be large, as shown in Fig. 1.1(b). This results in additional cost and increases the height of the overall structure. The spherical dome scheme illustrated in Fig. 1.1(c) would tend to be the most expensive of the three schemes considered as it would involve curved shuttering to form the shape of the concrete during construction. Nevertheless, it may be chosen over the alternatives for its superior aesthetic qualities.

The form of a structure often emerges as a result of discussion between various parties and good solutions are the result of clear lines of communication as well as the skill of the individuals involved in the project.

1.2 Structural materials

Until recent times, the use of a material for construction depended heavily on its local availability. Nowadays, with improved technology and communications, there is a greater variety and availability of materials suitable for construction in most parts of the world. Hence, at an early stage in the design of a modern structure, the structural engineer must choose a suitable material or group of materials which will form the main structural elements. Owing to the range of construction materials available, it is often necessary to carry out approximate preliminary designs for a number of material options in order to determine the most appropriate for the project. More often, however, the choice of material is founded on a knowledge of the properties of alternative materials and on experience gained from previous design projects.

The principal structural materials

The principal **raw materials** of structural design are steel, concrete, timber and masonry bricks. Of these, steel and concrete are the most widely used in practice. The main advantage of steel over other construction materials is its great strength, both in tension and compression. The strength of concrete is dependent on the type, quality and relative proportions of its constituents. To grade the strength of concrete, the compressive strength of simple cylinder or cube samples at 28 days is generally used. Values of the compressive strength of concrete at 28 days can vary from 5 N/mm^2 to 90 N/mm^2 but typically range between 35 N/mm^2 and 55 N/mm^2. An important characteristic of a hardened concrete is that its tensile strength is much less than its strength in compression, generally being between 1 N/mm^2 and 3 N/mm^2. For simplicity, designers will often assume the tensile strength to be equal to zero.

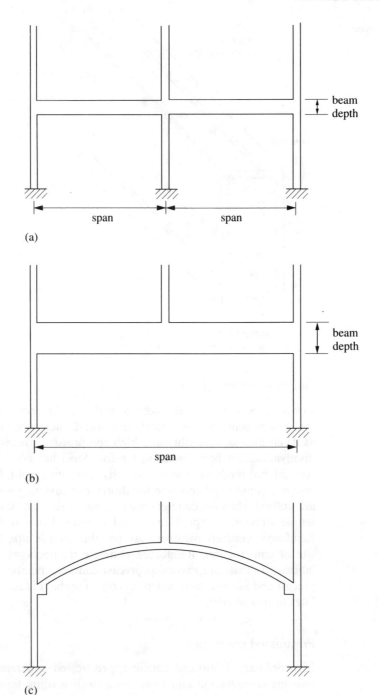

Fig. 1.1 Alternative structural schemes:
(a) two short spans;
(b) one long span;
(c) spherical dome

The raw materials of construction are often combined to form what are loosely referred to as **structural materials**. In this way, the distinctive properties of the different raw materials can be used to the greatest advantage. The principal structural materials are described in the following sections.

Fig. 1.2 Reinforced concrete

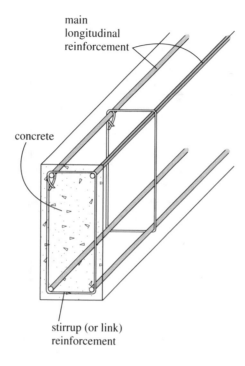

main longitudinal reinforcement

concrete

stirrup (or link) reinforcement

Ordinary reinforced concrete

Concrete reinforced with steel is perhaps the most widespread structural material presently in use around the world. Concrete has many advantages such as its cheapness, versatility and high compressive strength but it has the great disadvantage of being weak in tension. Steel has considerably higher tensile strength but tends to be more expensive per unit weight. In ordinary reinforced concrete (reinforced concrete for short), the advantages of both raw materials are utilized when the concrete resists compressive stresses while the steel resists tensile stresses. A typical reinforced concrete beam is illustrated in Fig. 1.2. Reinforced concrete members can be fabricated *in situ*, that is, directly at the site of construction. Reinforced concrete members which are prepared and fabricated offsite are known as **precast** concrete members. The choice between precast and *in situ* concrete depends on a number of factors which are discussed later in this section.

Prestressed concrete

Like ordinary reinforced concrete, prestressed concrete consists of concrete resisting compression and reinforcing steel resisting tension. However, unlike reinforced concrete, the concrete in prestressed concrete is compressed during construction and is held in this state throughout its design life by the reinforcing steel. The advantage of having the concrete in a compressed state is that tensile cracking is prevented thereby increasing the resistance of the steel to corrosion. In addition, prestressing of the concrete increases the overall stiffness of the member. A typical prestressed concrete member is illustrated in Fig. 1.3. Like

Fig. 1.3 Prestressed
concrete

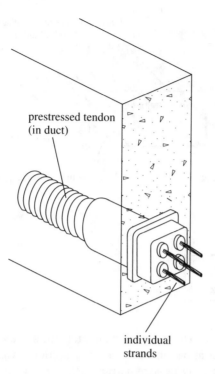

prestressed tendon
(in duct)

individual
strands

ordinary reinforced concrete, prestressed concrete members can be fabricated
in situ or as precast units.

Structural steel

Unlike concrete, steel can be used by itself as a structural material for most
types of member. Structural steel is available in many shapes, some illustrated
in Fig. 1.4, which are efficient in resisting bending and buckling. Structural steel

Fig. 1.4 Structural
steel sections

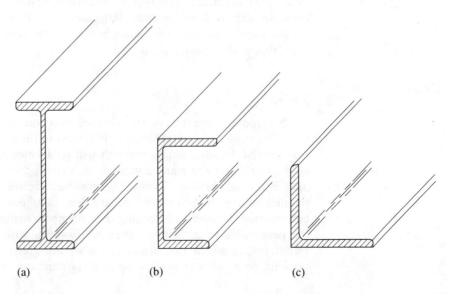

(a)   (b)   (c)

Fig. 1.5 Composite
construction

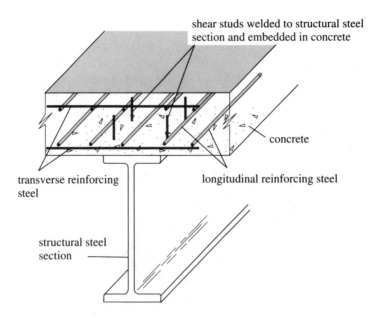

shear studs welded to structural steel
section and embedded in concrete

concrete

transverse reinforcing
steel

longitudinal reinforcing steel

structural steel
section

has the property of exhibiting approximately the same stress–strain relationship
in tension and compression and so steel sections which carry their loads in
bending will generally be symmetrical about the neutral axis. However, local
buckling due to large shear forces often places further restrictions on the
allowable compressive stresses in such members.

Composite construction

The advantages of reinforced concrete and structural steel can be combined
by what is known as composite construction. Figure 1.5 illustrates a typical
example in this increasingly popular structural material. The cheaper reinforced
concrete slab is used to span large areas to create floor space while a
combination of the structural steel beam and the concrete is used to support
the slab and the loads applied to it.

Timber

Timber from mature trees is one of the earliest construction materials used by
man. The strength of timber is directly related to the variety of tree from which
it is derived. In addition, its strength will be affected by its density, moisture
content, grain structure and a number of inherent defects such as cracks, knots
and insect infestations. Typical permissible stresses for softwoods loaded
parallel to the grain orientation are less than 6 N/mm^2 for members in
compression, tension and bending. For members loaded normal to the grain,
the permissible stress is even smaller. However, with the use of laminating
techniques, in which thin strips of timber are glued together to form hefty
sections, permissible stresses of up to 20 N/mm^2 can be achieved.

Masonry

Masonry members are made up of a combination of clay bricks or concrete blocks fixed together with mortar (cement mixture). Masonry structures are characterized by being strong in compression but weak in tension. In compression, among the factors which affect strength are the strength and shape of the units (bricks or blocks), the composition and thickness of the mortar joint and the bond between the mortar and the unit. With a strong clay brick and a mortar with a high cement content, the compressive strength of masonry can reach $15 \, N/mm^2$ or more. However, high variability in the quality of both manufacture and construction results in safe design strengths much less than this. In flexure, the strength of masonry is limited by the low tensile strength. However, in most instances this can be overcome by the provision of steel reinforcement in the bed joints and/or prestressing.

Factors affecting choice of structural material

The principal criteria which influence the choice of structural material are:
(a) strength
(b) durability (resistance to corrosion)
(c) architectural requirements
(d) versatility
(e) safety
(f) speed of erection
(g) maintenance
(h) cost
 (i) craneage

The properties of reinforced and prestressed concrete are compared below with the properties of structural steel, timber and masonry under each of these nine headings. It should be noted that only one or two structural materials tend to be used in any given construction project. This is to minimize the diversity of skills required in the workforce.

Strength

The relative strengths of the six main structural materials have already been discussed above. However, it should also be noted that the ability of a material to sustain external loads is dependent on the mechanisms by which the loads are carried in a member. For example, members which are in pure compression or tension will carry their loads more efficiently than members in bending since the stress is evenly distributed across the section (this will be seen in the following section). For this reason, the available strength of a structural material depends as much on the method of load transfer as its characteristic strength. Nevertheless, it can in general be stated that reinforced and prestressed concrete and structural steel are strong materials. Relative to these, timber and masonry are generally rather weak and are more suitable for short spans and/or light loads.

Durability

The durability of a material can be defined as its ability to resist deterioration under the action of the environment for the period of its design life. Of the four raw materials used in construction, steel has by far the least resistance to such corrosion (or 'rusting' as it is more commonly known), particularly in aggressive humid environments. Hence, the durability of a structural material which is wholly or partly made from steel will largely be governed by how well the steel is protected.

A significant advantage of reinforced and prestressed concrete over other structural materials is their superior durability. The durability of the concrete itself is related to the proportions of its constituents, the methods of curing and the level of workmanship in the mixing and placing of the wet concrete. The composition of a concrete mix can be adjusted so that its durability specifically suits the particular environment. The protection of the steel in reinforced and prestressed concrete against the external environment is also dependent on the concrete properties, especially the porosity. However, its resistance to corrosion is also proportional to the amount of surrounding concrete, known as the **cover**, and the widths to which cracks open under day-to-day service loads.

Structural steel, like concrete, is considered to be very durable against the agents of wear and physical weathering (such as abrasion). However, one of its greatest drawbacks is its lack of resistance to corrosion. Severe rusting of steel members will result in a loss in strength and, eventually, to collapse. The detrimental effect of rusting is found to be negligible when the relative humidity of the atmosphere is less than approximately 70 per cent and therefore protection is only required in unheated temperate environments. Where corrosion is likely to be a problem, it can often be prevented by protective paints. Although protective paints are very effective in preventing corrosion, they do add significantly to the maintenance costs (unlike concrete for which maintenance costs are minimal).

For timber to be sufficiently durable in most environments, it must be able to resist the natural elements, insect infestation, fungal attack (wet and dry rot) and extremes in temperature. Some timbers, such as cedar and oak, possess natural resistance against deterioration owing to their density and the presence of natural oils and resins. However, for the types of timber most commonly used in construction, namely softwoods, some form of preservative is required to increase their durability. When suitably treated, timber exhibits excellent properties of durability.

Masonry, like concrete, can also be adapted to suit specific environments by selecting more resistant types of blocks/bricks for harsh environments. Unreinforced masonry is particularly durable and can last well beyond the typical 50 year design life.

Architectural requirements

The appearance of a completed structure is the most significant architectural feature pertinent to material choice since the aesthetic quality of a completed structure is largely determined by the finish on the external faces. For concrete, this final appearance is dependent on the standards of placement and compaction and the quality of the formwork. Badly finished concrete faces, with little or no variation in colour or texture over large areas, can form the most unsightly views. Concrete is a versatile material, however, and when properly placed, it is possible to produce structures with a wide variety of visually appealing finishes. In the case of precast concrete, an excellent finished appearance can usually be assured since manufacture is carried out in a controlled environment.

Exposed structural steel in buildings is displeasing to the eye in many settings and must be covered in cladding in order to provide an acceptable finish. An exception to this is the use of brightly painted, closed, hollow, circular or rectangular sections.

Timber and masonry structures will generally have an excellent finished appearance, providing a high quality of workmanship is achieved. Masonry also offers a sense of scale and is available in a wide variety of colours, textures and shapes.

In addition to their aesthetic qualities, concrete and masonry structures also have the advantage of possessing good sound and thermal insulation properties.

Versatility

The versatility of a material is defined as its ability (a) to be fabricated in diverse forms and shapes and (b) to undergo substantial last-minute alterations on site without detriment to the overall design. Steel can easily be worked into many efficient shapes on fabrication but is only readily available from suppliers in standard sections. Concrete is far more versatile in this respect as it can readily be formed by moulds into very complex shapes. Timber is the most limited as it is only available from suppliers in a limited number of standard sizes. Laminated timber, on the other hand, can be profiled and bent into complex shapes. Masonry can be quite versatile since the dimensions of walls and columns can readily be changed at any time up to construction. The disadvantage of steel, timber and precast concrete construction is their lack of versatility on site compared with *in situ* reinforced concrete and masonry to which substantial last-minute changes can be made. *In situ* prestressed concrete is not very versatile as changes can require substantial rechecking of stresses.

Safety

The raw material of concrete is very brittle and failure at its ultimate strength can often occur with little or no warning. Steel, being a very ductile material, will undergo large plastic deformations before collapse, thus giving adequate warning of failure. The safety of reinforced concrete structures can be increased by providing 'under-reinforced' concrete members (the concepts of under-reinforced and over-reinforced concrete are discussed in Chapter 7). In such members, the ductile steel reinforcement effectively fails in tension before the concrete fails in compression, and there is considerable deformation of the member before complete failure. Although timber is a purely elastic material, it has a very low stiffness (approximately 1/20th that of steel) and hence, like steel, it will generally undergo considerable deflection before collapse.

An equally important aspect of safety is the resistance of structures to fire. Steel loses its strength rapidly as its temperature increases and so steel members must be protected from fire to prevent collapse before the occupants of the structure have time to escape. For structural steel, protection in the form of intumescent paints, spray-applied cement-binded fibres or encasing systems, is expensive and can often be unsightly. Concrete and masonry possess fire-resisting properties far superior to most materials. In reinforced and prestressed concrete members, the concrete acts as a protective barrier to the reinforcement, provided there is sufficient cover. Hence, concrete members can retain their strength in a fire for sufficient time to allow the occupants to escape safely from a building. Timber, although combustible, does not ignite spontaneously below a temperature of approximately 500 °C. At lower temperatures, timber is only charred by direct contact with flames. The charcoal layer which builds up on the surface of timber during a fire protects the underlying wood from further deterioration and the structural properties of this 'residual' timber remain unchanged.

Speed of erection

In many projects, the speed at which the structure can be erected is often of paramount importance due to restrictions on access to the site or completion deadlines. In such circumstances, the preparation and fabrication of units offsite will significantly reduce the erection time. Thus, where precast concrete (reinforced and/or prestressed) and structural steel are used regularly, the construction tends to be very fast. Complex timber units, such as laminated members and roof trusses, can also be fabricated offsite and quickly erected. The construction of *in situ* concrete structures requires the fixing of reinforcement, the erection of shuttering, and the casting, compaction and curing of the concrete. The shutters can only be removed or 'struck' when the concrete has achieved sufficient strength to sustain its self-weight. During the period before the shutters can be struck, which can be several days, very little other construction work can take place (on that part of the structure) and hence the overall erection time of the complete structure tends to be slow. Masonry construction, although labour intensive, can be erected very rapidly and the structure can often be built on after as little as a day.

Maintenance

Less durable structural materials such as structural steel and timber require treatment to prevent deterioration. The fact that the treatment must be repeated at intervals during the life of the structure means that there is a maintenance requirement associated with these materials. In fact, for some of the very large exposed steel structures, protective paints must be applied on a continuous basis. Most concrete and masonry structures require virtually no maintenance. An exception to this is structures in particularly harsh environments, such as coastal regions and areas where de-icing salts are used (bridges supporting roads). In such cases, regular inspections of reinforced and prestressed concrete members are now becoming a standard part of many maintenance programmes.

Cost

The cost of structural material is of primary interest when choosing a suitable material for construction. The relative cost per unit volume of the main construction materials will vary between countries. However, the overall cost of a construction project is not solely a function of the unit cost of the material. For example, although concrete is cheaper per unit volume than structural steel, reinforced concrete members generally require a greater volume than their equivalent structural steel members because of the lower strength of concrete. As a consequence, reinforced concrete can become the more expensive structural material. If reinforced concrete members are cast *in situ*, construction costs tend to be greater than for the steel structure because of the longer erection time and the intensive labour requirements. However, the high cost of structural steel and its protection from corrosion and fire counteract any initial saving with the result that either material can be more cost effective. In general, it is only by comparing the complete cost of a project that the most favourable material can be determined. As a general guide, however, it can be said that reinforced concrete and structural steel will incur approximately the same costs, masonry will often prove cheaper than both where it is feasible while the cost of timber is very variable.

Craneage

In certain circumstances, the choice of structural material and construction method may be determined by the availability of craneage. For example, in a small project, it may be possible to avoid the need for cranes by the use of load-bearing masonry walls and timber floors. Depending on their weight and size, structural steel and precast concrete units may require substantial craneage and it is often the limit on available craneage that dictates the size of such units. In general, *in situ* concrete requires little craneage although cranes, when available, can be used for moving large shutters.

Table 1.1 serves as a summary of the relative advantages and disadvantages of the four types of structural material under the categories discussed above. At this stage, it should be appreciated that the choice of any structural material is heavily dependent on the particular structure and the conditions under which it is constructed. The following examples briefly illustrate the process of selecting an appropriate structural material.

Example 1.1   Multi-storey warehouse

Problem   Your client requires a multi-storey warehouse in an industrial estate. In order to have it operational for the Christmas rush, construction time must be kept to a minimum.

Solution   From the location and function of the proposed building, appearance is assumed to be non-critical. To ensure a minimum erection time, structural steel is used for the main structural members. For fire resistance, the structural steel is sprayed with cement-binded fibre.

Table 1.1 Comparison of the structural properties of concrete (reinforced and prestressed), structural steel, timber and masonry

	Reinforced and prestressed concrete	Structural steel	Timber	Masonry
Strength	Excellent	Excellent	Fair	Good except in tension
Durability	Excellent	Poor against corrosion*	Poor*	Excellent
Appearance	Fair	Fair	Excellent	Excellent
Safety	Excellent	Poor fire resistance*	Good	Excellent
Speed of erection	Slow for *in situ*	Very fast	Very fast	Very fast but labour intensive
On-site versatility	Excellent for *in situ* reinforced concrete, poor otherwise	Poor	Fair	Very good

* Unless protected.

Example 1.2 Grandstand

Problem Your client requires a grandstand to be constructed between rugby seasons. This is to be a prestigious structure and so its appearance is of primary concern. Adequate safety, especially fire resistance, is also a primary concern.

Solution Precast concrete is selected for the main structural members of the grandstand because it is fast to erect and efficient in carrying the loads. In addition, it has a high natural fire resistance and good appearance. Any members which are too large for precasting are constructed *in situ*. Structural steel is chosen for the roof which is to cover the stands because of its low self-weight and high strength.

Example 1.3 Office building

Problem Your client requires a four-storey office building to be constructed in the centre of a town. Appearance and a minimum running maintenance are the governing design constraints.

Solution Masonry is chosen as the main structural material since it requires the minimum of maintenance and (for external façades) has an excellent appearance. In addition, blockwork (for internal walls) is inexpensive compared with other materials and requires little or no craneage (this may be a factor on such a constrained site).

1.3 Basic structural members

A complete structure is essentially a combination of members which can be categorized by their main function. Some structures can be broken down into sub-systems in which groups of these members act together to perform a specific function. However, before such complex systems are considered, it is necessary first to review the five basic mechanisms of load transfer which arise in members.

Tension

When a member is being stretched by forces parallel to its axis, as illustrated in Fig. 1.6(a), the stress produced is known as tension. Members used primarily to resist tension (such as ropes) need not have a capacity to resist transverse forces or bending moments. Homogeneous members in tension only have a uniform stress distribution over their cross-section as illustrated in Fig. 1.6(b) and hence can utilize the available material strength most efficiently. In reinforced concrete, the concrete cracks under the smallest tensile force. Once this occurs, the tensile force is carried solely by the reinforcement crossing the cracks, as illustrated in Fig. 1.6(c).

Fig. 1.6 (a) Tension member; (b) stress distribution at section X–X; (c) stress distribution in reinforced concrete after cracking

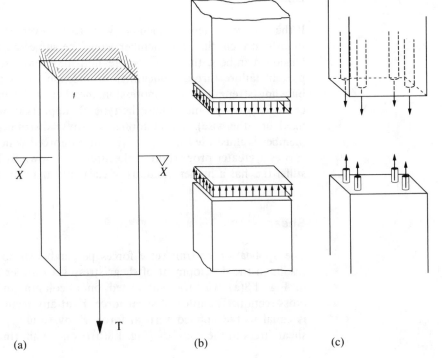

(a)   (b)   (c)

Fig. 1.7 (a) Compression member; (b) stress distribution – uniform section; (c) stress distribution in reinforced concrete member

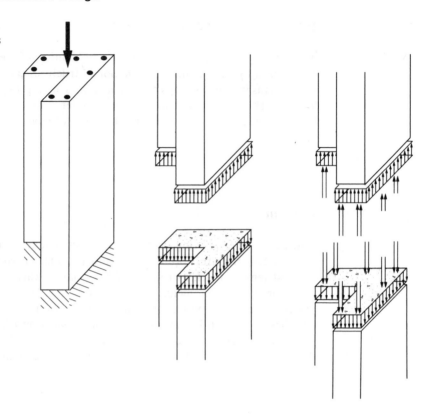

Compression

If the loads on a tension member were to be reversed so as to squeeze rather than to stretch, then the member would be subjected to compression. Unlike tension members, those in compression must have some flexural rigidity to prevent failure through buckling. In addition to its material properties, the buckling strength of a compression member is dependent on its length, its cross-sectional geometry and the type of supports in which it is held (pinned, fixed or otherwise). The compressive stress distribution in a homogeneous member is illustrated in Fig. 1.7(b). In reinforced concrete, however, the steel carries a greater proportion of the stress, as shown in Fig. 1.7(c), because it is stiffer (i.e. has a higher modulus of elasticity) than the surrounding concrete.

Shear

The application of transverse forces perpendicular to the axis of a member results in the development of shear stresses. Consider the horizontal member in Fig. 1.8(a) which is supported on a column and has a homogeneous cross-section. The internal shear force, V, at any section X–X along its length is equal to the applied vertical force, P, by equilibrium. The distribution of shear stress in the member (Fig. 1.8(b)) is not uniform over its cross-sectional

Fig. 1.8 Member in shear: (a) geometry and loading; (b) shear stress at section X–X; (c) shear in reinforced or prestressed concrete

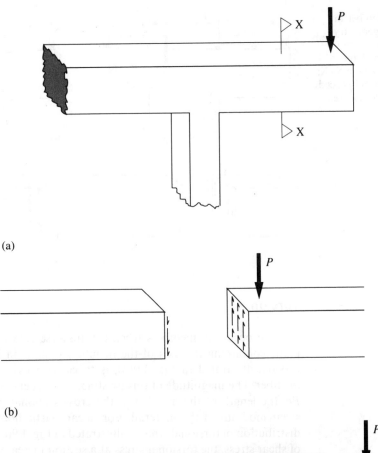

(a)

(b)

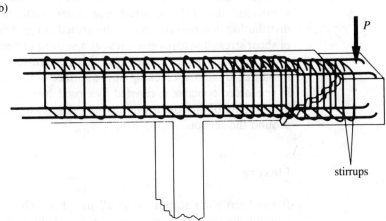

stirrups

(c)

area but, for linear elastic materials, varies parabolically from zero at the top and bottom surfaces to a maximum at the centre. However, reinforced and prestressed concrete are not homogeneous and when failing in shear are neither linear nor elastic. The shear failure of a typical reinforced concrete member is illustrated in Fig. 1.8(c). The inclined tensile cracks which are formed in the concrete are held closed by vertical shear reinforcement known as **stirrups** or **links**.

Fig. 1.9 Member in
torsion: (a) geometry
and loading;
(b) deformed shape;
(c) stress at section X–X;
(d) torsion in reinforced
or prestressed concrete

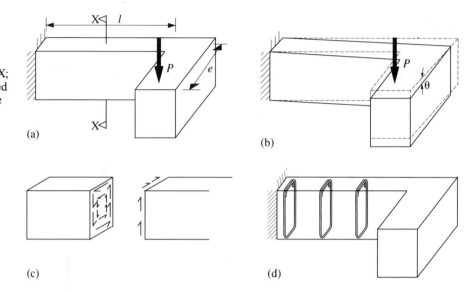

Torsion

Torsion occurs in members when a transverse external force acts outside the
plane containing the axis of the member, such as in Fig. 1.9(a). The effect of
torsion, illustrated in Fig. 1.9(b), is to cause a twisting action in the loaded
member. The magnitude of this twisting, θ, is dependent on the applied torque,
Pe, the length of the member, l, the cross-sectional geometry and the elastic
shear modulus of the material. For linear elastic homogeneous members, the
distribution of torsional stress is illustrated in Fig. 1.9(c). Unlike the distribution
of shear stress, the torsional stress at a section increases from zero at the centre
to a maximum at the edges. For this reason, torsional failure in concrete is
initialized by tensile cracking at the surface of the member. Torsional cracking
in reinforced concrete is resisted by closed stirrups, as illustrated in Fig. 1.9(d).
The lapping of the stirrups in this way ensures a continuity of reinforcement all
around the section.

Flexure

It has been shown above that all members which transmit transverse loads
laterally to one or more supports do so, at least partially, by shear force
mechanisms. If the loads are applied outside the plane of the member's axis
then the loads are also transferred by torsion. In addition to these two
mechanisms, transversely loaded members also transmit their loads by bending
action (flexure). The central point load in Fig. 1.10(a) exerts a bending moment
and causes the member to sag. In a linear elastic homogeneous member, the
longitudinal fibres at the top become shorter due to the bending and are,
therefore, stressed in compression, while the fibres at the bottom face become
longer and are so stressed in tension (apart from timber, most materials are
not fibrous and the concept of fibres in a bending member is only used as a

Fig. 1.10 Homogeneous linear elastic member in flexure: (a) geometry, loading and deflected shape; (b) stress distribution at section X–X

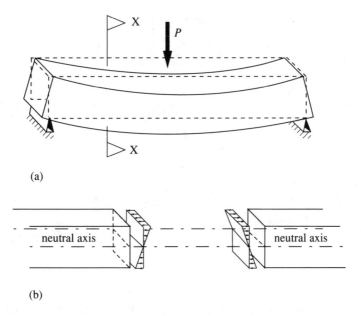

(a)

(b)

helpful analogy for the behaviour of the material). From Fig. 1.10(b), it can be seen that the outer fibres in both tension and compression will extend or shorten more than the internal fibres, and for an elastic material the stress distribution will be triangular, as shown. The location within the member where the bending stress is zero, between the tensile and compressive zones, is known as the neutral axis of the member. Of course, in three-dimensional structures this is a surface, as illustrated in the figure, not an axis. For homogeneous members (remember that concrete with reinforcement is not homogeneous) the neutral axis passes through the centroid of the cross-section.

Owing to its minimal tensile strength, concrete is assumed to crack under the smallest of tensile stresses. Thus, for a reinforced concrete member in bending, cracks extend through the tension zone to the neutral axis, as illustrated in Fig. 1.11(a). Failure of the member is prevented by the longitudinal steel which traverses the cracks and resists the tensile forces. Reinforced concrete only behaves as an elastic material under everyday loads. Under the much larger ultimate loads for which sections are designed, the stress distribution becomes non-linear, as illustrated in Fig. 1.11(b).

Member nomenclature

It has been seen above that flexure, shear and torsion often occur simultaneously in many transversely loaded members. The historical name for one-dimensional members in this category is **cantilever** for a member which transfers load to only one support and **beam** for a member which carries load between two or more supports. The two-dimensional equivalents of cantilevers and beams, where the applied loads are carried by bending, shear, etc., in two directions, are historically known as **slabs**. Members such as beams which carry their loads

Fig. 1.11 Reinforced concrete member in flexure: (a) geometry, loading and deflected shape; (b) stress distribution at section X–X under ultimate loading

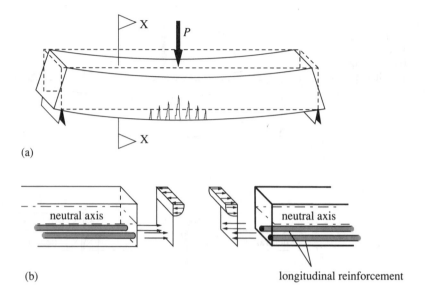

(a)

(b)   longitudinal reinforcement

by bending and shear may also be subjected to either tensile or compressive forces. An example of such a member is the beam–column, which is commonly found in multi-storey frames having moment-resisting connections between members (see Fig. 1.12). The two-dimensional equivalent of a column is known as a **wall**. The Eurocode for concrete, EC2, suggests that a column be classed as a wall if the larger side dimension is greater than four times the smaller side dimension.

By increasing the depth of a beam while keeping the span length constant, as illustrated in Fig. 1.13(a), the member becomes very stiff in the plane of bending. If the depth is increased substantially, the member becomes so stiff that the applied load is effectively carried through tension and compression zones, as shown in Fig. 1.13(b), rather than by bending and shear. This can be referred to as **membrane action** although historically such members are aptly named **deep beams**. EC2, the Eurocode for concrete design, defines a deep beam as one in which the span is less than twice the depth. A cantilever with a particularly deep section might also resist load by membrane action and be termed a **deep cantilever**. Deep cantilevers in buildings, such as that illustrated in Fig. 1.14, are more commonly known as **shear walls**.

For the purposes of design, it is often convenient to label individual members by the mechanism of load transfer which governs the design of the member, rather than by their historical names. This mechanism is termed the 'primary' mechanism of load transfer and the other, less critical mechanisms are termed 'secondary' mechanisms. For instance, beams and slabs where bending, say, is the primary mechanism are sometimes termed flexural members.

Example 1.4   Flexural members I

Problem   Figure 1.15(a) illustrates a reinforced concrete beam. For the applied loads illustrated determine the primary methods of load transfer and suggest how the member should be reinforced.

Fig. 1.12 Column with
shear, flexure and axial
force: (a) frame;
(b) column from frame

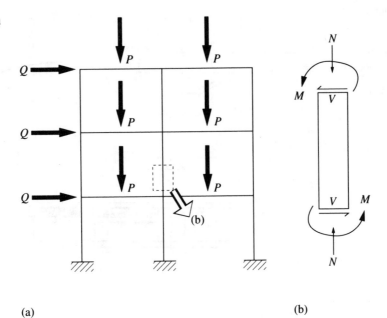

(a)

(b)

Fig. 1.13 Deep beam
member: (a) geometry
and loading; (b) tension
and compression zones

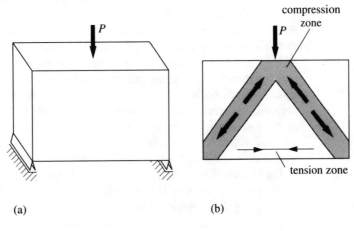

(a)

(b)

Fig. 1.14 Shear wall

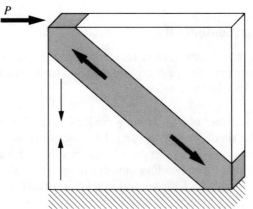

Fig. 1.15 Reinforced
concrete beam:
(a) geometry;
(b) deflected shape;
(c) shear force diagram;
(d) bending moment
diagram; (e) beam
reinforcement

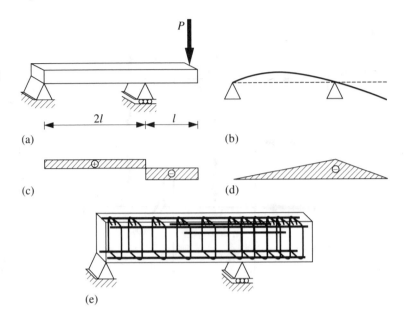

(a)

(b)

(c)

(d)

(e)

Solution As the span lengths are much greater than the depth of the member, the primary
methods of load transfer for this example are bending and shear. Figures
1.15(b)–(d) illustrate the deflected shape and the shear force and bending
moment diagrams. The shear force is constant in each span and is greater in
the shorter right-hand span. The bending moment varies across each span and
reaches a maximum over the internal support. Note from the deflected shape
that the member is hogging (as opposed to sagging) throughout its length. This
results in tension in the top of the member. Thus, to prevent flexural cracking,
longitudinal steel reinforcement must be provided along the top of the member
in both spans, as illustrated in Fig. 1.15(e). More steel is provided over the
internal support since the bending moment is greatest here. For practical
purposes, some longitudinal reinforcement would also be placed in the bottom
of the section. Links are provided in each span to prevent shear cracking. The
spacing of the links in the right-hand span is less than that in the left-hand
span as the shear force is higher.

Example 1.5 Flexural members II

Problem Figure 1.16(a) illustrates a reinforced concrete frame. For the applied horizontal
load, determine the primary methods of load transfer and suggest how the
member should be reinforced.

Solution For the member of Fig. 1.16(a), the load is carried by shear through AB and
CD and a small amount is carried by compression through BC but primarily
the load is carried by bending in all members. Figure 1.16(b) illustrates the
deflected shape and bending moment diagram for the member. Steel reinforcement is provided where tension cracks are likely to occur in the member, as

Fig. 1.16 Reinforced
concrete frame:
(a) geometry;
(b) deflected shape and
moment diagram;
(c) longitudinal
reinforcement

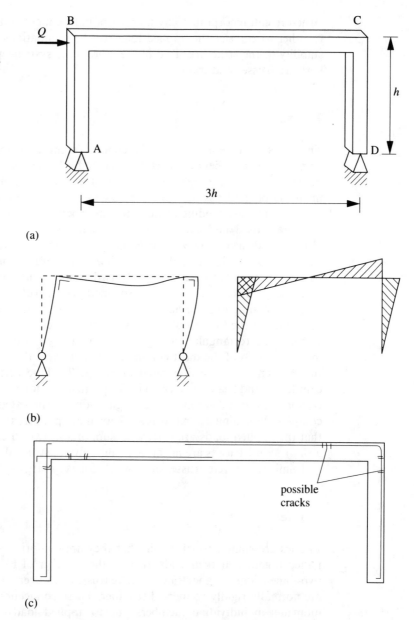

(a)

(b)

possible
cracks

(c)

shown in Fig. 1.16(c). In addition, some links should be provided to resist the
shear forces in AB and CD.

1.4 Structural systems

Structural systems are defined as groups of structural members (as defined in
section 1.3) which act together to perform a specific function. A complete

structure can incorporate any number of independent systems all of which act together to transfer the applied loads to the foundations and provide overall stability to the structure. Two of the more basic systems which are considered here are trusses and frames.

Trusses

The truss is a simple structural concept comprised solely of tension and compression members connected together by hinges or pins. The role of the hinges is to prevent the transmission of any bending moment through the members of the truss. In their simplest form, trusses are only loaded at their joints to preclude bending of individual members. In practice, hinge connections are rarely used and even steel truss members are in fact welded together. However, idealized hinge connections give fairly accurate results and for the purposes of simplified analyses such pin-joints are assumed in the design of trusses. Truss systems are commonly used as an alternative to flexural members where large distances need to be spanned. Thus, truss systems transmit their load by bending overall, but are comprised of members acting together which only transfer the load axially. For stability, trusses are built up in a triangular rather than rectangular configuration, with each triangle comprising three members. Two of the most common forms of truss, the 'N' or Pratt truss and the Warren truss, are illustrated in Fig. 1.17. These perform the function of cantilevers and beams, respectively, for particularly long spans. The tension and compression members in each configuration are indicated in the figure. More complex three-dimensional trusses, known as space trusses, can be built up so that the applied loads are carried to supports in two directions (not unlike in a slab). While trusses are far more commonly constructed using structural steel and timber, concrete trusses are feasible for very large spans.

Frames

Frames are similar to trusses in that they are comprised of straight members joined together at their ends. In fact, the member of Fig. 1.16(a) is a simple two-dimensional (plane) frame. Unlike trusses, however, the members in a frame are normally rigidly connected together. These connections allow the bending moments in individual members due to applied loads to be transferred to adjoining members, as can be seen from Fig. 1.16(b). The strength of the overall frame is thus derived from the bending and shear resistance of its members. Axial forces are also present in frames but their influence is usually small compared with the moments and shear forces and so frames are generally considered as systems of flexural members.

The frame of Fig. 1.18 consists of ten flexural members. The vertical forces at J and K are transmitted by bending and shear in members CF and FI, respectively, to the top of the vertical members. The loads are carried from there to the supports by compression in the vertical members and bending/shear in all members. The horizontal loads at B and C are transmitted to the supports

Fig. 1.17 Truss systems: (a) cantilevered 'N' truss; (b) Warren truss

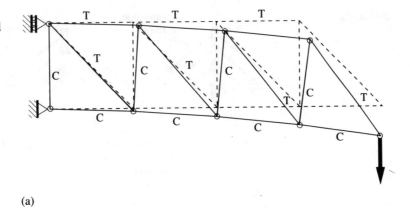

(a)

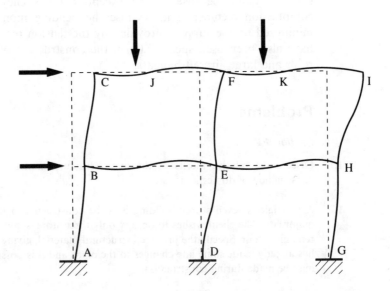

(b)

Fig. 1.18 Frame

Fig. 1.19 Cable-stayed
system

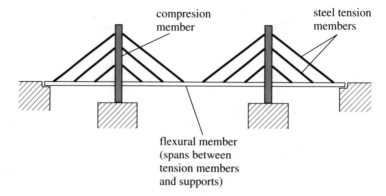

compresion
member

steel tension
members

flexural member
(spans between
tension members
and supports)

by bending action in all of the members. Although the vertical point loads exert compressive forces on the vertical members and the horizontal forces cause compression in the horizontal members, the section requirements of all members in a frame are generally governed by the bending and shear capacity.

Multiple systems

Structural systems which incorporate two or more types of members acting together are called multiple systems. A common example of a multiple system is the cable-supported structure which is a combination of tension, compression and bending members. Such systems are used to span very large distances where the use of flexural members is unfeasible owing to extremely high bending moments and the distances are too great for truss systems. Figure 1.19 is a simple diagrammatic representation of a cable-supported structure which is frequently adopted in the design of long bridges. The main cables form the tension members, compression occurs in the two towers and small bending moments are developed in the deck members. Great spans can be achieved with such a system because of the exceptional efficiency of the tension and compression members and because the bending moments in the span are minimized by the support provided by the tension members. Similar systems have also been used successfully in the construction of cantilevered stadium roofs and large aircraft hangars.

Problems

Section 1.2

1.1   Discuss the relative merits of the main structural materials in the context of the construction of domestic houses.

1.2   A prestigious seven storey building is to be constructed in the business district of a major city. The client wishes to occupy only three storeys and to secure tenants for the remaining four. Specify the preferred structural material, giving your reasons. The client has already made some late changes to the plans and it is possible that further changes may be made during construction.

1.3 A jetty is to be constructed in a marine environment for industrial use. Discuss the merits of alternative structural materials.

1.4 A bridge is to be constructed on level ground to carry a minor road over a proposed new motorway. Recommend a suitable structural material.

1.5 A pedestrian bridge is to be constructed over an existing motorway. Suggest a suitable construction material.

1.6 A large single storey supermarket is to be constructed in which there are to be very long clear spans with no internal obstructions. Suggest a suitable structural material.

Section 1.3

1.7 For the member illustrated in Fig. 1.20, how are the loads transferred to the support? If this member is made from reinforced concrete, where would you expect cracking to be most severe:

(a) if $H \gg P$;
(b) if $P \gg H$.

Fig. 1.20 Member of Problem 1.7

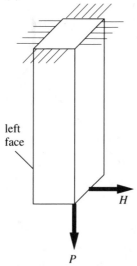

1.8 How is the load transferred to the supports in the reinforced concrete member illustrated in Fig. 1.21?

Fig. 1.21 Member of Problem 1.8

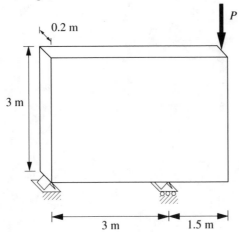

27

Section 1.4

1.9 What are the mechanisms of load transfer in each of the structures illustrated in Fig.1.22?

Fig. 1.22
Structures for
Problem 1.9

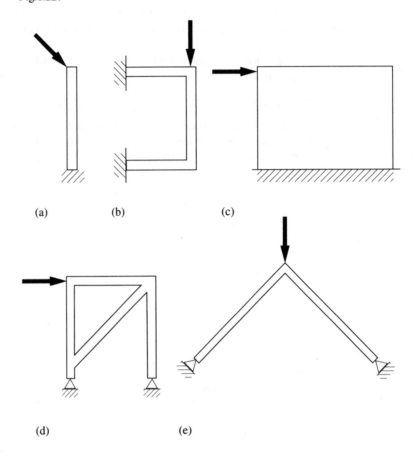

(a)   (b)   (c)

(d)   (e)

2

Basic Layout of Concrete Structures

2.1 Identification of load paths in structures

Load paths are the routes by which external actions (loads) are carried through the members of a structure to its foundations. An ability to recognize these paths and the mechanisms by which loads are transmitted through individual members is central to the development of a qualitative understanding of the behaviour of a complete three-dimensional structure. The external loads applied to a structure can usually be resolved into horizontal and vertical force components which are resisted by structural members acting in bending, tension, compression, torsion, shear or some combination of these mechanisms. Identification of the primary load paths and transfer mechanisms in a structure provides information on the precise function that each member plays in carrying the external loads to the foundations.

Load paths are most easily identified by a consideration of force equilibrium and the concept of relative stiffness of members carrying the loads. The concept of relative stiffness is that applied loads in statically indeterminate structures tend to be distributed between adjacent structural members in proportion to their relative stiffnesses, with the stiffer members tending to carry the larger proportion of the load. The stiffness, k, of a member can be defined as either (a) the force which is required to cause a unit displacement or (b) the moment required to cause a unit rotation; that is:

$$k = P/\delta \tag{2.1}$$

or:

$$k = M/\theta \tag{2.2}$$

For a particular member, stiffness clearly depends on where the force/moment is applied among other things. Stiffnesses for some of the more common arrangements of load on members are given in Appendix A.

The examples below illustrate the techniques, based on the concepts of force equilibrium and relative stiffness, which are used to identify the load paths and, hence, the primary functions of members in a variety of structures.

Fig. 2.1 Geometry and loading

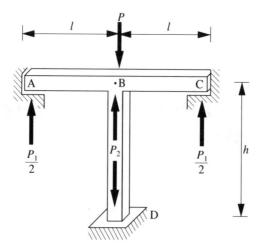

Example 2.1 Concept of relative stiffness

Problem   For the structure of Fig. 2.1, determine the portions of the total load, P, carried by the two members, beam ABC and column BD. Assume a constant Young's modulus, E, throughout.

Solution   In the structure of Fig. 2.1, part of the load, P, is carried by beam ABC to the supports at A and C and the rest is carried by column BD to the support at D. From Appendix A, No. 6, the beam stiffness is:

$$k_1 = \frac{48EI}{(2l)^3} = \frac{6EI}{l^3}$$

where I is the second moment of area of the beam. From Appendix A, No. 7, the column stiffness is:

$$k_2 = \frac{AE}{h}$$

where A is the cross-sectional area of the column. For typical values of A, E, I, h and L, k_2 is much greater in magnitude than k_1. Thus, the column will generally carry the greater portion of the load (in fact the column will carry over 90 per cent of P in most practical cases). By relative stiffness, the precise portion of the load carried by compression in the column is P_2 where:

$$P_2 = \left(\frac{k_2}{k_1 + k_2} \right) P$$

Similarly, the portion of the load carried by bending and shear in the beam is:

$$P_1 = \left(\frac{k_1}{k_1 + k_2} \right) P$$

Fig. 2.2 Rigidly constructed portal frame with masonry shear wall buttress

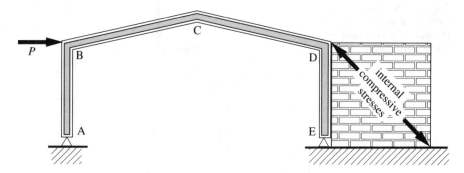

Example 2.2 Portal frame

Problem In the frame of Fig. 2.2, by what primary mechanism is the horizontal force carried to the supports?

Solution The solution is best determined by a consideration of the relative stiffnesses of the members of the structure. In this case, the horizontal thrust on the structure is transferred to the foundation principally by deep beam action in the shear wall rather than by bending in the frame, ABCDE. This is because of the greater stiffness of the shear wall against horizontal force.

Example 2.3 Sports stadium roof loads

Problem In the reinforced concrete structure of Fig. 2.3, by what mechanisms are the forces, F_1 and F_2, transferred to the ground?

Solution Figure 2.3 illustrates a section through a sport stadium. Members AD, DF, FH, BE, EG, GI, DE and FG are rigidly connected at their ends. All other members are effectively pinned at their ends and hence do not transfer bending moment.

The answer to the problem is approached initially using the basic principles of equilibrium and force resolution. In Fig. 2.4(a), the force, F_1, is resolved into components parallel and perpendicular to the neutral axis of member ABC (the curve through the arrow representing F_1 indicates that this is to be replaced with the two other arrows). It is clear from the figure that the perpendicular component, F_{1y}, is greater in magnitude than F_{1x}. This component is carried to points A and B by flexure and shear mechanisms. The smaller parallel component, F_{1x}, is transferred to the supports by compression of member ABC. The fact that the perpendicular component, F_{1y}, is larger, combined with the fact that reinforced concrete members are less strong in flexure than in compression, results in the primary mechanism of transfer of F_1 to A and B being deemed to be by bending and shear in member ABC.

The reactions at the pinned connections, A and B, due to F_1 are illustrated in Fig. 2.4(b). These reactions can readily be determined by static equilibrium of the external forces. Alternatively, resolution of the internal forces from their local coordinates (i.e. their components parallel and perpendicular to the neutral axis of the member) to global horizontal and vertical components at points A and B will yield the same results. It can be seen that the total force, F_1, tends to cause a clockwise rotation of member ABC about point B with the

Fig. 2.3 Section through a sports stadium showing roof and stand loads

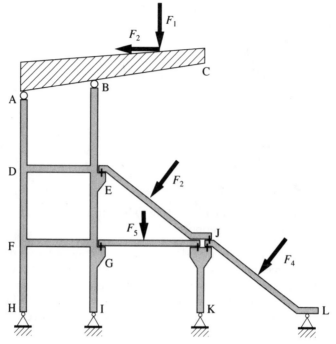

Fig. 2.4 Effects of force F_1: (a) resolution of F_1 into components; (b) reactions due to F_1

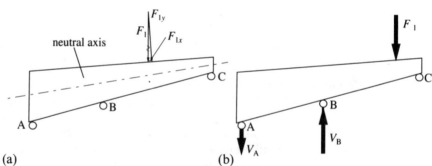

(a) (b)

effect of generating compression in member BE and tension in member AD. In fact, because the connections are pinned, F_1 is transferred purely by tension and compression from points A and B down through the vertical members to the supports at H and I.

The applied horizontal force, F_2, on member ABC can similarly be resolved into components parallel and perpendicular to the neutral axis as illustrated in Fig. 2.5. The component of force perpendicular to the neutral axis is clearly smaller than that parallel to it. However, the capacity of a reinforced concrete member to resist bending is also smaller. Hence, in this case, either axial compression or bending/shear could be the principal mechanism of load transfer.

The horizontal force, F_2, also tends to cause rotation of member ABC about point B but in an anti-clockwise direction and this results in a small tension in member BE and compression in member AD. However, the primary effect of the load, F_2, is to cause bending in members AD and BE as illustrated in

Fig. 2.5 Resolution of force F_2 into components

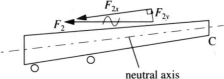

Fig. 2.6 Bending in members AD and BE due to load F_2

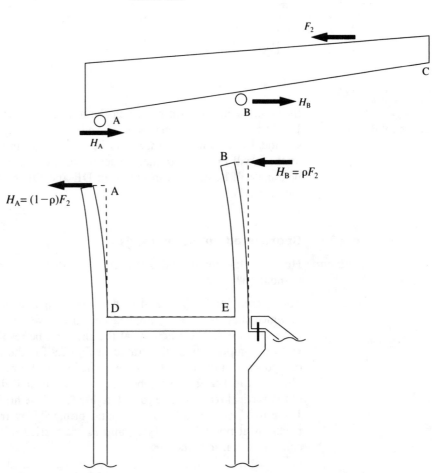

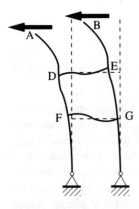

Fig. 2.7 Frame action due to load F_2

Fig. 2.6. The ratio in which the load is carried by the two members is determined by their relative stiffnesses where the stiffness of a cantilevered member in bending by definition is inversely proportional to the cube of the length of the member (see Appendix A, No. 4). In this case, member AD is shorter than member BE which makes it significantly stiffer. Hence, member AD carries the larger proportion of the load, F_2, by bending and shear.

The internal forces due to the horizontal load, F_2, may be transferred from points D and E to the foundations by the 'frame' action illustrated in Fig. 2.7 which primarily involves bending of the members. However, concrete members of normal proportions are stiffer against axial force than against flexure. Thus, the greater proportion of the horizontal load is transmitted from points D and E to the foundations by the 'truss' action shown in Fig. 2.8. Under this action,

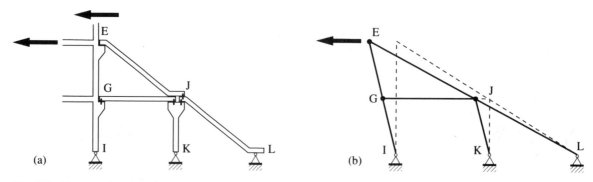

Fig. 2.8 Truss action due to load F_2: (a) geometry and loading; (b) deflected shape

the primary mechanism by which the applied load, F_2, is carried from points D and E is tension in members EJ and JL and compression in members EG, GI and JK. In addition, there will be some bending in members DE and EG as these will take the moment reaction to the cantilever, BE. Similarly, some bending will be induced in members DE and DF by the moment reaction to member AD.

Example 2.4 Sports stadium stand loads

Problem How are the forces, F_3, F_4 and F_5, illustrated in Fig. 2.3, carried to the foundation?

Solution The external loads, F_3, F_4 and F_5, cause bending and shear force in the members to which they are directly applied and in this way are transferred to the ends of the members. The reactions at the ends can be resolved into horizontal and vertical components as illustrated in Fig. 2.9 for the load, F_3. The horizontal components of these reactions are carried to the foundation by the truss action illustrated in Fig. 2.8. Some bending will occur at E due to the eccentricity, e, of the vertical reaction due to F_3 (see Fig. 2.9). The horizontal reaction at point J due to F_4 is also carried to ground primarily by truss action. The vertical reactions at point J from F_3, F_4 and F_5 are carried to the ground primarily by compression in member JK.

Example 2.5 Ramp and wall

Problem How are the forces, F_1, F_2 and F_3, in the *in situ* reinforced concrete structure of Fig. 2.10 carried to ground?

Solution The ramp and wall system in Fig. 2.10 is a simple, but interesting, three-dimensional concrete structure. The system is comprised of three members which, because they are cast *in situ*, are fully connected to form a continuous structure. It is rigidly supported at all points along its base. The mechanisms of load transfer in three-dimensional structures are not much more complex than in two-dimensional structures and the load paths in this system can be identified using the same techniques as were employed for the previous examples.

Fig. 2.9 Transfer of F_3 into grandstand structure

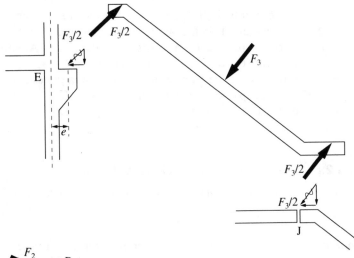

Fig. 2.10 Ramp and walls structure

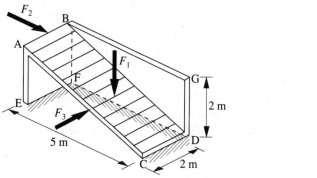

Consider first the external point load, F_1, which is applied at the mid-point of member ABCD. This load is transferred principally by bending and shear force mechanisms to the three supported sides of the slab member. Bending in two directions is taking place simultaneously in this member as the load spans between AB and CD but is also supported by cantilever action from BD. The vertical force reaction at AB is transferred to the foundations by compression in ABEF. There is also bending in ABEF as the rigid connection at AB will result in a moment reaction. Similarly, the force and moment reactions along BD result in compression and bending in BFD. The rigid connection between members ABEF and BGFD reduces the bending action in each member near this support. Thus, in ABEF, the moment near A is considerably higher than at B. This tends to cause twisting between A and B. Further details of plate theory are given in Chapter 5 and in specialist texts such as that of Timoshenko and Woinowski-Krieger (1970).

Resistance to the applied force, F_2, is provided by member BGFD acting as a deep cantilever and member ABCD acting in compression combined with ABEF acting in tension. As it is rigid at EF, member ABEF can also act as a cantilever. However, the other mechanisms are far stiffer than member ABEF in bending and so the force, F_2, is principally carried to the ground by membrane action (i.e. axial force).

Member ABCD also acts in bending and shear to transfer the applied load,

F_3, to points A, B, C and D. However, the member does not act as a slab in this case and is in fact acting as a simple beam (note that it is not a deep beam since its span/depth ratio is greater than two). At end AB the reaction due to F_3 is carried to the ground by member ABEF acting as a deep cantilever (shear wall) and so the tension and compression zones generated in the member are the principal load-carrying mechanisms. Member ABEF will also tend to warp owing to the moment reaction from ABCD.

Example 2.6   Three-dimensional skeletal frame

Problem   How does the frame illustrated in Fig. 2.11 deform under the applied horizontal load, F_1?

Solution   Figure 2.11 represents a two-storey skeletal concrete structure which has rigidly connected members. Under the external horizontal load, F_1, which is applied at joint I, the space frame can be thought of as three interconnected single-bay plane frames like frame ABCDEF. The horizontal force is principally transmitted to the foundations of the structure by a 'frame' action in each of the three plane frames. The largest portion of the force, F_1, is transferred directly through the central frame, GHIJKL, rather than being dispersed by transverse bending (bending in the horizontal plane) through members such as CI and IO to the two outer frames. This is because a frame at the point of application of a load provides more stiffness than frames connected to that point by flexible members.

Fig. 2.11   Two-storey concrete frame with rigid internal connections and pinned supports

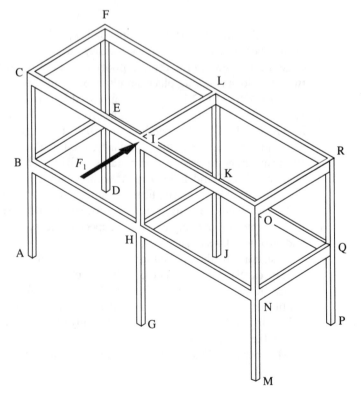

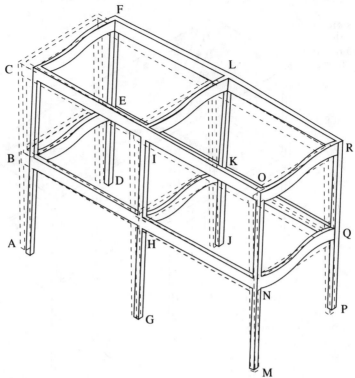

Fig. 2.12 Deflected shape of frame of Fig. 2.11

The central frame, therefore, deforms more than the outer frames as can be seen in Fig. 2.12.

Example 2.7 **Frame stiffening**

Problem How can the deformations in the frame in Fig. 2.11 be reduced?

Solution The introduction of a continuous floor system to the top level of the frame, as illustrated in Fig. 2.13, has the effect of increasing the rigidity of the entire structure against the horizontal load, F_1. The slab effectively acts as a deep beam and forces each frame to deform by the same amount. Hence the applied load is distributed in equal proportions to the three plane frames. Thus, the deformation of the entire structure, illustrated in Fig. 2.14, is uniform at any horizontal cross-section and is significantly smaller than the deformations in the structure of Fig. 2.11. If the external frames were replaced by continuous panels, such as in Fig. 2.15, then the primary mechanisms by which the external horizontal load, F_1, is resisted would not involve frame action. Instead, most of the load would be transferred by the deep beam to the stiffer external panels which would then act as shear walls (cantilevers) in carrying the load to the foundations. Such a system is far more rigid than the frames in Figs 2.11 and 2.13 and entails very little deformation of the members.

Example 2.8 **Elevator core**

Problem By what mechanisms are the loads, F_1 and F_2, in the structure of Fig. 2.16 carried to the foundations?

Fig. 2.13 Two-storey concrete frame with second floor slab rigidly connected to the frame

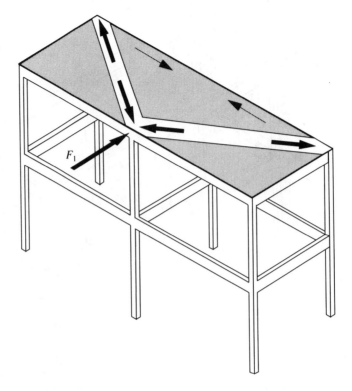

Fig. 2.14 Deflected shape of frame of Fig. 2.13

Fig. 2.15 Frame with rigid external panels

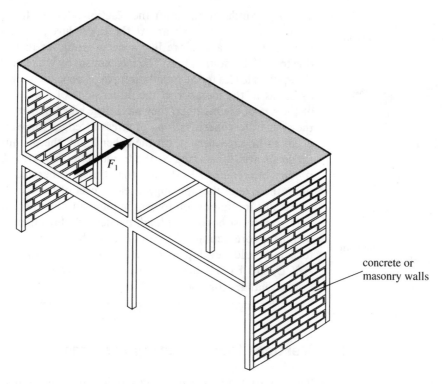

concrete or masonry walls

Solution The continuous structure illustrated in Fig. 2.16 represents a simple model of a reinforced concrete elevator core as commonly found in multi-storey buildings. In this example, horizontal loads are applied at the two levels where the floor slabs of the structure meet the core. The two applied forces, F_1 and F_2, are transferred by two-way bending and shear in member FBHD to members ABCD and EFGH. The greater portion of each force is carried to the stiffer solid member, ABCD, and this could cause torsion of the core. However, if the building as a whole is symmetrical and is joined together with horizontal floor slabs, both members will deflect by the same amount and, consequently, any twisting of the core due to the unequal stiffness of these vertical members will be prevented. For the dimensions of the core given in Fig. 2.16, member ABCD

Fig. 2.16 Elevator core

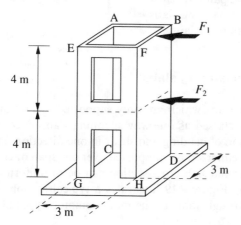

acts as a cantilever beam for the reactions due to the load, F_1, and carries the force by bending and shear. At the lower level, however, the member, being short but deep, acts more like a deep cantilever in transmitting the reaction due to F_2 to the foundation. The mechanism by which member EFGH transmits the applied forces to the foundation pad is dependent on the size of the lift access openings in the member. If the dimensions of the openings are small relative to the size of the member, they have little effect on the behaviour except in that they generate concentrations of stress at edges for which extra reinforcement needs to be provided. If, on the other hand, the openings are relatively large, the loads are transmitted by the frame action illustrated in Fig. 2.17. Thus, the rigidity of member EFGH and the stresses generated within it are dependent upon the size of the openings at each level of the core.

Finally, it is important to bear in mind another mechanism by which elevator cores can resist horizontal load, particularly for taller buildings. This is by the complete core acting as a cantilever of hollow box section. In this example, tension would develop in FBHD and compression in EAGC with the walls ABCD and EFGH acting as webs of the cantilever which join the top and bottom flanges together.

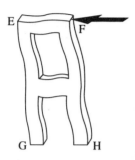

Fig. 2.17 Frame action of member EFGH when the openings are large

2.2 Vertical-load-resisting systems

The gravity loads which act on a building are normally applied directly to the floors and roof. For a multi-storey structure, the loads are transferred from the roof and floors by a system of compression and bending/shear members to the foundations. The precise vertical-load-resisting system used is dependent on the specific function of the structure, its layout and the magnitude of the vertical loads.

Many multi-storey structures will have reinforced or prestressed concrete floor slabs because of the excellent fire resistance and sound insulation properties of concrete. Concrete floor slabs can be divided into the following three general categories:

(a) one-way spanning slabs
(b) two-way spanning slabs
(c) flat (beamless) slabs

One-way spanning slabs

Rectangular slabs which are only supported at two opposite edges by beams or walls are classed as one-way spanning slabs since the loads are carried by a combination of bending and shear in one direction only (Fig. 2.18). One-way slabs can be either simply supported over one span or continuous over a number of spans and the slab cross-section can be of uniform solid, voided or ribbed construction. Figure 2.19 illustrates a typical section, such as section X–X in Fig. 2.18, through some of the more common types of slab. It is important to note that for voided and ribbed slabs the voids/ribs run from one supported end to the other.

Two-way spanning slabs

Slabs which are supported along all four edges by beams or walls are known as two-way spanning slabs since the applied loads are effectively transferred in

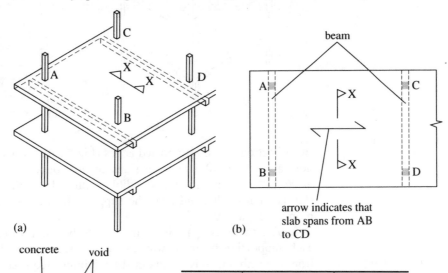

Fig. 2.18 One-way spanning slabs: (a) geometry of structure; (b) plan view of slab ABCD

beam

arrow indicates that slab spans from AB to CD

(a)

(b)

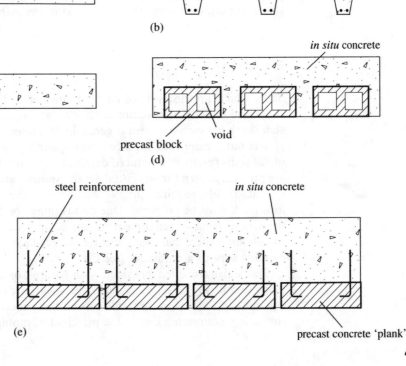

Fig. 2.19 Alternative sections through slab (section X–X in Fig. 2.18): (a) voided (usually precast and prestressed); (b) ribbed; (c) solid; (d) ribbed with block infill; (e) precast planks with *in situ* infill

concrete   void

(a)

(b)

in situ concrete

(c)

precast block   void

(d)

steel reinforcement   *in situ* concrete

(e)

precast concrete 'plank'

41

Fig. 2.20 Two-way
spanning slabs:
(a) geometry of
structure; (b) plan view
of two-way slab;
(c) waffle slab (view
from below)

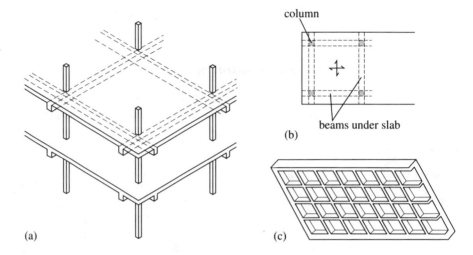

(a)

(b)

(c)

column

beams under slab

two directions to the supported edges (Fig. 2.20). Two-way spanning slabs are normally either solid uniform slabs or, for longer spans, 'waffle' slabs (Fig. 2.20(c)) of a shape not unlike edible waffles. In waffle slab construction, the slab is often made solid near the supports to increase the shear and bending moment capacity.

Square two-way spanning slabs tend to be the most economical shape since each supporting beam or wall carries the same proportion of the total load from the slab and this results in the minimum required slab and beam depth. It is a basic principle relating to slabs that load tends to be transferred to the nearest support (more in section 5.6). Thus, in a rectangular slab, the greater proportion of the load spans across the shortest distance. If the length of the slab is significantly greater than its width, the slab effectively spans one way.

Flat slabs

Slabs which are not supported by beams or walls along the edges but are supported directly by columns are known as flat slabs (Fig. 2.21). The required slab depth for such a system is generally less than for a one-way spanning slab system but greater than for a two-way spanning slab system. However, the use of flat slabs results in a reduced depth of structure overall as beams, when they are present, govern the structural depth. Another great advantage of this system is that it only requires simple shuttering – there are no beams for which formwork must be prepared. One disadvantage of flat slab construction is that the arrangement of reinforcement can be very complex, particularly adjacent to columns where punching shear reinforcement is often required if the slab depth is kept to a minimum (see Chapter 10).

Flat slabs are often provided with 'drops' or enlarged column heads, as illustrated in Fig. 2.22, to increase the shear strength of the slab around the column supports. However, this substantially increases the complexity of the shuttering, countering one of the principal advantages of the system.

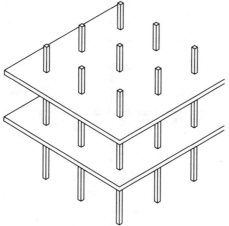

Fig. 2.21 Flat slab construction

Fig. 2.22 Flat slab construction details:
(a) no column head;
(b) flared column head;
(c) slab with drop panel;
(d) flared column head and drop panel

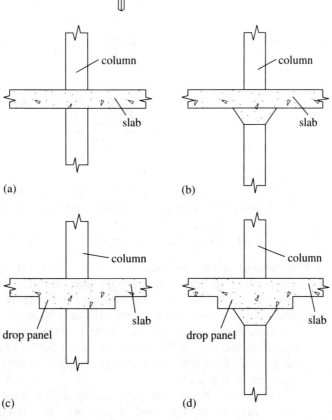

Example 2.9 Floor slabs

Problem Figure 2.23 represents the plan of a typical floor in a five-storey office building. The office space is to have an open plan with non-permanent partition walls and a minimum number of internal load-bearing walls. Determine an appropriate floor system for the building if:

(a) The main service ducts are to run through the corridors between a suspended ceiling and the floor above.

43

Fig. 2.23 Layout of proposed office block

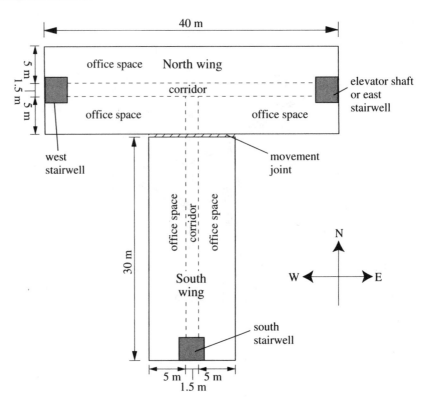

(b) The floor-to-floor height is to be kept to a minimum.

(c) A movement joint is to be positioned between the north and south wings of the building.

(d) A single floor scheme is to be used throughout.

Solution Movement joints are used to divide a large structure into a number of independent sections. They should pass through the entire structure above ground level in one plane. In reinforced concrete frame structures, movement joints should be provided at approximately 50 m centres in both directions. In addition, movement joints should be provided where there is any significant change in the type of foundation or of the height of the structure and should be located, if possible, at a change in plan geometry (as is the case for the structure of Fig. 2.23).

For the plan dimensions given in Fig. 2.23, a two-way spanning slab system would need to be supported in two bays, that is on three rows of columns (suggested preliminary span/depth ratios for slabs are given in Chapter 6). The internal columns are located along one wall of the corridor in each wing, as illustrated in Fig. 2.24, to minimize internal obstructions. However, the scheme illustrated in Fig. 2.24 is not a very effective solution because the service ducts would have to run under the lateral beams, which increases the total floor height over that for other schemes.

An alternative scheme of flat slab construction is illustrated in Fig. 2.25. The maximum span/depth ratios for flat slab construction are somewhat less than

Fig. 2.24 Two-way spanning slabs

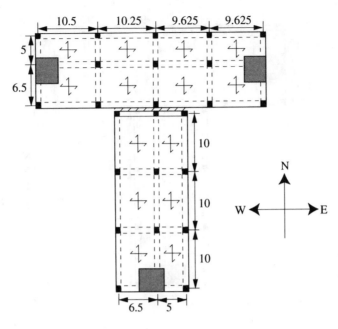

Fig. 2.25 Flat slab scheme

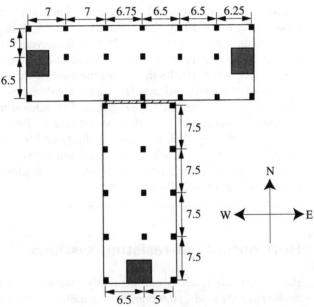

for two-way spanning slabs. For this reason, it is necessary to reduce the spacing of the columns in both wings of the building as shown. However, the overall storey depth of the flat slab scheme is less and this proves advantageous where the services are to be fixed below the floor.

A second alternative scheme, this time with one-way spanning slabs, is illustrated in Fig. 2.26. The allowable span/depth ratios for one-way spanning slabs are less than those for both flat slabs and two-way spanning slabs. For the building layout given in Fig. 2.23, a scheme of one-way spanning slabs

Fig. 2.26 One-way
spanning slab scheme

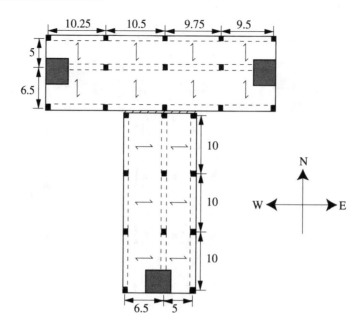

would usually be supported in two bays. The slabs are to span in an east–west direction in the south wing, as illustrated in Fig. 2.26, to minimize obstruction of the main service ducts by the beams. However, the beams still obstruct the secondary service ducts leading from the main ducts in the corridors into the office space which results in a small increase in storey height.

A third alternative scheme is to use a combination of flat slab and one-way spanning slabs, as illustrated in Fig. 2.27. This scheme minimizes the obstruction to both the main and secondary service ducts. For this scheme, the load on regions of the slab, such as region A illustrated in Fig. 2.27(b), spans to other strips of slab, B and C, which, in turn, span between columns (D, E and F, G respectively). Thus the strips of slab over the corridor between the columns act as beams and are heavily reinforced.

2.3 Horizontal-load-resisting systems

Horizontal wind forces are generally smaller in magnitude than vertically applied gravity loads. However, the resistance of structural systems to horizontal load is often considerably less than to gravity loads. A qualitative appreciation of the alternative methods of resisting horizontal load enables the engineer to select the most suitable structural layout for the structure under consideration.

One adverse effect of horizontal forces is their tendency to cause overturning of the entire structure, which can occur with tall, slender structures. For example, overturning occurs in the structure of Fig. 2.28(a) if the overturning moment, Fh, due to the horizontal force, is greater in magnitude than the stabilizing or restoring moment, Wb, where W is the self-weight of the structure and its foundations. Overturning is prevented in practice either by tying the

Fig. 2.27 Combined flat slab and one-way spanning slab scheme

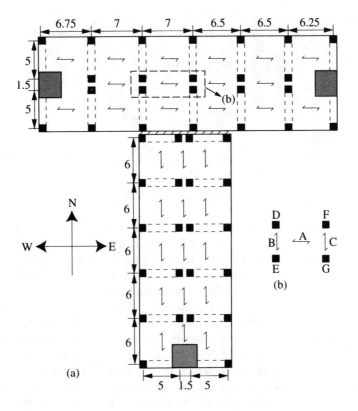

(a)

(b)

Fig. 2.28 Overturning of a structure due to a horizontal load: (a) geometry and loading; (b) heavy foundation; (c) wide foundation

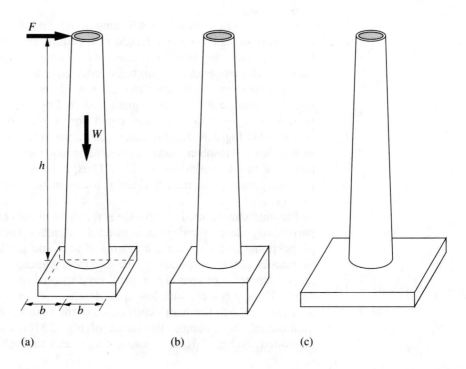

(a)   (b)   (c)

47

Fig. 2.29 Collapse of a dowel-jointed precast concrete frame

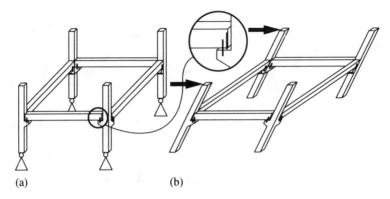

(a)   (b)

structure down to heavy foundations (Fig. 2.28(b)) or by providing the structure with a more expansive foundation pad (Fig. 2.28(c)). The restoring moment in the former method is increased by the added weight of the heavy foundations, and in the latter method the lever arm of the restoring moment is increased.

More critical for most concrete structures is the prevention of collapse by the 'racking' effects of horizontal forces, illustrated for a precast concrete frame in Fig. 2.29(a). A relatively small applied horizontal force causes rotation at the dowel connections and can lead to collapse of the structure (Fig. 2.29(b)). This mode of collapse in such a structure is known as a 'mechanism'. The racking effect of loads in a practical structure is opposed by effectively increasing its rigidity against horizontal forces. The actual method by which a structure is stiffened is normally dependent on its height (and hence the magnitude of the horizontal forces) and on the material from which it is constructed.

One common method of stiffening a reinforced concrete frame against horizontal forces is to provide rigid connections between the members of the frame to prevent any relative rotation of the connected members. Theoretically, only a small number of rigid joints are required in a frame to provide horizontal stability. However, in practice most, if not all, members are rigidly connected so that resistance to horizontal loads is shared by all members of the frame. A typical rigid joint construction detail for a reinforced concrete frame is illustrated in Fig. 2.30. Rigidly connected frames are quite effective in supporting vertical loads. However, frames of typical proportions are a relatively inefficient means of providing lateral stability. Thus, it is unusual to find such frames providing stability against horizontal forces in buildings of more than a few storeys.

Framing can be used to provide stability to significantly taller buildings if particularly deep members are utilized. Member stiffnesses in frames are proportional to their second moments of area and are inversely proportional to the cube of their length, as can be seen from Appendix A, Nos 4 and 5. Gross second moments of area are, in turn, proportional to the cube of the member depth. Thus, members with low span/depth ratios offer substantial resistance to rotation. When the span/depth ratios of the beams are of similar magnitude to those of the columns, the frame of Fig. 2.31(a) deflects in the manner illustrated in Fig. 2.31(b). Now, if the beams in such a frame are greatly

Fig. 2.30 Typical detail for a rigid reinforced concrete connection

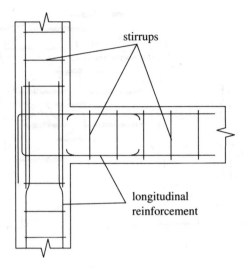

Fig. 2.31 Frame action: (a) simple frame; (b) deflected shape for beams of typical stiffness; (c) deflected shape when beams are stiff

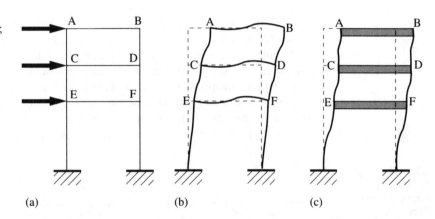

deepened, the beam members become very stiff and restrict the rotation at the joints (A, B, etc.). This substantially reduces the horizontal deflection as it restricts the rotation at the ends of the columns as illustrated in Fig. 2.31(c). Often, the depth of internal beams is restricted and hence such deep members can only readily be provided on the perimeters. In these structures, load is transferred transversely across the frame through the floor slabs to the perimeter from where it is taken by frame action to the foundations. When both beams and columns on the perimeter of a building have very low span/depth ratios and the openings between them become small (see Fig. 2.32), the overall behaviour becomes more like that of a solid cantilever member than that of a frame. In three-dimensional structures of this type, the perimeter of the building can act like a hollow tube to resist horizontal load as illustrated in Fig. 2.32(b). This principle is used to provide lateral stability in particularly tall buildings such as the World Trade Center in New York. Further details on the design of tall buildings are given in the excellent book on the subject edited by Beedle (1983).

An alternative to providing horizontal stability through rigid connections

Fig. 2.32 Hollow tube structure: (a) short and deep beams and columns; (b) very short and deep beams and columns

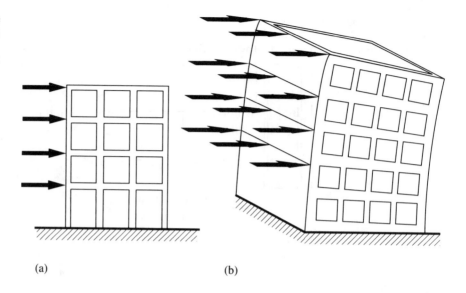

(a)   (b)

and/or deep beams is to incorporate diagonal or 'cross' bracing in skeletal frames to carry the bulk of the loads. In fact, bracing can be used along with rigid joints to reduce greatly the magnitude of frame deformations in taller structures. The use of conventional diagonal bracing is limited to panels of a frame where openings, such as windows, are not required. Steel bracing is usually provided only by members acting in tension, such as in Fig. 2.33(a), because of the possibility of buckling in compression members (Fig. 2.33(b)). In order to provide stability in both directions, steel tension members are often

Fig. 2.33 Horizontal stability through bracing: (a) tension members; (b) compression members buckle; (c) tension members for load in either direction; (d) K-bracing

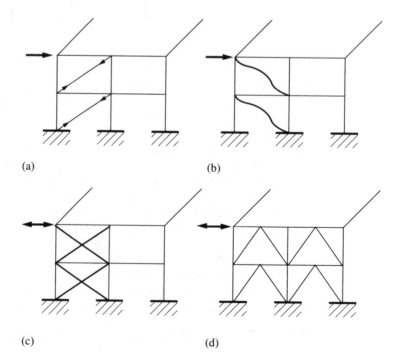

(a)   (b)

(c)   (d)

provided on both diagonals as illustrated in Fig. 2.33(c). A special type of bracing, known as K-bracing (Fig. 2.33(d)), can be used where openings are required. However, this form is structurally less efficient at resisting horizontal loads than the conventional diagonal type. If the frame is fitted with rigid floors, bracing is often only required in the perimeter of the structure or around the stairwells and elevator cores.

Bracing is not a commonly employed method of providing stability against horizontal loads in concrete buildings because of the unsuitability of ordinary reinforced concrete for tension members and the high cost of prestressed concrete. While it is not very common to mix structural materials on site, it is perfectly conceivable to provide steel bracing members in a concrete frame.

Shear walls, in reinforced concrete or masonry, are perhaps the most popular method of providing lateral stability in concrete structures. The precise effect of introducing such rigid panels into a skeletal frame has been discussed in Example 2.7. Unlike most braced frames, small openings can be provided in concrete shear walls, often with little detriment to their strength. Concrete shear walls can be used effectively to provide horizontal stability for structures of up to 20 storeys. Masonry shear walls are also good but cannot be used to the same height because low tensile strength leads to diagonal cracking in taller masonry structures as illustrated in Fig. 2.34. However, masonry used as an infill for a skeletal frame, as illustrated in Fig. 2.35, can be used successfully in taller structures since the frame confines the masonry and reduces the stresses which would cause diagonal cracking. Disadvantages of infill masonry panels are an increase in construction time and a sensitivity of the panels to openings (such as for doors and windows). Shear walls can be provided either internally, typically as walls to the stairway, or on the perimeter, where they form the outer panels of the structure. However, external panels are not always acceptable aesthetically.

Another system commonly used to provide lateral stability in concrete structures is the reinforced concrete core, an example of which is illustrated in Fig. 2.36. Core systems are conveniently provided in multi-storey structures as

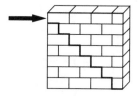

Fig. 2.34 Diagonal cracking of masonry shear wall

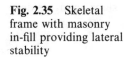

Fig. 2.35 Skeletal frame with masonry in-fill providing lateral stability

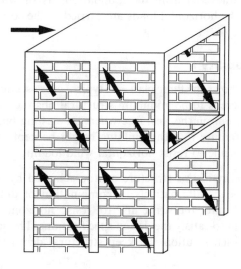

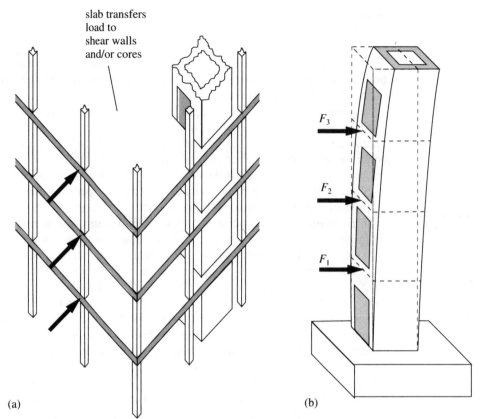

slab transfers
load to
shear walls
and/or cores

F_3

F_2

F_1

(a)

(b)

Fig. 2.36 Stability
using concrete core:
(a) portion of structure;
(b) deflected shape of
core

enclosures for lift shafts and stairwells. In such a capacity, the core not only
provides lateral stability to the structure but also serves as a fire-resisting shell
in the case of staircases and as a rigid supporting tube in the case of lifts. The
core illustrated in Fig. 2.36(a) does not simply act as four shear walls providing
stability in two perpendicular directions – it acts as a rigid hollow box in
bending, cantilevered from the foundations as shown in Fig. 2.36(b). As for
shear walls, the horizontal loads are carried to the core by slabs acting as deep
beams.

Example 2.10   Lateral stability I

Problem   Figure 2.37 shows the floor plan and typical cross-section for a two-storey office
building. There is to be minimum structural obstruction internally and on the
perimeter except on the west face where an existing building adjoins. Planning
restrictions require that the building height be kept to a minimum. Devise a
structural scheme to resist horizontal wind loading.

Solution   Lateral stability of this structure can be achieved by any one of the methods
described above. Alternatively, a combination of different methods may be
incorporated into the structure to resist horizontal loads from different
directions and to satisfy the design specifications. The scheme illustrated in Fig.
2.38 is one such solution. To keep structural depth to a minimum, flat slab

Fig. 2.37 Two-storey office building: (a) floor plan; (b) typical section X–X

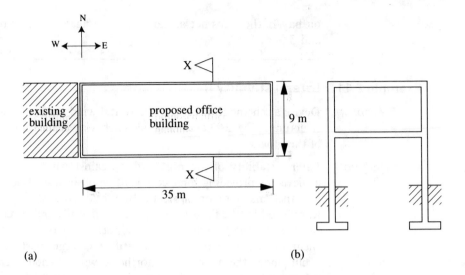

(a)   (b)

Fig. 2.38 Structural scheme for the provision of lateral stability

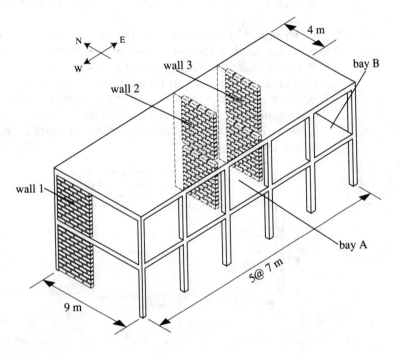

construction will be used. To resist E–W load with the minimum of internal obstruction, rigidly connected frames are provided in which slabs form the horizontal members. As the slabs are not very stiff, columns are provided at 7 m intervals along the north and south perimeters. The 7 m × 9 m grid is also acceptable for a flat slab of reasonable depth. To resist N–S wind, the cheaper alternative of masonry infill shear walls is selected. Wall 1 is located on the west perimeter as no windows are possible on this face. Walls 2 and 3 are placed on either side of the stairs. For wind from the south on bay A, say, the slabs at each level act as beams to transfer the load to walls 1, 2 and 3. For wind on bay B, the slabs act as cantilevers to again transfer the load to walls 1, 2 and 3.

Example 2.11   **Lateral stability II**

Problem   Devise a scheme to resist the horizontal wind loads on the five-storey concrete structure of Fig. 2.23 assuming that each wing of the building acts independently of the other.

Solution   Lateral stability of the north wing against wind from all directions is best achieved by providing some combination of shear walls and/or cores constructed in either masonry or concrete. In the scheme of Fig. 2.39, wind loads are transferred by the floor slabs to the shear walls and/or cores which then carry the load to the foundations in proportion to their relative stiffnesses. If the shear walls/cores are of unequal stiffness, some twisting of the structure will occur under the action of a northerly wind. When such twisting effects are substantial, they must be considered and allowed for in the design (see Section 5.7).

To resist horizontal loads in the south wing, a reinforced concrete core is provided around the south stairwell. However, to prevent twisting of the structure under the action of wind from the east or west, some form of lateral support is also required at the northern end of the south wing. The provision of masonry or concrete shear walls at the position illustrated in Fig. 2.39 satisfies this requirement but does cause significant internal obstruction. Alternatively, frame action could be utilized to give the necessary E–W stability at the north end of the south wing.

Fig. 2.39 Provision of lateral stability in North and South wings

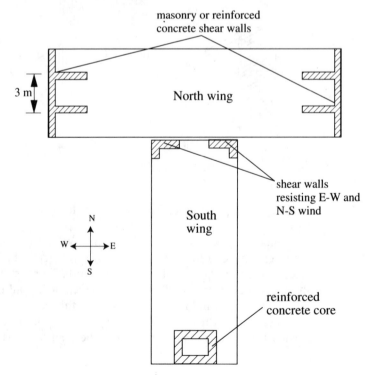

Note: The Manual for the Design of Reinforced Concrete Building Structures, published by the Institution of Structural and Civil Engineers (1985) (henceforth referred to as the ISE manual), recommends that joints be provided in buildings where there are significant changes in plan geometry such as in this example. However, joints are best avoided if at all possible as they are difficult to construct and complicate the cladding requirements in the immediate vicinity of the joint. Further, as can be seen above, the provision of the joint makes it more difficult to provide stability against horizontal loads. For these reasons, it may sometimes be appropriate not to follow exactly the recommendations of the ISE manual in the interests of a better design overall.

Connections

In the provision of lateral stability it is of vital importance to ensure that the connections between the adjoining members in a structure are sufficiently adequate to transfer safely the applied horizontal loads. For rigidly connected frames, the moment capacity of the connections, such as that illustrated in Fig. 2.30, must be of sufficient strength to resist the bending moment induced by frame action. In shear wall and beam–slab details, each member, where appropriate, should be adequately anchored or tied to its adjoining member.

2.4 Resistance of three-dimensional structures to incremental collapse

The term 'incremental collapse' is used to describe the behaviour of a complete structure when accidental failure of a single member causes the collapse of neighbouring members and, in certain cases, the entire structure. In the case of such a 'domino' effect, collapse of the initial member is known as primary damage and collapse of adjoining members is called secondary damage. Incremental collapse is often described as disproportionate collapse because the extent of the overall damage is out of proportion to the magnitude of the initial source of damage. Probably the best documented case of incremental collapse was the disaster at the Ronan Point flats, England, in 1968, in which the failure of one vertical concrete panel due to a localized gas explosion resulted in extensive secondary damage to the whole high-rise structure. The vertical panel had been supporting the floor slab above at one corner and its failure led to collapse of that floor. The external wall slabs of the floors above collapsed in turn and the weight of their impact caused a progressive collapse of the floor and wall panels in one corner of the block right down to ground level.

The integrity of a structure against incremental collapse due to accidental loads, otherwise known as its robustness, is dependent on several factors, including design methods, structural layout, the type of connections between members and the nature of the accidental load. The current trends of reducing factors of safety in design and making more efficient use of materials have led to a reduction in the reserve capacity of structures to accommodate the abnormal load conditions which lead to incremental collapse. Therefore, while a formal consideration of robustness is not necessary for many types of structure, it is wise always to bear the principles in mind in the preliminary stages of design.

The best approach to take when making a qualitative assessment of the robustness of a structure is to determine the effect on the stability of the structure of removing each member in turn. Statically determinate structures, such as that illustrated in Fig. 2.40, are inherently non-robust because the primary damage of any one member by accidental loads will generally result

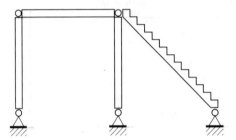

Fig. 2.40 Removal of any member of the pin-jointed structure will result in disproportionate collapse

in disproportionate collapse. Statically indeterminate or 'redundant' structures, however, are robust as they contain more support restraints and/or internal members than are required for stability. This allows a redistribution of force through secondary load paths in members adjacent to the damaged member and the structure is often capable of absorbing the accidental load without the occurrence of significant secondary damage. For instance, if member JL is removed from the structure in Fig. 2.3, the horizontal reactions from force F_3 are taken by frame action in DEFGHI. However, the failure of the inclined member in the determinate frame of Fig. 2.40 would lead to collapse of the entire structure.

Example 2.12   Multi-storey plane frame

As an example of qualitative assessment of the robustness of a redundant frame, consider the two-dimensional concrete structure in Fig. 2.41(a). The elimination of the external column, DH, results in the deformation of the frame illustrated in Fig. 2.41(b). Specifically, the removal of this member causes a substantial change in the distribution of bending moments and axial forces within the frames, particularly in members GH, KL, OP, PL and LH which now transfer the vertical load into the rest of the structure by frame action. Thus, incremental collapse of the structure is prevented only if these members and their connections are sufficiently strong to resist the new distribution of force/moment.

Similarly, the removal of member BF, as illustrated in Fig. 2.41(c), results in complete structural failure if members AE and CG are unable to carry the extra compressive force and members EFG, IJK and MNO are not strong enough to carry the increased bending moment. However, in this case when deformations are very large, members EFG, IJK and MNO could still act as tension members even if they have yielded in bending.

By the hypothetical removal of other individual members, the requirements

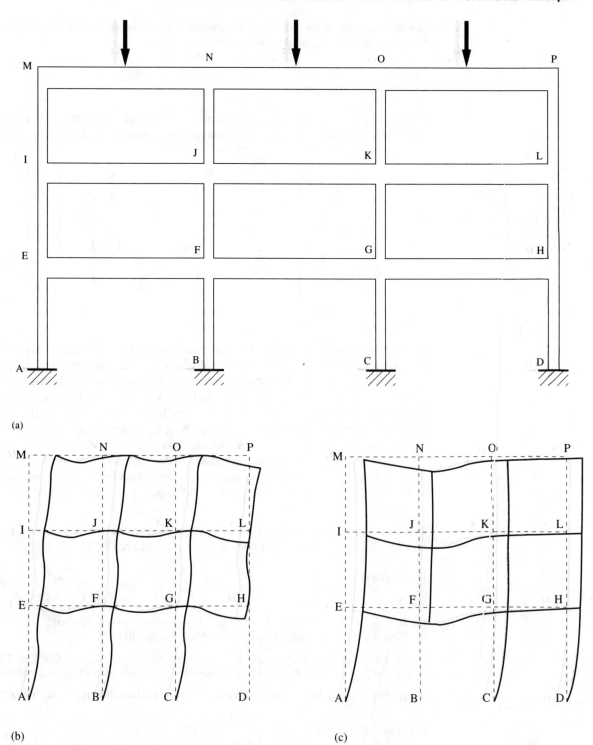

Fig. 2.41. Robustness of rigid frame: (a) original frame; (b) effect of removing member DH; (c) effect of removing member BF

of each member in the structure to prevent total collapse due to the failure of a single member can be determined.

Continuous members

Continuous three-dimensional structures, such as that illustrated in Fig. 2.42, are generally more robust than skeletal frames as they have a superior ability

Fig. 2.42 Resistance of continuous structures to accidental loads by arching of plate members: (a) load taken by arch action; (b) load taken by flexure

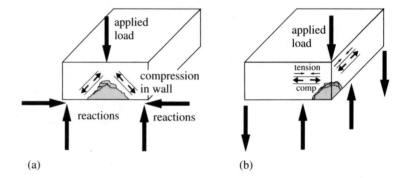

(a)   (b)

to resist incremental collapse. In a properly connected continuous structure, the extent of damage due to accidental loads is often limited to individual members. It is because of their continuous nature that these members are able to transfer the forces around the damaged area with little detrimental effect to the adjacent members. It is of vital importance, however, to emphasize the need for adequate connections between the panels of a precast continuous structure to prevent secondary damage under accidental loads. All interconnecting vertical and horizontal concrete members should be securely fixed together using reinforcing ties or protected steel. Failure to provide adequate connections will leave the structure with little residual capacity to prevent incremental collapse.

Further information on incremental collapse is given in the papers of Gross and McGuire (1983) and Ellingwood and Leyendecker (1978).

Problems

Section 2.1

2.1 (a) How is the force, F, illustrated in Fig. 2.43, transferred to the supports?
(b) Find the axial force and bending moment in member BD.

2.2 By what mechanisms of load transfer is the horizontal force illustrated in Fig. 2.44 transferred to the supports? Where is the bending moment due to this force a maximum?

2.3 What percentage of the total vertical force, F, in the structure in Fig. 2.45, is carried by the member AC?

Sections 2.2 and 2.3

2.4 A single storey factory is to be designed in which the structure may be exposed to corrosive gases. Propose a preliminary structural solution given that the dimensions in plan are 15 m × 40 m and that there are to be no internal obstructions.

Fig. 2.43 Structure for Problem 2.1

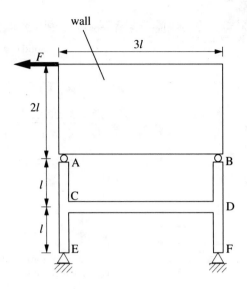

Fig. 2.44 Structure for Problem 2.2

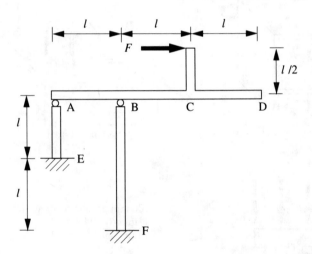

Fig. 2.45 Structure for Problem 2.3

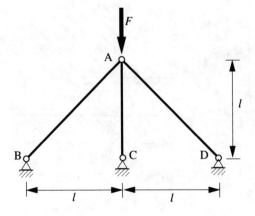

2.5 State, giving reasons, whether the proposals in Fig. 2.46 are structurally stable and robust.

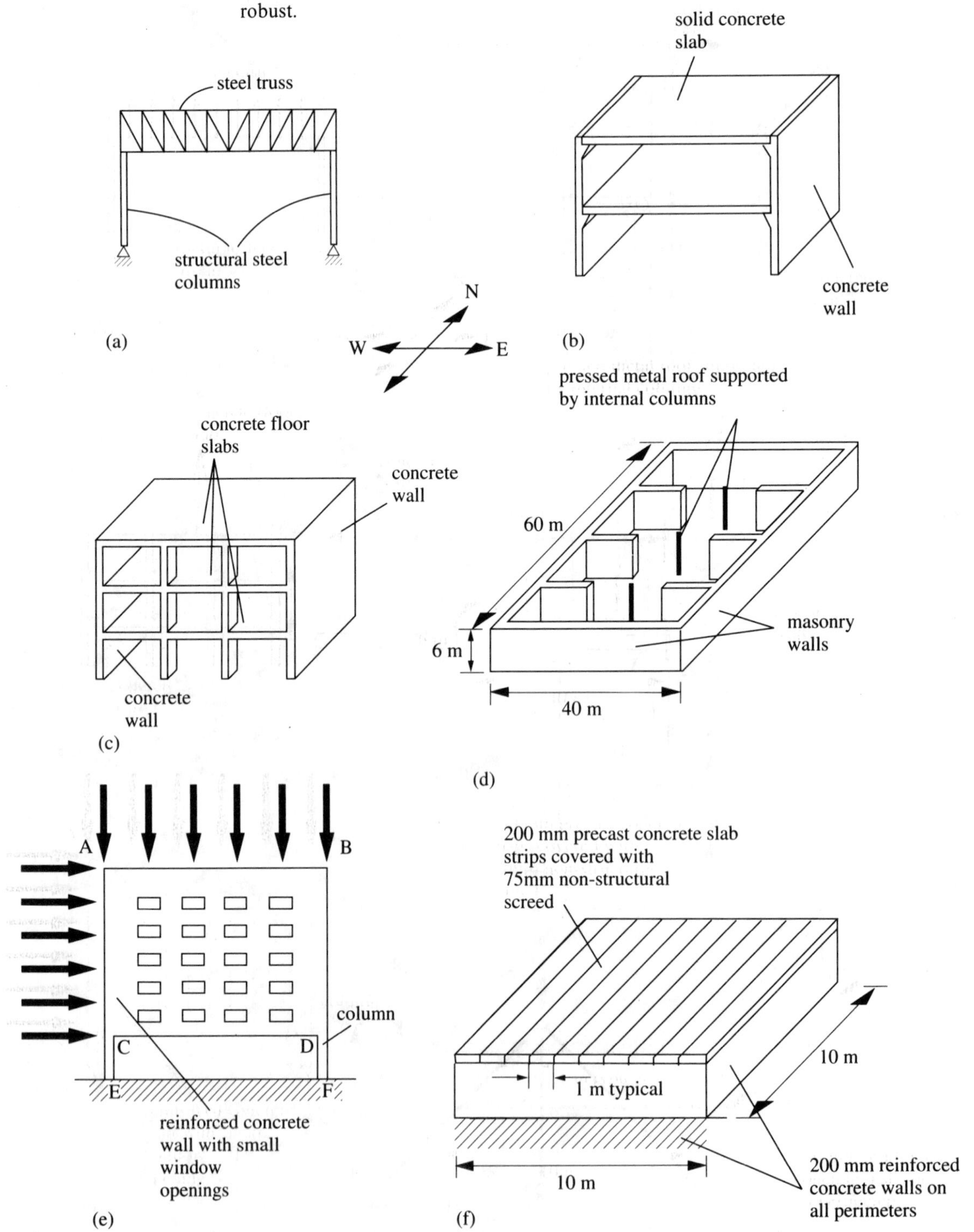

(a)

(b)

(c)

(d)

(e)

(f)

Fig. 2.46. Structures for Problem 2.5

3

Loads and Load Effects

3.1 Introduction

The aim of structural design is to ensure that, with an acceptably high probability, a structure will remain fully functional during its intended life. The expected lifetime of a structure is formally known as its **design life** and is usually a period of at least 50 years. During its design life, a structure must be capable of safely sustaining all applied loads and other stress-inducing actions that might reasonably be expected to occur. Thus, it is necessary to identify and, more importantly, quantify the various types of load which act on its members. The different classes of loads which commonly act on structures and must be considered in design are the subject of this chapter. Owing to the variable nature of most loads and materials, structural design philosophy relies heavily on the use of statistical principles. The application of these principles to ensure structural safety is discussed in section 3.7.

It is worth pointing out at this stage that in contrast to construction procedures, in which the structure is erected from the ground upwards, building structures are generally designed from the top downwards. This is because the magnitudes of the internal gravity forces in members at any level depend on the forces being transferred from the levels above as well as the loads applied directly at that level.

3.2 Classification of loads

The term **action** is used in the Eurocodes collectively to describe forces and environmental effects on structures. An action can be defined as anything that gives rise to internal stresses in a structure and can be direct or indirect in nature. **Direct actions** are forces applied to the structure by external agents such as wind or vehicular traffic whereas **indirect actions** are imposed deformations in the structure which do not result from external forces. Temperature variation, settlement of supports and shrinkage of concrete members are examples of indirect actions which give rise to internal stresses in structures.

Actions can be classified further by their variation with time. Those which have little or no variation in magnitude over the design life of the structure are known as **permanent actions** (sometimes called dead loads). On the other hand, actions which are unlikely to maintain a constant magnitude over the design life are termed **variable actions**. Two important types of variable action are those due to wind and the gravity loading due to the occupants of the structure, known as occupancy load, or **imposed load**. In a typical building, the self-weight of the floor slab is an example of a permanent action while the weight of people and furniture is a variable action. **Accidental actions**, such as the impact of a vehicle against a column, are also time variant. However, in contrast to variable actions, accidental actions are those which are unlikely to occur very often in significant magnitude and duration. It is worth noting that indirect actions are also classified by their variation (or lack of variation) with time. Shrinkage of a beam can induce stress in the columns to which it is attached. As this is long term in nature, shrinkage is classified as an (indirect) permanent action.

Direct actions can also be classified by their spatial variation, that is variation in the area of application. **Fixed actions** are direct actions which have no freedom of movement within or on the structure. Clearly, structural self-weight falls into this category. **Free actions**, on the other hand, are direct actions which can occur at arbitrary locations on or within the structure. Furniture, for example, is a free action.

Two other terms which are applied to the classification of loads are **static** and **dynamic**. Direct actions are known as 'dynamic' actions if they cause significant acceleration or vibration of the structure to which they are applied, and as 'static' actions if they cause no such significant acceleration. In sufficiently slender structures many different types of action can be dynamic. For example, the surge due to acceleration in a gantry crane can be dynamic. Also, in slender towers, wind can be dynamic although for most concrete structures it is essentially a static action.

Actions which have magnitudes which are exactly or approximately known and which are closely maintained during the design life of the structure are called **closely bounded actions**. An example of a closely bounded permanent action is the self-weight of structural concrete or steel.

Table 3.1 summarizes the different load classifications, giving some common examples of each type of action. It is important to realize that most actions can be categorized under each class. For example, self-weight can be classified as a direct action which is permanent, fixed, closely bounded and static.

In the following three sections, the more familiar types of direct actions, namely permanent gravity loads, variable gravity loads and the static effect of wind loads, are treated in greater detail. More specialized actions, such as seismic loads and vehicle loads, are not considered further.

3.3 Tributary areas

When a uniformly loaded floor slab is simply supported on beams on two opposite sides, half of the load is carried by each beam. The slabs of Figs 3.1(a) and (b) are simply supported on beams AD, BE and CF. Thus, all of the load on the shaded area in Fig. 3.1(a), known as the tributary area of beam BE, is

Table 3.1 Classification of loads

		Examples	
Class	Action	Direct	Indirect
Time variation	Permanent	Soil pressure, self-weight of structure and fixed equipment	Settlement, shrinkage, creep (results from direct permanent actions)
	Variable	People, wind, furniture, snow, traffic, construction loads	Temperature effects
	Accidental	Explosion, vehicular impact	Temperature rise during fire
Spatial variation	Fixed	Self-weight (generally), trains (fixed in direction normal to rails)	–
	Free	Persons, office furniture vehicles	–
Static/ dynamic	Static	All gravity loads	–
	Dynamic	Engines, turbines, wind on slender structures	–
Others	Closely bounded	Water pressure, self-weight	–
	Not closely bounded	Snow, people	–

supported by that beam. This follows from the fact that the reaction at each end of a uniformly loaded simply supported beam is half the total load (see Appendix C, No. 1). If the slab of Fig. 3.1(a) were continuous over beam BE, as illustrated in Fig. 3.1(c), the reaction on BE due to a uniformly distributed load throughout the floor would be greater with this beam carrying a greater portion of the total load. It can be seen from Appendix C, No. 4, that, in this situation, beam BE would take $\frac{5}{4}$ of the total load and the tributary area would be $\frac{5}{4}(6) \times 10$ m^2.

When loading is not uniformly distributed on all spans, as is the case for variable gravity load, the reaction and hence the tributary length is affected. For example, if the slab of Fig. 3.1(a) is continuous over beam BE, as illustrated in Fig. 3.1(c), and is loaded uniformly on the panel, ABDE only, then the tributary length is calculated from Appendix C, No. 6. The tributary length for support BE is then:

$$l_b = \frac{l(4k + 1)}{8k}$$

where $l = 6$ m and $k = 1$. Hence, $l_b = \frac{5}{8}(6) = 3.75$ m and the tributary area for beam BE is 3.75×10 m^2.

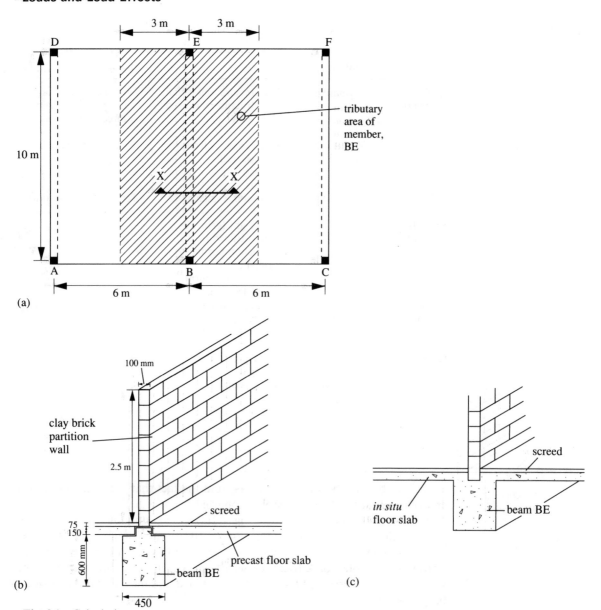

Fig. 3.1 Calculation of loads: (a) floor plan; (b) section X–X; (c) alternative section X–X if slab were continuous over BE

Tributary areas for slab panels

For two-way spanning slabs, the determination of tributary areas for the supporting beams is a little more subjective. Different engineers make different assumptions and can arrive at significantly different results. When selecting a method, it is useful to bear in mind that actual variable loading due to people, furniture, etc., is represented by equivalent uniformly distributed loading. In view of this, excessive levels of refinement in calculating tributary areas would seem inappropriate. One approach, which is sufficiently accurate for most purposes, is illustrated in the following example.

Example 3.1   **Tributary areas in two-way spanning slabs**

Problem   A floor system, illustrated in Fig. 3.2, consists of two-way spanning continuous slabs supported by continuous beams which, in turn, are supported by columns. Determine the loading for beam abcd which will give maximum moment at b if the permanent gravity load is 10 kN/m² and the variable gravity load varies from 0 to 7 kN/m².

Solution   As the slab spans both N–S and E–W, portions of the load applied to each panel will be carried by each of the four beams around its perimeter. Thus, the pattern of tributary areas will be of the type illustrated in Fig. 3.2(b). Many designers assume that the lengths, L_1 and L_2, of Fig. 3.2(b) are equal to half the span, that is – they ignore the effect of continuity of the slab. The following approach is, perhaps, more accurate.

Permanent gravity load

The strip of slab at X–X in Fig. 3.2(b) spans over three unequal spans, as illustrated in Fig. 3.3(a). As permanent gravity load is uniform over all spans, the tributary lengths for this beam are as given in Appendix C, No. 8. Hence, the length L_1 illustrated in Figs 3.2(b) and 3.3(a) is:

$$L_1 = \frac{L(6k^4 + 15k^3 + 6k^2 - 2k - 1)}{4k(4k^2 + 8k + 3)}$$

where $k = \frac{6}{8} = 0.75$ and $L = 8$ m. Thus:

$$L_1 = 2.157 \text{ m}$$

The length, L_3, also illustrated in Figs 3.2(b) and 3.3(a) is half the span length, that is 4 m. Similarly, considering section Y–Y of Fig. 3.2(b), illustrated in Fig. 3.3(b), we find from Appendix C, No. 5:

$$L_5 = L\left(\frac{3k^2 + k - 1}{8k}\right)$$

where, again (by coincidence), $k = \frac{6}{8} = 0.75$ and $L = 8$ m. Thus:

$$L_5 = 1.917 \text{ m}$$

A sensible pattern of tributary areas might then be that illustrated in Fig. 3.4(a). This has been constructed by adjusting the angles, θ_1, θ_2, etc., from 45° to reflect the bias towards the simply supported ends of the slab. On this basis, θ_1 is defined by:

$$\tan \theta_1 = \frac{L_5/(L_5 + L_6)}{L_1/(L_1 + L_2)} = \frac{1.917/6}{2.157/6}$$

$$= 0.889$$

$$\Rightarrow \qquad \theta_1 = 42°$$

65

Fig. 3.2 Two-way
spanning slab on
continuous beams:
(a) geometry; (b) pattern
of tributary areas (plan)

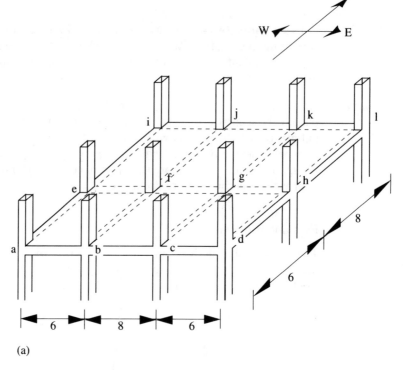

(a)

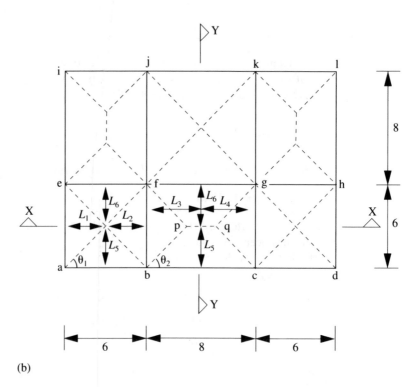

(b)

Fig. 3.3 Tributary lengths: (a) section X–X of Fig. 3.2(b); (b) section Y–Y of Fig. 3.2(b)

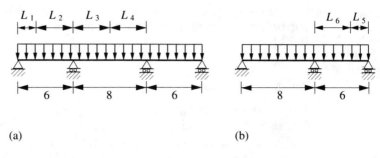

(a)   (b)

Similarly:

$$\theta_2 = \tan^{-1}\left(\frac{L_5/(L_5 + L_6)}{L_3/(L_3 + L_4)}\right) = \tan^{-1}\left(\frac{1.917/6}{4/8}\right)$$

$$= 33°$$

In practice, the tributary area for beam abcd can often be constructed without calculation of such angles. This is done by first constructing the tributary areas on the assumption that all slabs are simply supported (as illustrated in Fig. 3.2(b)) and then editing this pattern. For example, points p and q (Figs 3.2(b) and 3.4(a)) are adjusted in the ratio $L_3/(L_3 + L_4)$:0.5 for the E–W direction and $L_5/(L_5 + L_6)$:0.5 for the N–S direction. The result of this is the pattern illustrated in Fig. 3.4(a).

Having determined the pattern of tributary areas, the loading on each beam can be found. Thus the permanent gravity loading on beam abcd varies linearly in span ab from zero at a to a maximum intensity of $(10L_5 =)$ 19.2 kN/m and

Fig. 3.4 Permanent gravity load on beam abcd: (a) tributary areas (part plan view); (b) applied loading on beam

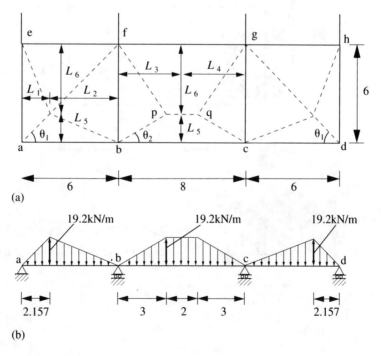

back to zero again at b as illustrated in Fig. 3.4(b). Similarly, there is a trapezoidal distribution of load on span bc with a maximum intensity of 19.2 kN/m.

Variable gravity load

The variable gravity load on this floor can have any intensity of loading from zero to 7 kN/m². Load in spans ab and bc will cause hog moment at b while load in span cd will actually reduce that moment. Thus, the load case which maximizes the moment at b will include the full loading on panels abef and bcfg and zero loading on panel cdgh. Load on other panels will not greatly influence the moment at b. For such panels, it is reasonable (and allowed by the Eurocode) to assume the minimum level of variable gravity loading (i.e. zero). Thus, variable gravity loading for maximum moment at b is taken to be that illustrated in Fig. 3.5(a). The strip of slab at section Y–Y is now loaded

Fig. 3.5 Tributary areas for variable gravity load: (a) loaded panels and tributrary areas; (b) applied load on beam abcd

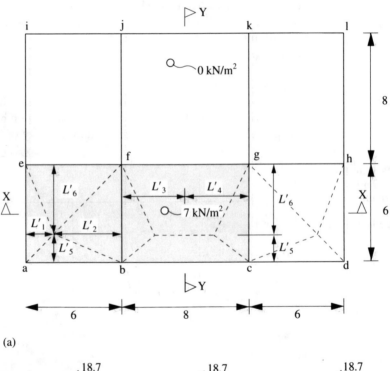

(a)

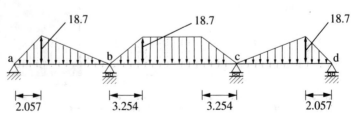

(b)

on one span only and the tributary lengths are calculated from the formulae in Appendix C, No. 6. Evaluation of the formulae in the appendix with $k = 8/6$ gives reactions of 45 and 59 per cent of the applied load at supports A and B respectively (refer to figure in appendix). The fact that the sum of reactions equals 104 per cent is explained by a downward reaction at the third support equal to 4 per cent of the applied load. The tributary lengths, l_A and l_B, are 45 and 55 per cent of the span length. While a complete assigning of the loading in these proportions would underestimate the reaction at B, it is reasonably accurate for the reaction at A which is all that is of interest for this load case. Hence:

$$L'_5 = (0.45)(6) = 2.678 \text{ m}$$

$$L'_6 = (0.55)(6) = 3.322 \text{ m}$$

Similarly, from Appendix C, No. 9:

$$L'_1 = 2.057 \text{ m} \qquad L'_2 = 3.943 \text{ m}$$

$$L'_3 = 4.338 \text{ m} \quad \text{and} \quad L'_4 = 3.662 \text{ m}$$

Thus, the variable gravity loading on beam abcd is, from Fig. 3.5(a), as illustrated in Fig. 3.5(b).

The total loading on a beam such as that considered in Example 3.1 is the sum of the loadings due to permanent and variable gravity load. As an alternative to adding the loading intensities, it is sometimes more convenient to analyse separately to find the moment due to permanent and variable loading and to add the results.

3.4 Permanent gravity loads on structures

Permanent gravity loads consist of the self-weight of the structural members plus the weight of objects permanently fixed to the structure over its design life. These materials include fixed partitions and walls, windows, ceilings, cladding, heating ducts, plumbing and electrical installations and any other essentially unmovable equipment.

The self-weight of individual structural members is normally determined from their dimensions and the weight density of the materials from which they are composed. The weight of non-structural objects is estimated either from their approximate dimensions and the density of their constituent materials (given in EC1, Part 2.1) or from manufacturers' specifications. The characteristic weight densities for various materials, which can be used to determine the magnitude of permanent gravity loads in a structure, are listed in Table 3.2.

Example 3.2   **Permanent gravity loads**

Problem   For the structure of Fig. 3.1(a) and (b), determine the permanent gravity load acting on the support at point B.

69

Table 3.2 Recommended values of weight density for common construction materials (from EC1, Part 2.1)

Construction material	Weight density (kN/m^3)
Concrete	
reinforced or prestressed	25*
unreinforced (natural aggregate)	24*
cement mortar	19–23
Masonry	
medium-weight concrete block	12–19
medium-weight clay brick	18–20
stone masonry	
granite	27–30
limestone	20–29
Metals	
aluminium	27
cast iron	71
lead	112
steel	77
Timber	
softwood	4–6†
hardwood	6–10†
chipboard	8
plywood	4–6

* These values may vary, depending on locally available materials.
† Not confirmed at time of going to press (June 1994).

Solution   The permanent gravity load, W, acting on the member, BE, is the combined weight of the permanent partition wall, a portion of the slab and screed and the self-weight of the member itself. The precast floor slab units are supported solely by the three members, AD, BE and CF, in two simply supported spans. The weight on each span is distributed equally between these beams, with each member carrying the weight of 3 m by 10 m of the slab. The total tributary area supported by the member BE is shown in Fig. 3.1(a). The total permanent gravity load acting on the member BE is (assuming the weight density of screed is 23 kN/m^3):

$$W = \text{weight of partition wall}$$
$$+ \text{ weight of slab and screed}$$
$$+ \text{ self-weight of beam BE}$$

$\Rightarrow \quad W = 20 \times 0.1 \times 2.5 \times 10 + 2(25 \times 0.15 \times 3 \times 10)$
$$+ 2(23 \times 0.075 \times 3 \times 10)$$
$$+ 25 \times 10 \times (0.6 \times 0.45) \text{ kN}$$

(note that the top 0.15 m of the beam has been allowed for in the calculation of the slab weight)

$\Rightarrow \quad W = 446 \text{ kN}$

Since member BE is itself simply supported and the load is uniform, half the total load is supported on each column. Therefore, the total permanent gravity load at point B is:

$$W/2 = 223 \text{ kN}$$

Permanent gravity loads for initial sizing of members

The magnitudes of the permanent gravity loads on a structure are dependent on the size of the structural members which are, in turn, governed by the applied loads. Thus, in order to determine the permanent gravity loads acting on a structure, it is necessary to obtain initial estimates for the member sizes. Methods for determining initial estimates for the sizes of structural concrete members are given in Chapter 6. In checking the adequacy of such initial designs, the ISE manual recommends the following guidelines for complete components:

(a) For floors where the finish is not specified, add 1.8 kN/m^2 to the weight of the slab to allow for a screed finish of about 75 mm depth (density of screed taken as 23 kN/m^2).
(b) The weight of ceilings and services for normal use can be taken initially as 0.5 kN/m^2, acting on the entire plan area of the floor.
(c) Often at the preliminary design stage, the precise location of permanent partitions is not known. In such circumstances, a nominal load can be smeared across the total floor area. Specifically, the weight of solid blockwork partitions can be represented by a load intensity of 2.5 kN/m^2 on the entire floor area.

3.5 Variable gravity loads on structures

Variable gravity loads which act on the floors and roof of a structure are those which vary in magnitude during the design life of the structure. They include the weight of persons, furniture, movable partitions, stored material, snow, vehicles and any other objects which are removable.

Imposed loads on floors

Imposed loads on floors can be classified as those which relate to the occupancy of a structure. They may be caused by people, furniture, movable partitions, stored material or vehicles. As for permanent partitions, when lightweight non-permanent partitions are to be present but their location is not known, they can be represented by a load intensity of 1.0 kN/m^2 acting on the entire plan area of the floor (from the ISE manual).

The magnitude of the variable gravity loads due to occupancy is determined by the specific functions for which different areas of the building are used. In the determination of imposed loads on floors, a distinction can be made between the following types of areas in a building:

- Assembly areas
- Commercial and administration areas
- Escape routes
- Parking areas
- Maintenance areas
- Residential areas
- Storage and production areas

The spatial and time variation of occupancy floor loads is generally different for each type of use. For instance, the imposed load in a hospital waiting room is greater than that in a hospital ward because the former is classed as an assembly area which is susceptible to overcrowding.

In recent years, probabilistic models have been developed for the derivation of suitable intensities of imposed loads which take account of their time-varying and space-varying nature. For the variation in time, the imposed load is separated into two components: a sustained (quasi-permanent) component and an extraordinary (short-term) component (Fig. 3.6). The sustained component represents the mean loading for normal day-to-day use (persons, miscellaneous stored material, furniture, etc.). This changes, for example, with the tenant using the floor area – some tenants will use heavier furniture than others. The extraordinary component represents unusual extremes in loading such as concentrations of persons (emergency situations, parties, ceremonies) and short-term stacking of furniture as might occur during redecoration work.

Fig. 3.6 Variable gravity loads: (a) sustained load; (b) extraordinary load; (c) total load (sum of (a) and (b))

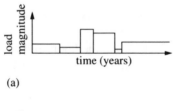

(a)

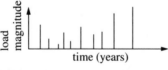

(b)

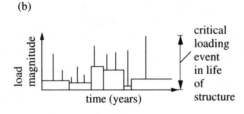

(c)

It is impractical to represent all possible spatial arrangements of occupancy loads at all times. For design purposes, an equivalent uniformly distributed load (EUDL) is used. Using probabilistic load models, values of load obtained from load surveys and engineering judgement, intensities of EUDL have been derived for a wide range of building types and uses. It is intended that the bending moment, shear force, etc., due to the EUDLs will equal the values that would actually occur in buildings under the same use with an acceptably remote probability of occurrence. Values for different occupancies are given in Part 2.1 of the draft Eurocode for actions on structures, EC1 (1993). Some of these values are given in Table 3.3. The figures in Table 3.3 represent nominal values which are adequate for most structures. However, where it is likely that these values are exceeded, such as in storage areas, precise figures should be obtained

Table 3.3
Recommended values for imposed floor loads (from EC1, Part 2.1)

Category	Description	Examples	Intensity of EUDL (kN/m^2)
A	Areas for general domestic & residential activities	Rooms in houses, rooms & wards in hospitals, bedrooms in hotels, kitchens and toilets	2.0
	Stairs		3.0
	Balconies		4.0
B	Office areas		3.0
C	Areas where people may congregate		
	C1 Areas with tables	In schools, restaurants, reading rooms	3.0
	C2 Areas with fixed seats	In churches, theatres, lecture halls, waiting rooms	4.0
	C3 Areas for moving people	In museums. Access areas in hotels and public buildings	5.0
	C4 Areas for physical activities	Dance halls, stages, gymnasia	5.0
	C5 Areas susceptible to overcrowding	Concert halls, grandstands	5.0
D	Shopping areas		
	D1 Retail shops		5.0
	D2 Department stores	Areas in warehouses, office stores	5.0
E	Areas susceptible to accumulation of goods	Areas for storage use, libraries	6.0

from the weight densities of the stored material. An extensive list of recommended values for the weight densities of stored material is given in Part 2.1 of EC1.

Example 3.3 Imposed loads on floors

Problem The two-storey structure illustrated in Fig. 3.7 is intended for use as a library. The ground floor and second-floor levels are to be used as reading rooms only while the first floor is to be used for reading and book storage. Lightweight partitions are to be present at unspecified locations. The total variable gravity load on the roof (snow plus imposed) is 1.2 kN/m^2. Determine the total variable gravity load in column E due to the two floors and the roof. The floors are of flat slab (beamless) construction.

Solution From Table 3.3, the characteristic variable gravity load on each floor is:

$$\text{Ground floor (reading room only)} \quad = 3 \text{ kN/m}^2$$
$$\text{First floor (reading room + stacks)} \quad = 6 \text{ kN/m}^2$$
$$\text{Second floor (reading room only)} \quad = 3 \text{ kN/m}^2$$
$$\text{Roof} \quad = 1.2 \text{ kN/m}^2$$

The partition loading consists of a further 1 kN/m^2 on each of the ground, first and second floors. The tributary areas are found by assuming two-span continuous beam behaviour in both directions. Hence, from Appendix C, No. 4, column E supports an area of $\frac{5}{4}(8) \times \frac{5}{4}(6) = 75 \text{ m}^2$. The tributary areas of all

Fig. 3.7 Calculation of imposed gravity loads: (a) floor plan; (b) end elevation; (c) tributary areas for columns

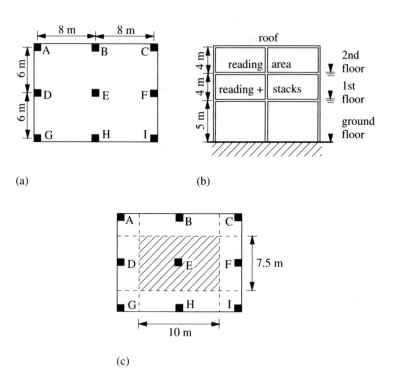

(a)

(b)

(c)

columns are illustrated in Fig. 3.7(c). The total variable gravity load in column E is the sum of the loads acting on the roof and the first and second floors (the load from the ground floor is assumed to be transferred directly to the foundations), that is:

$$P_E = 75[(6 + 1) + (3 + 1) + (1.2)]$$

$$P_E = 915 \, \text{kN}$$

Under certain circumstances, it is possible to reduce the recommended value of the variable gravity loads for buildings over two storeys in height and/or buildings with a large floor area. This is allowed because, in such conditions, it is unlikely that an extraordinary loading event will occur simultaneously throughout the structure. EC1 allows a reduction in variable gravity load as a function of the number of storeys and the loaded area.

Variable gravity loads on roofs

Roofs can be classified as being either accessible or non-accessible except for maintenance and repair. In the case of an accessible roof, the intensity of the variable gravity load due to occupancy is taken from the appropriate class in Table 3.3. When access is restricted, that is when no ready means exists to access the roof, the load intensity on flat roofs is taken as $0.75 \, \text{kN/m}^2$.

In addition to the variable gravity loads associated with occupancy, roofs are also exposed to gravity snow loads which must be considered in design. The deposition and accumulation of snow on roofs is dependent on a number of factors, including roof geometry and surface properties, the proximity of adjacent buildings, local climate and local surface terrain.

In Part 2.1 of EC1, design values for the intensity of snow loads are determined using the formula:

$$s = \mu_i C_e C_t s_k \quad [\text{kN/m}^2] \tag{3.1}$$

where C_e and C_t are exposure and thermal coefficients respectively, and usually have values of unity. The parameter s_k is the characteristic snow load, and its intensity depends on geographic location and the altitude of the site. Methods for the calculation of s_k vary between countries. For the United Kingdom, it is calculated as follows:

$$\left.\begin{array}{ll} \text{for } A \leq 100; & s_k = s_b \\ \text{for } 100 < A \leq 500; & s_k = s_b + (0.09 + 0.1s_b)(A - 100)/100 \end{array}\right\} \tag{3.2}$$

where A is the altitude of the site in metres above mean sea level and s_b is the basic snow load given in Fig. 3.8.

The parameter μ_i is a snow load shape coefficient which is dependent on the geometry of the structure and the number of consecutive snow falls before melting occurs. It is calculated using tables and formulae. For example, for a duopitch roof (see Fig. 3.9) the shape coefficients are as given in Table 3.4. In

Fig. 3.8 Basic snow load, s_b for UK (from EC2)

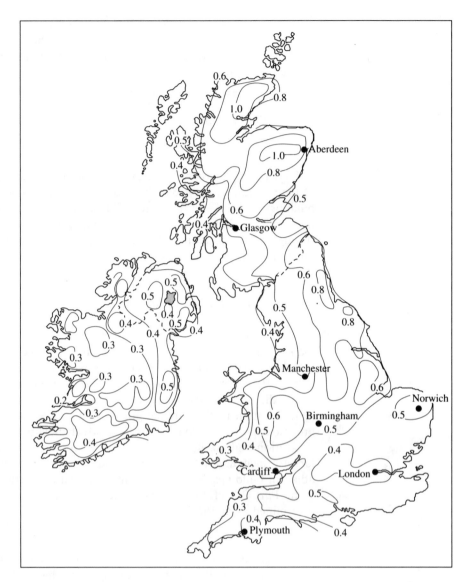

such structures, EC1 recommends that two load cases be considered. These are snow load on both roof pitches simultaneously, and snow load on one roof pitch only, as illustrated in Fig. 3.9. When snow is present on both pitches, drifting may result in there being more on one pitch than the other. For this reason, EC1 specifies two different values for μ_i, namely, μ_1 and μ_2 as given in Table 3.4. When snow is only present in one pitch, the lesser shape coefficient of $0.5\mu_1$ is used.

Example 3.4 Snow load on roofs

Problem Determine the snow load for the duopitch roof of a single bay building (pitch = 20°) which is located at sea level in the city of Norwich. Exposure and thermal coefficients of unity may be assumed.

Fig. 3.9 Snow load cases for single bay pitched roofs: (a) snow on both roof pitches; (b) snow on one roof pitch only

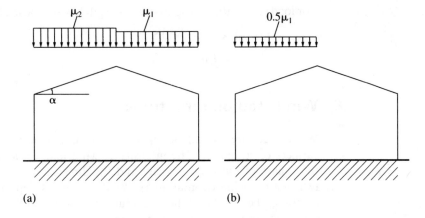

(a)   (b)

Table 3.4 Shape coefficients, μ_i, for symmetrical duopitch roofs (from EC1, part 2.1)

Roof pitch α	Both pitches loaded		One pitch loaded
	μ_1	μ_2	$0.5\mu_1$
0° to 15°	0.8	0.8	0.4
16° to 30°	0.8	$0.8 + 0.6(\alpha - 15)/30$	0.4
31° to 59°	$0.8(60 - \alpha)/30$	$1.1(60 - \alpha)/30$	$0.4(60 - \alpha)/30$
> 60°	0	0	0

Solution   From Fig. 3.8, Norwich has a basic snow load of 0.5. From Eq. 3.2 ($A = 0$), we have:

$$s_k = s_b = 0.5 \text{ kN/m}^2$$

The shape coefficients are taken from Table 3.4. For load case 1:

$$\mu_1 = 0.8$$

and

$$\mu_2 = 0.8 + 0.6(20 - 15)/30$$
$$= 0.9$$

The shape coefficient for load case 2 is:

$$0.5\mu_1 = 0.4$$

From equation (3.1), the load intensities for load case 1 are:

$$s = \mu_1 C_e C_t s_k$$
$$= 0.8 \times 1 \times 1 \times 0.5$$
$$= 0.4 \text{ kN/m}^2$$

and

$$s = 0.9 \times 1 \times 1 \times 0.5$$
$$= 0.45 \text{ kN/m}^2$$

For load case 2, there is no load on one pitch while the load on the other pitch is:

$$s = 0.4 \times 1 \times 1 \times 0.5$$
$$= 0.2 \text{ kN/m}^2$$

3.6 Wind load on structures

Wind forces are variable loads which act directly on the internal and external surfaces of structures. The intensity of wind load on a structure is related to the square of the wind velocity and the dimensions of the members that are resisting the wind (frontal area). Wind velocity is dependent on geographical location, the height of the structure, the topography of the area and the roughness of the surrounding terrain.

The response of a structure to the variable action of wind can be separated into two components, a background component and a resonant component. The background component involves static deflection of the structure under the wind pressure. The resonant component, on the other hand, involves dynamic vibration of the structure in response to changes in wind pressure. In most structures, the resonant component is relatively small and structural response to wind forces is treated using static methods of analysis alone. However, for tall or otherwise flexible structures, the resonant component of wind should be calculated using dynamic methods of analysis. Such structures are not considered further here.

Static effects of wind load on buildings

Reference wind velocity

The reference wind velocity for a locality is defined as the mean wind velocity at 10 m above farmland averaged over a period of 10 minutes with a return period of 50 years. It is calculated using:

$$v_{ref} = c_{DIR} c_{TEM} c_{ALT} v_{ref,0} \tag{3.3}$$

where $v_{ref,0}$ is the basic reference wind velocity 10 m above sea level and c_{DIR}, c_{TEM} and c_{ALT} are factors relating to direction, seasonal variations in temporary structures and altitude respectively. Typical values of $v_{ref,0}$ for Europe are given in Fig. 3.10. Exact values for individual countries are still being drafted at time of printing.

The factors c_{DIR}, c_{TEM} and c_{ALT} will be specified for local conditions by individual countries and these too are still being drafted at time of printing. For each of these factors, a value of unity may be assumed unless otherwise specified for a particular region. The direction factor, c_{DIR}, allows for the orientation of the structure in relation to the direction of the prevailing wind. The seasonal variation factor, c_{TEM}, may be applied to structures of a temporary nature which are exposed to wind for only part of a given year. It reflects the fact that storm winds are less likely in the summer months in most European countries. (Temporary structures are subjected to a reduced risk of exposure

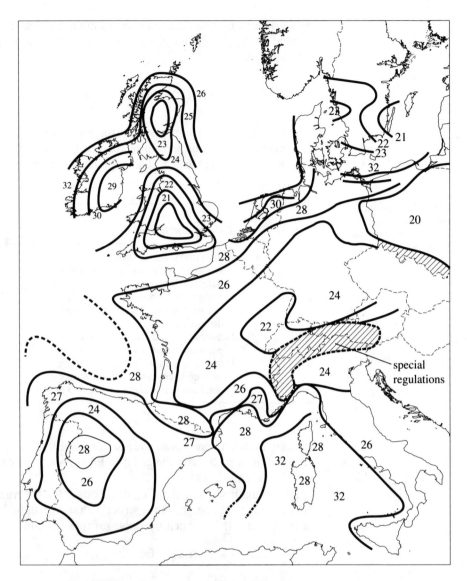

Fig. 3.10 Basic reference wind velocities, $V_{ref,0}$, for Europe

to strong winds simply by virtue of their reduced design life. This phenomenon can be allowed for by means of a separate adjustment to the wind reference velocity.) The altitude factor, c_{ALT}, allows for the altitude of the site on which the structure is located. Wind speeds tend to be greater in sites located at high altitudes.

Exposure coefficient

Wind velocity tends to decrease near ground level owing to frictional forces between the wind and the ground. If the terrain is rugged, the decrease in velocity can be quite substantial. The exposure coefficient takes account of the variation from the reference wind velocity due to the ground roughness around the structure, the local topography and the height of the structure above ground level. EC1 defines the exposure coefficient at height z metres, using the relationship:

$$c_e(z) = c_r^2(z)c_t^2(z)\left[1 + \frac{7k_r}{c_r(z)c_t(z)}\right]$$

(3.4)

where c_r and c_t are roughness and topography coefficients respectively and k_r is a terrain factor. The terrain factor is a function of the nature of the terrain and is given in Table 3.5. The topography coefficient, c_t, accounts for the

Table 3.5 Ground roughness categories and parameter values (from EC1, Part 2.3)

Category	Terrain description	k_r	z_0 (m)	z_{min} (m)
1	Rough open sea. Lakeshore with ≥ 5 km fetch upwind and smooth flat country without obstacles	0.17	0.01	2
2	Farmland with boundary hedges, occasional small farm structures, houses or trees	0.19	0.05	4
3	Suburban or industrial areas and permanent forests	0.22	0.3	8
4	Urban areas in which $\geq 15\%$ of the surface is covered with buildings and their average height exceeds 15 m	0.24	1	16

increase in mean wind speed over isolated hills and escarpments. Details for its calculation in such cases are given in EC1, Part 2.3. For all other situations, c_t may be taken as unity.

The roughness coefficient, $c_r(z)$, accounts for the variability of mean wind velocity due to the height of the structure above ground level and the roughness of the terrain. It is defined by the logarithmic relationship:

$$\left.\begin{array}{ll} c_r(z) = k_r \, \text{Ln}(z/z_0) & \text{for } z \geq z_{min} \\ c_r(z) = c_r(z_{min}) & \text{for } z < z_{min} \end{array}\right\}$$

(3.5)

where z_0 is the roughness length and z_{min} is the minimum height. Both z_0 and z_{min} are dependent on the ground roughness and are given in Table 3.5.

External wind pressure

The wind pressure acting on the external surface of a structure is a function of the reference wind pressure which is given by:

$$q_{ref} = \tfrac{1}{2}\rho v_{ref}^2 \quad [\text{N/m}^2]$$

(3.6)

where ρ = air density (kg/m^3)
v_{ref} = reference wind velocity (m/s) calculated using equation (3.3)

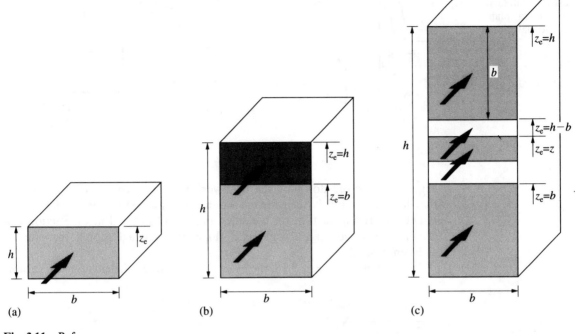

Fig. 3.11 Reference height, z_e, for rectangular buildings: (a) $h < b$; (b) $b < h < 2b$; (c) $h > 2b$

The density of air varies with temperature, elevation and the expected atmospheric pressure in the region during a storm. EC1 gives a recommended design value of $\rho = 1.25\,\text{kg/m}^3$. More accurate values may be specified by individual countries.

In order to determine the contact pressure on the outside of a structure or part of a structure, the reference pressure, q_{ref}, of the wind must be multiplied by an external pressure coefficient, c_{pe}, and an exposure coefficient. Thus, the external pressure is:

$$w_e = c_e(z_e)c_{pe}q_{ref} \qquad (3.7)$$

where $c_e(z_e)$ is the exposure coefficient evaluated at a reference height, z_e. Reference heights for the calculation of external pressure coefficients depend on the breadth to height ratio of the structure. For rectangular buildings whose breadth, b, is less than their height, h, as illustrated in Fig. 3.11(a), the reference height equals the actual height. When h exceeds b but is less than $2b$, the building is considered in the two parts illustrated in Fig. 3.11(b). When h exceeds $2b$, the building is considered in multiple parts. A lower part extends upwards from the ground a distance b. An upper part extends downwards from the top a distance b. The rest of the building can be divided into any number of parts, with the reference height in each case calculated as the distance from the ground to the top of the part.

The external pressure coefficient, c_{pe}, accounts for the variation in dynamic pressure on different zones of the structure due to its geometry, area and proximity to other structures. For instance, the wind acting on the structure in

Fig. 3.12 Wind flow
past a rectangular
building: (a) plan;
(b) end elevation

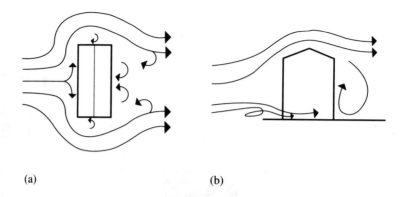

(a)   (b)

Fig. 3.12 is slowed down by the windward face and generates a pressure on
that face. The wind is then forced around the sides and over the top of the
structure, causing suction on the sides and on all leeward faces. Suction can
also be generated on the windward slope of a pitched roof if the pitch is
sufficiently small.

With reference to Fig. 3.13, the external pressure coefficients for the various

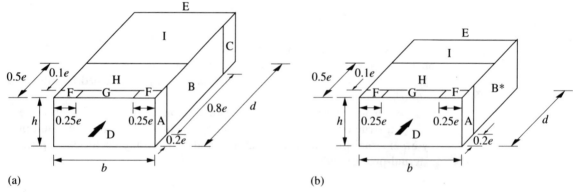

(a)   (b)

Fig. 3.13
External pressure
coefficient zones ($e =$
lesser of b and $2h$):
(a) $d > e$; (b) $d < e$

zones of the walls of a rectangular building are given in Table 3.6. Similar tables
are given in EC1 for other building shapes. The values in Table 3.6 are valid
for surface areas in excess of 10 m^2 only. Values for lesser surface areas are
given in the Eurocode. External pressure coefficients for the roof zones in a
flat-roofed building are given in Table 3.7. Other values are specified for areas
less than 10 m^2, or when parapets are present, or when the eaves are curved.
Pressure coefficients are considered positive when the pressure is acting on to
the surface of the structure and negative when the pressure is acting away from
that surface. Thus, the external pressure coefficient is positive when acting
inwards.

Internal wind pressure

Internal pressure arises due to openings, such as windows, doors and vents, in
the cladding. In general, if the windward panel has a greater proportion of

Table 3.6 External pressure coefficients for the walls of a rectangular building (from EC1, Part 2.3)

Zone (Fig. 3.13)	$d/h \leq 1$	$1 < d/h < 4$	$d/h \geq 4$
A	-1	-1	-1
B, B*	-0.8	-0.8	-0.8
C	-0.5	-0.5	-0.5
D	$+0.8$	$0.8 - 0.067(d/h - 1)$	$+0.6$
E	-0.3	-0.3	-0.3

Table 3.7 External pressure coefficients for a flat roof (from EC1, Part 2.3)

Zone (Fig. 3.13)	Coefficient
F	-1.8
G	-1.2
H	-0.7
I	± 0.2

openings than the leeward panel, then the interior of the structure is subjected to positive (outward) pressure as illustrated in Fig. 3.14(a). Conversely, if the leeward face has more openings, then the interior is subjected to a negative (inward) pressure as illustrated in Fig. 3.14(b). Like external pressure, internal

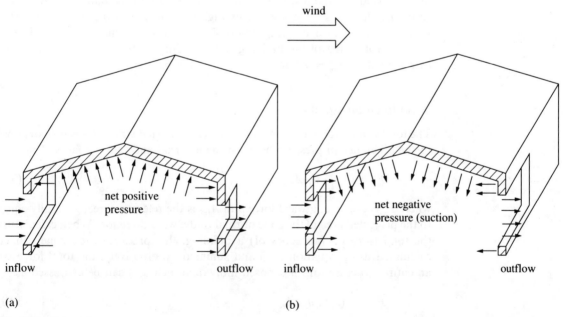

(a) (b)

Fig. 3.14 Internal pressures in structures

pressure is considered positive when acting on to the surface of the structure. Thus, internal pressure is positive when the pressure acts outwards.

Internal pressure on a building or panel is given by:

$$w_i = c_e(z_i) c_{pi} q_{ref} \tag{3.8}$$

83

Fig. 3.15 Internal pressure coefficients, c_{pi}, in buildings with openings (from EC1)

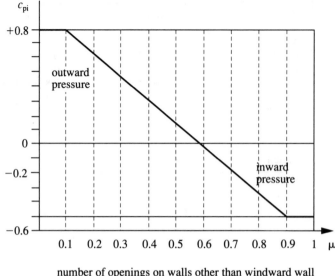

$$\mu = \frac{\text{number of openings on walls other than windward wall}}{\text{total number of openings}}$$

where z_i is the reference height for internal pressure equal to the mean height of the openings and c_{pi} is the internal pressure coefficient. The magnitude of c_{pi} depends on the distribution of openings around the building. The values recommended by EC1 are given in Fig. 3.15 for a building without internal partitions. In such a building, internal pressure is assumed to act uniformly over the total area of the building. For buildings with internal partitions the extreme values, $c_{pi} = 0.8$ and $c_{pi} = -0.5$, may be used.

Wind force on structures

The total wind force acting on individual zones of clad structures is proportional to the difference in pressure between the external and internal faces. That is:

$$F_w = (w_e - w_i)A_{ref} \tag{3.9}$$

where F_w is the total inward force and A_{ref} is the reference area, generally equal to the projected area of the zone normal to the wind direction. When calculating the total force on (all zones of) a building, the forces on each zone can be calculated using equation (3.9) and summed. Alternatively, the total force on an entire structure (or an exposed individual member) can be expressed as:

$$F_w = c_e(z_e)c_f q_{ref}A_{ref} \tag{3.10}$$

where c_f is a force coefficient. While, strictly speaking, the force coefficient is approximately equal to the algebraic sum of the external pressure coefficients on the windward and leeward faces, they are in fact slightly different owing to frictional effects on the side walls. EC1 provides tables of force coefficients for common forms of structures and sections used in structural frames.

Example 3.5 Wind loads

Problem The structure illustrated in Fig. 3.16 is to be located in the centre of Paris on a site surrounded by buildings of similar height. It is an apartment building

Fig. 3.16 Building of Example 3.5

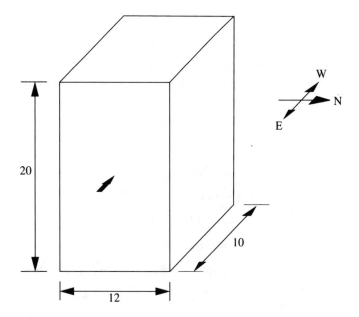

with internal partitions. Wind from the east and west is transmitted from the clad faces to the north and south masonry walls. Each external panel has opening windows equal in area to one tenth of the total wall area.

(a) Determine the total moment due to wind at the base of the north and south masonry walls.
(b) Calculate the maximum pressure on the east masonry wall.

Solution The reference pressure and exposure coefficient are first calculated.

Reference pressure

The basic reference wind velocity for Paris can be taken from the map of Fig. 3.10 and is 26 m/s. Assuming values of unity for c_{DIR}, c_{TEM} and c_{ALT}, the reference wind velocity is also (from equation (3.3)) 26 m/s. Hence the reference wind pressure is, from equation (3.6):

$$q_{ref} = \tfrac{1}{2}\rho v_{ref}^2$$
$$= \tfrac{1}{2}(1.25)(26)^2$$
$$= 423 \text{ N/m}^2$$

Exposure coefficient

As the height exceeds the breadth but is less than twice its value, the building is considered in two parts, as illustrated in Fig. 3.11(b). The reference heights for external pressure are thus:

$$z_e = h = 20 \text{ m}$$

and

$$z_e = b = 12 \text{ m}$$

As the building is located in an area of Roughness Category 4 (refer to Table 3.5), $k_r = 0.24$, $z_0 = 1$ m and $z_{min} = 16$ m. Equation (3.5) gives the roughness coefficients:

$$c_r(20) = k_r \text{ Ln}(20/z_0)$$
$$= 0.24 \text{ Ln}(20/1)$$
$$= 0.719$$

$$c_r(12) = c_r(z_{min})$$
$$= 0.24 \text{ Ln}(16/1)$$
$$= 0.666$$

Taking a topography coefficient of unity, equation (3.4) gives the exposure coefficients:

$$c_e(20) = c_r^2(20)c_t^2(20)\left[1 + \frac{7k_r}{c_r(20)c_t(20)}\right]$$
$$= (0.719)^2(1)^2\left[1 + \frac{7 \times 0.24}{0.719 \times 1}\right]$$
$$= 1.725$$

$$c_e(12) = c_r^2(12)c_t^2(12)\left[1 + \frac{7k_r}{c_r(12)c_t(12)}\right]$$
$$= (0.666)^2(1)^2\left[1 + \frac{7 \times 0.24}{0.666 \times 1}\right]$$
$$= 1.563$$

External pressure

It can be seen from Fig. 3.13 that only zones D and E are of interest in this example. The ratio d/h is $10/20 = 0.5$. Hence, from Table 3.6:

$$c_{pe}(\text{Zone D}) = +0.8$$

$$c_{pe}(\text{Zone E}) = -0.3$$

At the reference height of 20 m, the external pressure on zone D is (equation

(3.7)):

$$w_e = c_e(20)c_{pe}q_{ref}$$
$$= 1.725 \times 0.8 \times 0.423$$
$$= 0.584 \text{ kN/m}^2$$

The corresponding force on the upper part of zone D is $0.584(12 \times 8) = 56$ kN. At the reference height of 12 m, the external pressure on zone D is:

$$w_e = c_e(12)c_{pe}q_{ref}$$
$$= 1.563 \times 0.8 \times 0.423$$
$$= 0.529 \text{ kN/m}^2$$

and the corresponding force is $0.529(12 \times 12) = 76$ kN. The corresponding forces for zone E are -21 kN and -29 kN for the upper and lower parts respectively. These forces are illustrated in Fig. 3.17.

Fig. 3.17 Forces due to east wind

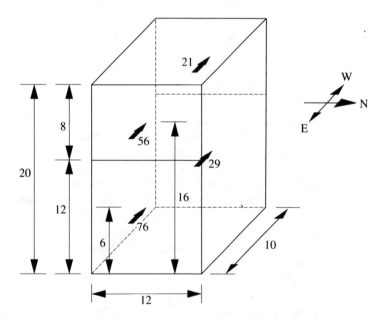

(a) Internal pressure within a structure is self equilibrating. Thus, while it can cause significant pressures on individual wall panels, it results in no net force on the structure overall. Accordingly, the overturning moment at the base of the north and south walls due to wind is unaffected by internal pressure and is given by:

$$\text{Moment} = (56 + 21)16 + (76 + 29)6$$
$$= 1862 \text{ kNm}$$

Of this, half will apply at the base of each of the two walls.

(b) To determine the total pressure on the east wall, it is necessary to calculate the internal as well as the external pressure. As there are internal partitions, the worst value for c_{pi} is assumed, that is, $c_{pi} = -0.5$. The maximum pressure will occur in the upper part of zone D. In this part of the building, the mean height of the windows will be assumed to equal the mean height of the part. Hence:

$$z_i = 16 \text{ m}$$

The exposure coefficient at this height is calculated as before and is 0.666. Thus:

$$w_i = c_e(z_i)c_{pi}q_{ref}$$
$$= 0.666(-0.5)(0.423)$$
$$= -0.141$$

The net pressure on the upper part of zone D is the difference between the external and the internal pressures, that is:

$$w_e - w_i = 0.584 - (-0.141)$$
$$= 0.725 \text{ kN/m}^2$$

3.7 Limit-state design

There is considerable random variation in the strength and stiffness of concrete structures and in the actions applied to them. Thus, a statistical analysis of the complete process would seem logical and indeed considerable progress has been made on research in this field. However, it is not at present practical to carry out a complete statistical analysis for the purposes of everyday design. What is current practice in most parts of the world is **load and resistance factor design** which is a deterministic design approach using factors to reflect the statistical variability in the parameters. In simple terms, load and resistance factor design consists of factoring up applied actions and factoring down the material resistances.

Limit-state theory

Limit-state theory is a philosophy of design under which structures are designed to fulfil a number of basic functions or conditions. A limit state is defined as a situation where the structure ceases to fulfil one of the specific functions or conditions for which it was originally designed.

For concrete structures, two main groups of limit state exist:

1. **The ultimate limit state (ULS)**
   This is reached when the structure, or part of the structure, collapses. The collapse may be due either to a loss of equilibrium or stability, or to failure by rupture of structural members.

2. The serviceability limit state (SLS)

This is reached when the structure, while remaining safe, becomes unfit for everyday use due to phenomena such as excessive deformation, cracking or vibration.

It is clear that the two groups of limit state have different degrees of importance. For example, the cracking of a concrete member affects its appearance and durability but is far less serious than collapse of the entire structure owing to the rupture of a critical member. Thus, in limit-state design, the factors applied for safety (ULS) are significantly greater than the factors applied for serviceability (SLS), reflecting the greater importance of preventing collapse of the structure.

Safety factors for loads and materials

A structural member will support its applied loads only if its ability to resist load is greater than the load applied. All loads and material strengths are variable and estimates of each must be made in design. Limit-state design uses the concept of 'characteristic' loads and strengths. For loads, these are the values which have a minimal probability (usually 2 per cent) of being exceeded even once in a given year (Fig. 3.18(a)). For material strengths, these are the values below which a minimum number (usually 5 per cent) of specimen strengths are expected to fall (Fig. 3.18(b)). To allow for the possibility that the

Fig. 3.18 Characteristic values of: (a) loads; (b) material strengths

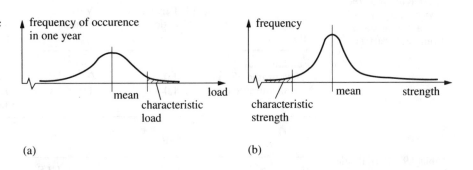

material resistance is less than its characteristic value and the load is greater than its characteristic value, partial safety factors are introduced to the characteristic strength and load before being applied as design values. Thus, the design load, F_d, is defined as:

$$F_d = \gamma_f F_k \tag{3.11}$$

where γ_f is the partial safety factor for load. For permanent loads, F_k is the characteristic value of the load, denoted G_k (throughout the Eurocodes, G is used to denote permanent load, Q is used to denote variable load and A is used to denote accidental load). In the case of variable loads, however, F_k is also known as the main representative value of the load. For accidental loads, the design value, denoted A_d, is normally specified directly.

The factor, γ_f, accounts for possible deviations of the load, inaccurate modelling of the load, uncertainty in the assessment of the load effects and uncertainty in the assessment of the limit state being considered.

Similarly, the design strength, X_d, is given by:

$$X_d = X_k/\gamma_m \tag{3.12}$$

where X_k is the characteristic stength and γ_m is the partial safety factor for material strength. The factor, γ_m, takes account of uncertainties in the stength and dimensions of the material along with any inaccuracy due to the methods of modelling member behaviour.

Values for γ_f and γ_m are recommended by the Eurocode for structural concrete, EC2, for two principal design situations, fundamental and accidental. Fundamental design situations correspond to normal conditions of use of the structure (known as persistent situations) and to transient situations, such as during construction or repair. Accidental design situations involve the consideration of accidental loads (vehicular impact, fire, explosion, etc.) which are generally greater than the fundamental permanent and variable.loads but shorter in duration. The recommended values for γ_f and γ_m for both design situations are given in Tables 3.8 and 3.9, respectively. In Table 3.8, γ_f becomes

Table 3.8 Partial safety factors (γ_f) for permanent loads and variable loads at the serviceability and ultimate limit states (from EC1, part 1)

Design situation	SLS		ULS	
	Permanent action (γ_g)	Variable action (γ_q)	Permanent action (γ_g)	Variable action (γ_q)
Fundamental				
favourable effect	1.0	0	1.0	0
unfavourable effect	1.0	1.0	1.35	1.5
Accidental	—	—	1.0	1.0

Table 3.9 Partial safety factors (γ_m) for the strength of concrete and steel reinforcement at the serviceability and ultimate limit states (from EC2)

Design situation	SLS		ULS	
	Concrete (γ_c)	Steel (γ_s)	Concrete (γ_c)	Steel (γ_s)
Fundamental	1.0	1.0	1.5	1.15
Accidental	—	—	1.3	1.0

γ_g or γ_q for permanent and variable actions, respectively. In Table 3.9, γ_m becomes γ_c or γ_s for concrete and steel, respectively.

In the case of safety factors for loads, the code also makes a distinction between the favourable and unfavourable effects of load for fundamental design situations. A load is favourable if its effect is to reduce the effect of other loads.

For instance, if permanent and imposed gravity loads act against the over-turning effect of wind load on a structure, their effects are considered favourable. However, only permanent loads can be applied favourably in design since the presence of variable loads cannot be depended upon. Hence, the safety factor for favourable variable loads is zero.

Representative values of variable loads

In the case of variable loads, portions of the characteristic values, known as **representative values**, are considered for various design situations. The representative values take account of the differing probabilities of a variable load reaching its full characteristic value in different circumstances. The characteristic value, Q_k, is known as the **main representative value**. Other representative values for combinations of loads are found by applying factors ψ_i (where $\psi_i < 1$) to the characteristic value, Q_k. The relevant design values used in EC2 are given by:

combination values: $\qquad Q_r = \psi_0 Q_k$

frequent values: $\qquad Q_r = \psi_1 Q_k$

quasi-permanent values: $\quad Q_r = \psi_2 Q_k$

where the terms 'combination', 'frequent' and 'quasi-permanent' have specific meanings as outlined below.

In a fundamental design situation, it is unlikely that all the variable loads (e.g. imposed gravity, wind and snow) will reach their characteristic values simultaneously. Therefore, in such situations, only the principal load (i.e. that which causes the greatest load effect) is applied at its main representative value, Q_k. The remaining variable loads are applied at their combination values, $\psi_0 Q_k$.

In design situations where an accidental load occurs in combination with other variable loads, it is unlikely that the other variable loads will be at their full combination values (accidental loading is very unlikely). Thus, in an accidental design situation, the accidental load is applied at its design (accidental load) value, A_d. Of the other variable loads, the principal one is applied at its frequent value, $\psi_1 Q_k$, and the remaining variable loads are applied at their quasi-permanent values, $\psi_2 Q_k$.

Table 3.10 gives the recommended values of the factors, ψ_0, ψ_1 and ψ_2, for variable gravity and wind loads to be applied at both the serviceability and the ultimate limit states.

Table 3.10 Representative load factors, ψ_0, ψ_1 and ψ_2 (from EC1) (see Table 3.3 for category description)

Action	ψ_0	ψ_1	ψ_2
Imposed loads			
category A, B	0.7	0.5	0.3
category C	0.7	0.7	0.6
category D	1.0	0.9	0.8
Wind	0.6*	0.5*	0*
Snow	0.6*	0.2*	0*

* Values may have to be modified for specific locations.

Example 3.6   Limit-state design

Problem   Determine the design ULS value of the internal bending moment for a beam in an apartment building given the following results from a computer analysis:

Internal moment (kN m)

Permanent gravity	200*
Imposed	150*
Wind	50*
Design accidental	100

(* characteristic value)

Solution   The design moment due to permanent gravity loading is $200\,\gamma_g$. For the fundamental design situation, imposed is the principal variable loading. Hence, the design moment due to variable loading is $\gamma_q(150 + 50\psi_0)$ and the total design moment is:

$$200\gamma_g + \gamma_q(150 + 50\psi_0)$$

$$= 200(1.35) + 1.5(150 + 50 \times 0.7)$$

$$= 548 \text{ kN m}$$

For the accidental design situation, the design moment due to variable loading is:

$$100 + \gamma_q(150\psi_1 + 50\psi_2)$$

Hence, the total design moment is:

$$200\gamma_g + 100 + \gamma_q(150\psi_1 + 50\psi_2)$$

$$= 200(1.0) + 100 + 1.0(150 \times 0.5 + 50 \times 0.3)$$

$$= 390 \text{ kN m}$$

Design procedure

For the design of reinforced concrete structures, it is normal practice to design first for the ultimate limit state as this tends to govern the design. Then, conditions of serviceability, which are usually less critical, are checked. For fully prestressed concrete structures, on the other hand, the serviceability limit state of stress usually governs the design and the normal practice is to design for this serviceability limit state and subsequently to check the other limit states.

The critical (maximum and minimum) effects of the applied loads on each structural member are established by a consideration of all possible realistic combinations of individual factored design loads. Taking the design values for γ_g and γ_q from Table 3.8, the design values of loads in combination at the ultimate limit state have been summarized in Table 3.11.

For design at the serviceability state, three combinations are specified for consideration by EC2, namely 'rare', 'frequent' and 'quasi-permanent'. The design values of loads for each combination are given in Table 3.12. The rare

Table 3.11 Design values for actions for use in combination with other actions at ultimate limit states

Design situation	Permanent actions	Accidental actions	Variable actions	
			Principal action	All other actions
Fundamental				
favourable	$1.0G_k$	–	0	0
unfavourable	$1.35G_k$	–	$1.5Q_k$	$1.5\psi_0 Q_k$
Accidental	$1.0G_k$	A_d	$1.0\psi_1 Q_k$	$1.0\psi_2 Q_k$

Table 3.12 Design values for actions for use in the three combinations at serviceability limit states

Combination	Permanent actions	Variable actions	
		Principal action	All other actions
Rare	$1.0G_k$	$1.0Q_k$	$1.0\psi_0 Q_k$
Frequent	$1.0G_k$	$1.0\psi_1 Q_k$	$1.0\psi_2 Q_k$
Quasi-permanent	$1.0G_k$	$1.0\psi_2 Q_k$	$1.0\psi_2 Q_k$

combination would be used in the calculation of maximum short-term deflections while the quasi-permanent combination would be used to determine the long-term deflections.

Example 3.7 Continuous beam

Problem The reinforced concrete beam illustrated in Fig. 3.19 is subject to its self-weight, a permanent gravity load of characteristic value 25 kN/m and an imposed

Fig. 3.19 Continuous beam

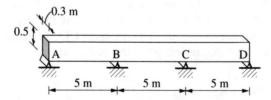

load with characteristic value 20 kN/m. Use a computer analysis program to determine the maximum and minimum ULS bending moments.

Solution Permanent load acting on continuous beams and slabs can have either a favourable or an unfavourable effect on the bending moments and shear forces within the member. Therefore, it is necessary to consider alternative configurations of applied load to determine the critical bending moments in each span and at the supports.

Critical moments between supports

To find the maximum bending moment in span AB, say, we put the maximum load (i.e. unfavourable permanent plus variable load) in spans AB and CD and

Fig. 3.20 Design loads for critical sag moments: (a) maximum moment in AB; (b) deflected shape for loading of (a); (c) maximum moment in BC

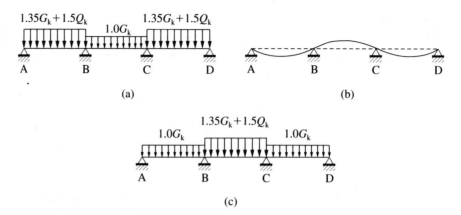

(a)

(b)

(c)

the minimum load (favourable permanent load only) in span BC, as shown in Fig. 3.20(a). It can be seen from the corresponding deflected shape (Fig. 3.20(b)) that the load in the centre span tends to reduce the curvature in span AB. As the moment is proportional to the curvature, the load in the centre span has the effect of reducing the bending moment in span AB. This is why minimum load is applied in the centre span. This configuration of load also gives the maximum moment in span CD and the minimum moment in span BC. In a similar manner it can be seen that the load configuration illustrated in Fig. 3.20(c) gives the maximum interior moment in the centre span and the minimum interior moments in the two outer spans. In this case the favourable design value of the permanent load is applied on the two outer spans.

Critical support moments

To get the maximum hog moment at support B (i.e. the minimum value of moment if hog is taken as negative), we put the maximum load in spans AB and BC and the minimum load in span CD, as illustrated in Fig. 3.21(a). A similar load configuration gives the minimum moment at support C.

Fig. 3.21 (a) Design load for maximum hogging moment in AB; (b) deflected shape for loading of (a)

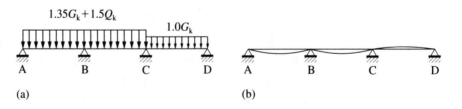

(a)

(b)

Thus, the total of four load cases must be analysed for this example. The self-weight is given by the product of concrete density and cross-sectional area. Thus, the total characteristic permanent load is given by:

$$G_k = 25 \times 0.3 \times 0.5 + 25$$

$$= 28.75 \text{ kN/m}$$

Fig. 3.22 Design loads in kN/m

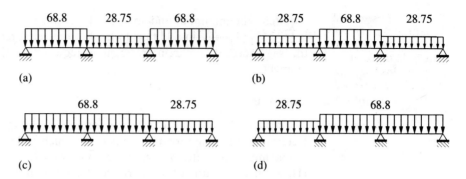

Design loads for each load case are given in Fig. 3.22. The beam is next analysed for these load cases to determine the bending moments. Alternative methods of analysis are discussed in Chapter 4 but typically this beam might be analysed using a program based on the stiffness method. The results of such an analysis are given in Fig. 3.23. Any of these four distributions of bending moment could

Fig. 3.23 Bending moment diagrams for load configurations of Fig. 3.22 in kNm

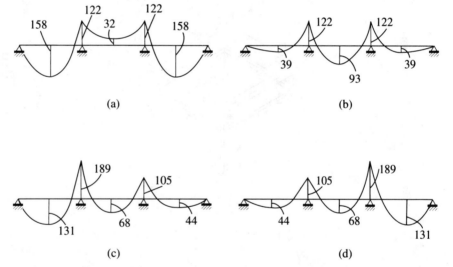

occur in the beam depending on the distribution of load and the beam must be capable of resisting the extremes of all four. The four diagrams of Fig. 3.23 can be superimposed to yield the 'bending moment envelope' of Fig. 3.24. This envelope gives the maximum and minimum applied bending moments at each

Fig. 3.24 Bending moment envelope (kNm)

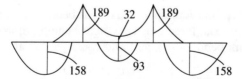

section of the beam. In span BC, for instance, the maximum (sag) mid-span bending moment is 93 kNm for which substantial bottom reinforcement is required. The minimum mid-span bending moment (i.e. greatest hog moment) is -32 kNm. Thus, in addition to the bottom reinforcement required to resist the sag moment, top reinforcement is required at this point to resist hog moment.

Example 3.8   Columns

Problem   The structural frame in Fig. 3.25 is to form the skeleton of a residential building. Determine the design load combinations which yield the worst combinations of maximum ultimate axial force and maximum bending moment in column HJ, given the characteristic permanent (G_k), imposed (Q_{ik}) and wind (Q_{wk}) loading shown.

Fig. 3.25 Skeletal frame: (a) geometry; (b) central frame

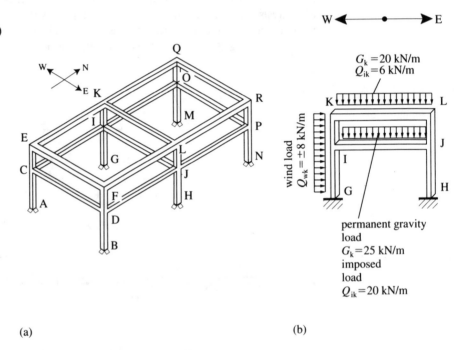

(a)   (b)

In reinforced concrete, the applied axial force and bending moments combine to cause the failure of columns. Any increase in the bending moment is always detrimental. However, the axial force component can sometimes be beneficial. For this reason, we must determine the combinations of applied load which yield the following conditions for column HJ:

(a) maximum axial force plus coexistent bending moment
(b) minimum axial force plus coexistent bending moment
(c) maximum bending moment and coexistent axial force

Maximum axial force

To determine the maximum axial (compressive) force in column HJ and the coexistent bending moment, the permanent and variable gravity loads on members IJ and KL are applied at their maximum unfavourable design values. The wind load on members GI and IK tends to cause the building to overturn, which results in tension in the windward columns and compression in the leeward columns. Thus, westerly wind load has the effect of increasing the axial compression in columns HJ and JL and is applied at its unfavourable design value. The design load combination for maximum axial compression in column HJ is, therefore, unfavourable permanent and variable gravity load plus unfavourable westerly wind load. If the variable gravity load is considered to have the greatest effect on the magnitude of the compressive force then, from Tables 3.10 and 3.11, the factored combination is:

$$1.35G_k + 1.5Q_{ik} + 1.5\psi_0 Q_{wk}$$
$$= 1.35G_k + 1.5Q_{ik} + 1.5(0.6)Q_{wk}$$

If, on the other hand, the wind load is considered to be the principal variable action, the factored combination is:

$$1.35G_k + 1.5\psi_0 Q_{ik} + 1.5Q_{wk}$$
$$= 1.35G_k + 1.5(0.7)Q_{ik} + 1.5Q_{wk}$$

The corresponding design loads are shown in Figs 3.26(a) and (b). The design maximum compressive force and coexistent moment in column HJ are taken from the results of the more onerous load case.

Fig. 3.26 Design loads

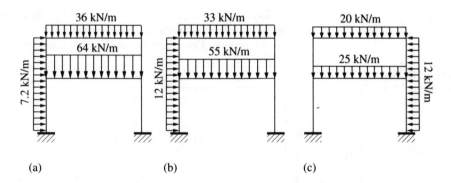

(a) (b) (c)

Minimum axial force

To determine the minimum axial force in column HJ and the coexistent bending moment, the permanent and variable gravity loads on members IJ and KL are applied at their favourable design values. Since easterly wind load will cause tension in column HJ it is applied at its unfavourable value. From Table 3.11, the combination of load which gives rise to the minimum compressive force in column HJ is $1.0G_k + 1.5Q_{wk}$ (design values for favourable variable action are zero). Figure 3.26(c) gives the corresponding design loads.

Fig. 3.27 Bending
moment diagrams: (a)
for gravity loads; (b) for
wind loads

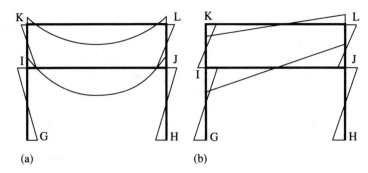

(a)   (b)

Maximum bending moment

Figure 3.27(a) illustrates the general shape of the bending moment diagram for
the frame resulting from the gravity loads. Similarly, Fig. 3.27(b) illustrates the
bending moment diagram for the frame resulting from the westerly wind load
only. It can be seen from the figure that, in column HJ, the bending moments
due to the gravity loads and the westerly wind load are additive. Therefore, the
design load combination which yields the maximum bending moment in column
HJ is unfavourable permanent and variable gravity load plus unfavourable wind
load, that is the same combination which is used to find the maximum axial
force in column HJ.

Problems

Section 3.3

3.1   The one-way spanning slab, illustrated in plan in Fig. 3.28, is continuous over the central
support, EB. The factored permanent and variable loading intensities are 9 kN/m² and
5 kN/m² respectively. Determine the maximum loadings on the supporting beams.

Fig. 3.28 Slab of
Problem 3.1

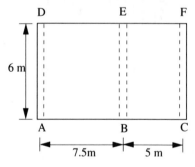

3.2   The two-way spanning continuous slab illustrated in Fig. 3.29 is subjected to factored
permanent and variable loads of 10 kN/m² and 6 kN/m² respectively. Determine the
maximum loadings on beam BE.

Sections 3.4 and 3.5

3.3   For preliminary design purposes, estimate the gravity loadings for a 200 mm thick
reinforced concrete floor in an office with non-permanent partitions.

Fig. 3.29 Slab of Problem 3.2

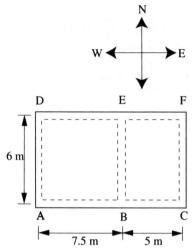

3.4 A hotel is to be constructed at a location near Birmingham, 150 m above sea level. The roof is to be used as a fire escape route. It is made of a reinforced concrete slab of thickness 175 mm which is covered in a lightweight sealant. The roof is sloped at 1:10 to facilitate the runoff of water. Calculate the permanent and variable gravity loading intensities.

3.5 For the hotel of Problem 3.4, it becomes apparent that designing the complete roof as an escape route results in an unduly large variable gravity loading. Accordingly, the escape route is limited to one portion of the roof, with access to the remaining area closed off by a masonry wall. Determine the gravity loadings for the latter part of the roof.

Section 3.6

3.6 The building illustrated in Fig. 3.30 is located in a suburban area outside London. In order to check the capacity of the structure to resist applied horizontal forces, check the wind forces in the N–S direction.

Fig. 3.30 Building of Problem 3.6

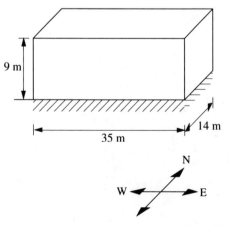

3.7 For the building of Problem 3.6, the interior is open-plan and windows and doors are located approximately uniformly around the perimeter. In order to check the capacity of the masonry wall panels, determine the maximum wind pressure on the South wall.

Section 3.7

3.8 For the floor of Problem 3.3, determine the maximum design SLS and ULS loading intensities for the fundamental design situation.

3.9 For the roof of Problem 3.4, determine the minimum and maximum ULS loading intensities for the fundamental design situation.

Part II

Preliminary Analysis and Design

Introduction

The conceptual design stage discussed in Part I may end with only one feasible solution which satisfies the design constraints being selected by the design team. More often, however, a handful of alternative schemes will be proposed which must be considered in greater detail before a final choice is made on the most appropriate solution. This second stage of the design process is known as preliminary analysis and design and it is carried out by the structural engineer in order to acquire more reliable estimates of structural actions and section requirements. The techniques involved at this stage are based on simple calculations which can be executed on a hand calculator or a slip of paper. The results provide the design team with more reliable information on the projected cost and structural efficiency of each proposed scheme. The results may also bring to light further design constraints which have to be checked before the scheme is provisionally accepted or finally rejected. Some of this part of the book deals with this stage in the design process. However, there are also two chapters on analysis as it applies to concrete structures. These are necessary for an understanding of the assumptions made in approximate and detailed analysis. In addition, they provide a sound basis for an understanding of a number of concepts used in detailed design (such as plastic moment redistribution).

In order to appreciate the assumptions commonly made in an approximate preliminary analysis, it is necessary to review the fundamentals of structural analysis. Thus, Chapter 4 deals with the analysis of determinate and indeterminate structures to find moment, shear and axial forces due to applied loads. In addition to linear elastic analysis, concepts of elastic–plastic and fully plastic analysis are dealt with as they are of particular relevance to the detailed design of ductile concrete structures such as beams and slabs. In Chapter 5, the applications of analysis fundamentals to concrete structures are considered and assumptions commonly made in both preliminary and detailed analysis are outlined. Of relevance to detailed design is the analysis of slabs which is also dealt with in this chapter. Shear wall systems are also a special application of an analysis technique to concrete buildings.

Finally, in Chapter 6, the rules of thumb used to determine preliminary sizes for concrete members are described and formulae are presented for the estimation of quantities of reinforcement.

4

Fundamentals of Structural Analysis

4.1 Introduction

Structural analysis, in the context of this book, is the process of calculating the effects of loads on a structure. The principal load effects of interest are the internal shear and axial forces, bending moments and torsion. In some cases, structures are also analysed to determine deflections. Other load effects are directly related to the principal effects. For example, crack widths in concrete members are related to internal bending moment and axial force.

Estimates of the load effects are used in the preliminary design stage to determine approximate values for the quantities of reinforcement required to resist them. For the same reason, accurate values for load effects are required at the detailed design stage to allow accurate calculation of reinforcement requirements. To perform an approximate analysis, the properties of the members, such as the second moment of area, must be known. These are calculated using preliminary member sizes which, in turn, are found from simple formulae and rules of thumb as described in Chapter 6. The preliminary sizes are also necessary for the calculation of structural self-weight. This chapter describes some of the fundamental concepts of analysis which are necessary for an understanding of concrete design. The principles underlying the more important methods of analysis are also outlined.

For the methods of analysis presented, a distinction is made between statically **determinate** and **indeterminate** structures. Statically determinate structures can be analysed solely by considerations of equilibrium. Statically indeterminate structures, on the other hand, require further information for their analysis and the solution process is more complex than for determinate structures. A distinction is also made between **linear elastic** and **non-linear** methods. Linear elastic methods of analysis are based on the assumption that deformation is proportional to the applied load (linear) and that deformation will dissipate if the load is removed (elastic). Linear elastic methods are very important for studying the performance of structures under relatively small loads – in practice, serviceability limit-state loads. Non-linear methods of analysis, on the other hand, consider the performance of structures in which some of the members have yielded under the external loads. Therefore, these methods are particularly important for studying the behaviour of structures under their ultimate limit-state loads.

Sign conventions and free-body diagrams

The distribution of internal shear force and bending moment in a structure, the results of analysis, are usually represented by diagrams superimposed on the structure. The sign convention adopted for internal moments throughout this book is that **moment is shown on the tension side** of members, that is, the side where cracks would form (see Fig. 4.1). For horizontal members, sag moment is taken as positive. For vertical members, a dotted line is shown on the right

Fig. 4.1 Sign convention for bending moment: (a) loading and deflected shape; (b) bending moment diagram; (c) internal moments

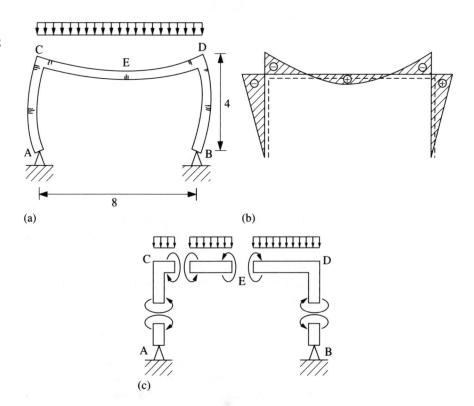

side of each member. If the moment causes tension on this side it is positive, otherwise it is negative (Fig. 4.1(b)). The internal moment at any point in a member can be represented by a pair of equal and opposite arrows. These **arrows should always emanate from the tension face** as illustrated in Fig. 4.1(c), that is, from the side on which the moment is drawn.

The sign convention for shear is illustrated for the same frame in Fig. 4.2. Shear force diagrams are drawn by following the directions of the arrows which represent the reactions and applied loads. The vertical reactions at A and B in this example are clearly both acting upwards. They are transmitted through compression in AC and BD to give vertically upward forces at each end of CD.

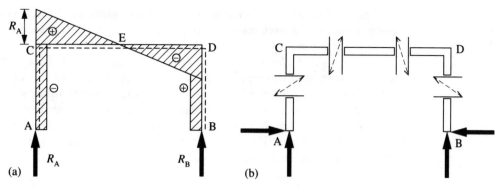

Fig. 4.2 Sign convention for shear forces: (a) shear force diagram; (b) reactions and internal shear forces

Starting at C (one should always start at the left-hand end or bottom) and drawing a line upwards a distance proportional to the magnitude of this vertical force gives the start of the diagram (Fig. 4.2(a)). From this elevated point, the diagram descends linearly in the direction of the applied uniformly distributed load by a total amount equal to the total load applied to this member, that is, by $8q$. At point D the upward reaction completes the shear force diagram on CD. From the deflected shape (Fig. 4.1(a)), it can be seen that the horizontal reactions at A and B are inward. Thus, the shear force in AC is drawn by moving to the right at A by a distance proportional to the horizontal reaction. As no load is applied to AC, the shear remains constant throughout the member length. Similarly, the shear force diagram for BD is drawn. Throughout this book, shear is positive if, when drawn in this way, it is on the **opposite** side to the dotted line illustrated in Fig. 4.2(a). As for moments, a shear force can be represented by a pair of equal and opposite arrows at any point. Negative shear in horizontal members is represented by a pair of arrows which, if joined as illustrated in Fig. 4.2(b), would form the shape of the letter 'N' (see right end of CD). Arrows representing negative shear in vertical members can be joined to form the letter 'Z' (see member AC). **Positive shear is represented by arrows which form a backward 'N' or a backward 'Z'.**

The sign convention for axial force is **positive tension** and **negative compression**. For the example of Fig. 4.1(a), all members are in compression. Arrows representing moment, shear and axial force are illustrated together in Fig. 4.3(a).

Fig. 4.3 Break-up of frame into free body diagrams: (a) complete frame showing internal forces and moments at break points; (b) typical free body diagrams which must satisfy equilibrium

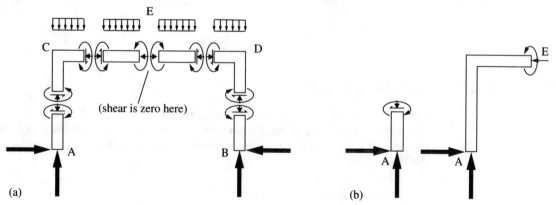

(shear is zero here)

The **pairs of arrows used to represent positive (tensile) axial force at a point should point towards each other**. The opposite rule applies to arrows representing negative (compressive) forces.

Moment, shear and axial force together represent all of the ways in which load is transmitted through two-dimensional frames such as that of Fig. 4.3. Hence, each of the segments of the frame of Fig. 4.3 constitutes a free-body diagram and must independently satisfy equilibrium. Regardless of where structures are 'broken' to form free-body diagrams, this principle stands. Thus, for each of the segments illustrated in Fig. 4.3(b), equilibrium of moments and forces must be satisfied.

Example 4.1 Frame example

Problem The bending moment diagram for a three-bay frame is illustrated in Fig. 4.4(a). From the bending moment diagram, calculate (a) the moment M_1 in the 6 m high column BF at B, and (b) the distribution of shear force in that column given that no horizontal load is applied to the member.

Fig. 4.4 Frame example: (a) bending moment diagram; (b) free body diagram; (c) moment equilibrium at B

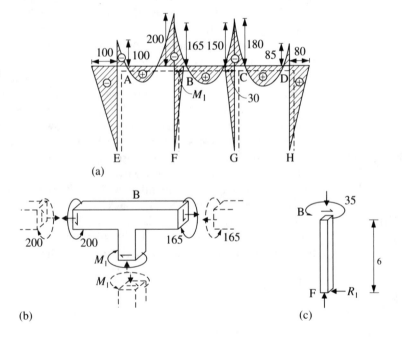

Solution (a) Calculation of M_1

The moments adjacent to joint B are represented by pairs of arrows in Fig. 4.4(b), remembering that the arrows must emanate from the side on which the moment is drawn on the bending moment diagram. Moment equilibrium at the joint gives:

$$M_1 = 200 - 165 = 35 \text{ kNm}$$

106

(b) Shear force distribution in BF

The free-body diagram for column BF is illustrated in Fig. 4.4(c). Taking moments about point B, we have:

$$R_1 \times 6 = 35$$

$$\Rightarrow \quad R_1 = 5.83 \text{ kN}$$

Hence, the distribution of shear force in column BF is uniform over its length and has a magnitude of 5.83 kN. As the arrow representing shear in Fig. 4.4(c) forms the lower part of a backward 'Z', shear in the member is positive.

4.2 Finding moment and shear in determinate linear elastic structures

In the majority of *in situ* reinforced concrete structures, the structural members, such as beams, columns and slabs, are cast as one, to give a continuous structure. Thus, rigid connnections between members are a principal characteristic of *in situ* reinforced concrete structures. To provide robustness in precast concrete structures, it is common practice to connect separate precast units together using rigid joints. One method of providing a rigid connection is the insertion of an *in situ* concrete strip between the ends of each unit, as illustrated in Fig. 4.5.

Fig. 4.5 *In situ* connection of precast members

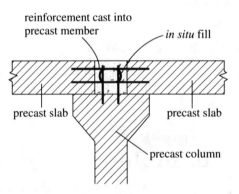

As a consequence of the methods of construction, there are few examples of concrete structures which are statically determinate. Possibly the only common examples of determinate concrete structures are the simply supported and cantilevered beam and slab. An example of a simply supported beam is illustrated in Fig. 4.6(a). The idealized model of this beam, which is used in its analysis, is illustrated in Fig. 4.6(b). It is common in such an idealized model to show only one roller support since, if both were on rollers, the structure would be unstable.

Fig. 4.6 Simply
supported beam:
(a) actual beam;
(b) idealized model

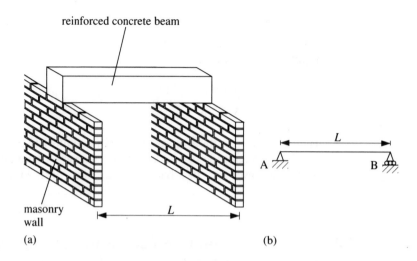

(a)

(b)

Example 4.2 Simply supported beam

Problem If the beam of Fig. 4.6 has an ultimate factored uniformly distributed load, ω, of 20 kN/m (included self-weight) and a span length, L, of 10 m, determine the distribution of internal bending moment and shear force in the member.

Solution By equilibrium, the sum of the vertical reaction R_1 at support A and the vertical reaction R_2 at support B must equal the total applied load, that is:

$$R_1 + R_2 = \omega L \tag{4.1}$$

By symmetry:

$$R_1 = R_2 = \omega L/2 \tag{4.2}$$

Therefore:

$$R_1 = R_2 = 20 \times 10/2 = 100 \text{ kN} \tag{4.3}$$

The free-body diagram for a length x of the beam is illustrated in Fig. 4.7(a). The internal bending moment, M_x $(0 < x \leq L)$, at a distance x from the left

Fig. 4.7 Uniformly
loaded, simply
supported beam: (a) free
body diagram;
(b) bending moment
diagram; (c) shear force
diagram

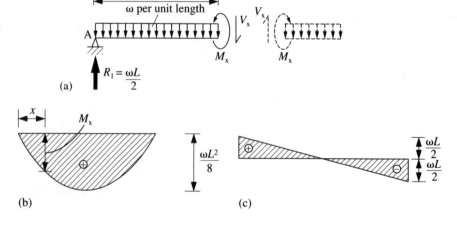

support is found by summing external bending moments about that point. Thus:

$$R_1(x) - \omega(x)(x/2) - M_x = 0$$

$$\Rightarrow \qquad M_x = R_1(x) - \omega x^2/2 \qquad (4.4)$$

$$\Rightarrow \qquad M_x = 100x - 10x^2 \qquad (4.5)$$

Using equation (4.5), the bending moment diagram for the beam can be drawn and is illustrated in Fig. 4.7(b). The maximum bending moment occurs at the centre ($x = 5$ m) and is equal to 250 kNm. It can be shown by substitution of equation (4.2) into equation (4.4) with $x = L/2$ that the maximum bending moment is given by:

$$M_{max} = \omega L^2/8 \qquad (4.6)$$

The shear force diagram for this beam is drawn by moving upwards from A by an amount proportional to R_1 ($=\omega L/2$), and then moving linearly downwards across the beam by a total vertical distance proportional to the total load, that is ωL. Finally, the diagram is completed at B by moving back up by a distance proportional to R_2 ($=\omega L/2$). The full shear force diagram is illustrated in Fig. 4.7(c).

Example 4.3   Balcony slab

Problem   Determine the distributions of bending moment and shear force in the one-way spanning slab of Fig. 4.8(a) for an applied uniformly distributed loading of ω kN/m².

Solution   If we take a 1 m wide strip through the slab, it will act like a continuous beam with an overhang. The idealized model for this strip is illustrated in Fig. 4.8(b) where the supports at A, B and C represent the support provided by the beams to the slab. Although rotation at supports A, B, and C is restrained by torsion in the beams, the torsional effects are small and are commonly ignored for the idealized model.

Fig. 4.8   Balcony slab: (a) actual slab; (b) idealized model (1 m wide strip)

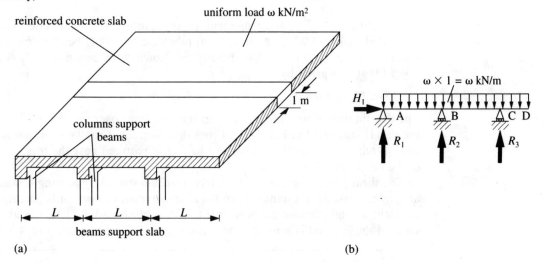

(a)   (b)

There are four unknown reactions (H_1, R_1, R_2 and R_3) from which it follows that the slab is indeterminate, with the degree of indeterminacy equal to one (three equations of equilibrium and four reactions). Thus, we are unable to determine the distribution of bending moment throughout the member by simple statics. However, the overhang portion of the slab acts as a cantilever and so this part is determinate. The free-body diagram for this portion of the member is illustrated in Fig. 4.9(a). By summing moments about a section a distance x from the free end of the cantilever, we have:

$$M_x + \omega x(x/2) = 0$$
$$\Rightarrow \quad M_x = -(\omega x^2/2) \tag{4.7}$$

Fig. 4.9 Bending moment and shear force diagrams: (a) free body diagram for cantilever; (b) bending moment diagram for one metre strip; (c) bending moment diagram for whole cantilever; (d) shear force diagram for one metre strip

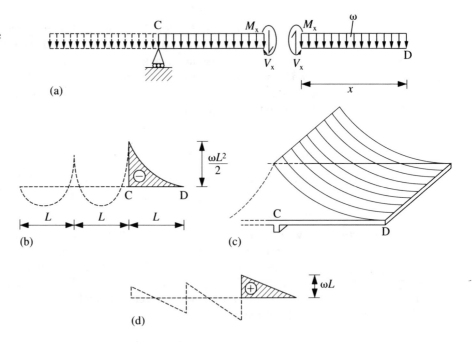

The distribution of bending moment in the cantilever is parabolic and is illustrated in Fig. 4.9(b). The maximum (absolute value of) bending moment occurs at support C and it can readily be shown from equation (4.7) (with $x = L$) that it is given by:

$$M_{max} = \omega L^2/2 \tag{4.8}$$

The distribution of bending moment in the cantilever over the entire width of the slab is illustrated in Fig. 4.9(c). From this diagram, it can be seen that steel tension reinforcement, aligned parallel to CD, is required near the top surface of the cantilever.

The shear force diagram for the 1 metre strip is started by moving upwards at C by an unknown distance. Then the diagram moves downwards linearly to the right a total distance proportional to the applied load, ωL. Hence, as the shear at the free end is zero, the shear just right of C must be ωL (Fig. 4.9(d)).

4.3 Finding internal bending moment in indeterminate linear elastic structures

There are many methods of linear elastic analysis of indeterminate structures that can be used to determine the internal bending moments. Two of these methods, the flexibility/force method and the stiffness/displacement method, reduce the problem to one of matrix algebra which can readily be solved by computer. Of these, the flexibility/force method consists of introducing 'releases' in the structure which reduce it to a familiar one. However, it is not always obvious (especially to a computer) where the releases should be introduced. For this reason, programmers tend to favour the stiffness/displacement method which can be readily applied to a greater range of structural geometries. Because of the widespread use of the stiffness method in computer programs, it is described briefly below.

Of the many hand methods for finding the internal bending moment in structures, moment distribution is by far the most popular method. An understanding of moment distribution also serves to better one's understanding of structural behaviour in general. For these reasons it too is described below.

Both the stiffness method and moment distribution are applicable to skeletal structures, that is structures in two or three dimensions made up of one-dimensional members such as beams or columns (Fig. 4.10(a)). Other methods are more suitable for the analysis of structures which incorporate two- or three-dimensional continuous members, such as slabs or walls (Fig. 4.10(b)). Of these, the computer-based finite element method (FEM) is the most popular and is becoming quite widely used in design offices for specialist applications. The finite element method may be viewed as an extension of the stiffness/displacement method to two and three dimensions. It can be used, among other things, to find the internal bending moments in slabs and walls.

Fig. 4.10 Skeletal and non-skeletal structures: (a) three-dimensional skeletal structure; (b) three-dimensional structure with one-dimensional members and a two-dimensional continuous member (slab)

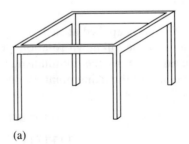
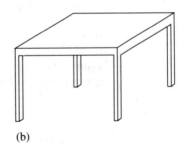

(a)   (b)

Stiffness method

The stiffness or displacement method of analysis is an extremely powerful method for the analysis of indeterminate frames to find internal bending moments. It is based on the use of a number of simple formulae and the application of the principle of superposition. The procedure is briefly explained here through the example illustrated in Fig. 4.11. Interested readers are referred

Fig. 4.11 Two-span beam

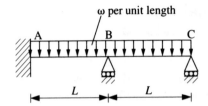

ω per unit length

to the book by Bhatt and Nelson (1990) for a more detailed exposition of the subject.

Example 4.4 Continuous beam

Problem For the beam of Fig. 4.11, use the stiffness method of analysis to find the internal bending moment at B and plot the bending moment diagram for the span BC.

Solution Fundamental to the stiffness method is a knowledge of the reactions and the distributions of the internal bending moment in simple members such as those of Appendix B. In addition, a knowledge of member stiffnesses (force/unit deflection or moment/unit rotation) is required (see Appendix A). Based on this knowledge, individual structural members are isolated by imposing **external** restraining moments to give a series of independent members. The structure is then equivalent to: a series of isolated structural members subjected to the applied loading (Fig. 4.12(b)) *plus* the effect of 'unfixing' the restraint(s) (Fig. 4.12(c)). For this equivalent structure, the moment at each fixing point must equal zero; that is, the sum of

(a) the moment reaction at the fixing points due to applied load (M_1 in Fig. 4.12(b)), and
(b) the external moment required to distort the structure into its final shape (M_2 in Fig. 4.12(c))

must equal zero. This principle is applied at each fixing point to establish a set of simultaneous equations in the unknown rotations. For this example, there is only one such equation and so the simultaneous solution of equations becomes trivial. The reaction at the fixing point, B, due to the applied load is,

Fig. 4.12 (a) Original structure; (b) fixed structure with applied loading; (c) imposed rotation to unfix structure. Basis of stiffness method is that (a) = (b) + (c)

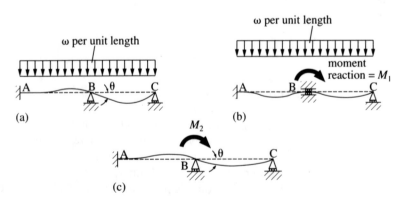

from Appendix B (Nos 6 and 8):

$$M_1 = \frac{\omega L^2}{12} - \frac{\omega L^2}{8} = -\frac{\omega L^2}{24}$$

It is important to remember here that the moment is an **external** reaction. Internal moments are represented by pairs of equal and opposite arrows and are positive if sagging. An external moment, on the other hand, is represented by a single arrow which here is taken as positive if it is clockwise.

The external moment required to distort the structure into its final shape is, from Appendix A (Nos 1 and 2):

$$M_2 = \frac{4EI\theta}{L} + \frac{3EI\theta}{L} = \frac{7EI\theta}{L}$$

where E is the modulus of elasticity and I is the second moment of area. The combined term, EI, is commonly known as the flexural rigidity of the member. As there is no external moment reaction at B:

$$M_1 + M_2 = 0$$

$$\Rightarrow \qquad -\frac{\omega L^2}{24} + \frac{7EI\theta}{L} = 0 \tag{4.9}$$

$$\Rightarrow \qquad \theta = \frac{\omega L^3}{168EI} \tag{4.10}$$

Having found the rotation(s) at the fixing point(s), the moment at B (or at any other point) is found by superimposing the effects of the rotation(s), shown in Fig. 4.13(b), on the moment in the fixed beam shown in Fig. 4.13(a). In the fixed beam:

$$M_B(\text{fixed}) = \frac{-\omega L^2}{8} \tag{4.11}$$

(This is negative as it is a hogging internal moment.) The effect of rotating point B through θ is:

$$M_B(\text{due to } \theta) = \frac{3EI}{L}\theta$$

$$= \frac{3EI}{L}\frac{\omega L^3}{168EI} = \frac{\omega L^2}{56} \tag{4.12}$$

Fig. 4.13 Superposition of moments: (a) fixed beam and associated bending moment diagram; (b) rotation at B and associated bending moment diagram

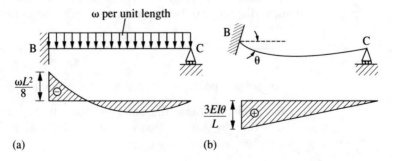

(a)   (b)

Fig. 4.14 Free body diagrams to determine distribution of internal moment in span BC: (a) internal moments just to the right of B; (b) free body diagram for span BC; (c) free body diagram for portion of span BC

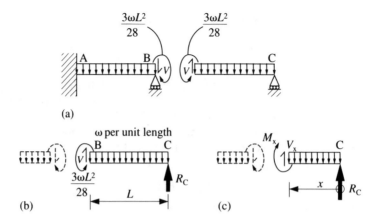

(a)

(b)

(c)

Summing equations (4.11) and (4.12) gives the internal moment at B due to all effects:

$$M_B(\text{total}) = \frac{-3\omega L^2}{28} \qquad (4.13)$$

The distribution of bending moment in span BC can now be found using equilibrium of the free-body diagrams illustrated in Figs 4.14(b) and (c). Taking moments about B in the free-body diagram of Fig. 4.14(b), the reaction at C is found to be:

$$R_C = \frac{\omega L(L/2) - M_B}{L} = \frac{\omega L}{2} - \frac{M_B}{L} \qquad (4.14)$$

Taking moments at a point a distance x to the left of C in the free-body diagram of Fig. 4.13(c) gives:

$$M_x = R_C x - \omega x^2/2$$
$$= \frac{\omega L x}{2} - \frac{M_B x}{L} - \frac{\omega x^2}{2} \qquad (4.15)$$

Substituting for M_B from equation (4.13), the bending moment diagram can be drawn and is illustrated in Fig. 4.15(a). It can be seen that this diagram is identical to the bending moment diagram in a simply supported beam (Fig. 4.15(b)) superimposed on a linearly decreasing bending moment diagram associated with M_B (Fig. 4.15(c)). This is indeed always true and the bending moment diagram can be drawn by superposition of these diagrams once M_B is known.

Two important points to note about this example are:

(a) the ability to draw the bending moment diagram for an indeterminate structure by superposition once the internal bending moments at the member ends are known; and

Fig. 4.15 (a) Bending moment diagram for span BC; (b) bending moment diagram if point B were simply supported; (c) bending moment diagram due to M_B. (a) = (b) + (c)

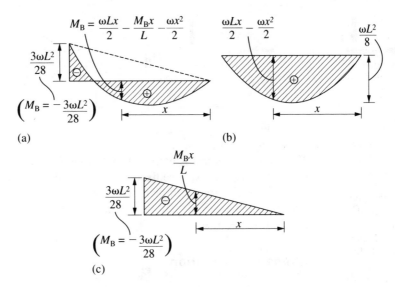

(a)

(b)

(c)

(b) the fact that, while the imposed rotation(s) are a function of EI, the bending moments are not; that is, for a member of constant flexural rigidity, EI, the final bending moment diagram is independent of the magnitude of E and I. However, the bending distribution is dependent on the **relative** magnitudes of E and I, as can be seen in the following example.

Example 4.5 Continuous beam with varying stiffness

Problem For the beam of Fig. 4.11, determine the internal bending moment at B if the flexural rigidity of span AB is $2EI$ and that of span BC is EI.

Solution In this case, the moment required to distort the structure into its final shape becomes:

$$M_2 = \frac{4(2EI)\theta}{L} + \frac{3EI\theta}{L} = \frac{11EI\theta}{L}$$

Adding this to M_1 and equating to zero gives:

$$\theta = \frac{\omega L^3}{264EI} \tag{4.16}$$

In a manner similar to Example 4.4, the combined internal moment is found by superimposing the effects of the rotation on the moment in the fixed beam. Summing the components as before gives the internal moment at B due to all effects:

$$M_B(\text{total}) = \frac{-5\omega L^2}{44} \tag{4.17}$$

Thus, while the internal moment is not proportional to the absolute magnitude of E and I, it is significantly affected by their relative magnitudes.

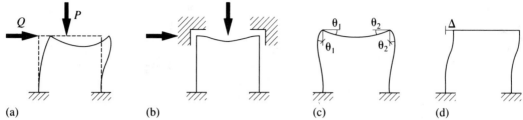

Fig. 4.16 Stiffness
method applied to
frame: (a) original
structure; (b) fixed
structure with applied
loads; (c) imposed
rotation to unfix
structure; (d) imposed
translation to unfix
structure.
(a) = (b) + (c) + (d)

The stiffness method is extremely versatile. In addition to being fixed against rotation, structures can be fixed against translation. This allows the method to be applied to more complex structures, such as the frame of Fig. 4.16. In addition, the method can be applied to three-dimensional structures using identical techniques.

Moment distribution

The moment distribution method is based on similar principles to the stiffness method. However, the moment distribution method, unlike the stiffness method, is particularly easy to apply by hand even if the number of fixing points is large. The method is often an iterative process and, as such, the solution can be derived to any required degree of accuracy.

Like the stiffness method, moment distribution is based on the principle of isolating individual members by fixing the joints of the structure. However, instead of solving simultaneously to find rotations and subsequently calculating internal moments, each joint is released in succession and the internal moments at the joints are 'distributed' and balanced until the joints have rotated to their correct orientations. The distribution of the internal moments is carried out in proportion to the relative rotational stiffnesses of adjacent members.

Rotational stiffness and carry-over factors

Rotational stiffness, in the context of moment distribution, is defined as the moment required to induce a unit rotation at one joint of a beam. The ratio of the moment induced at the other joint of the beam to this moment is defined as the carry-over factor. For the general cases of fixed- and pin-ended beams having uniform flexural rigidity, EI, the stiffness formulae are given in Appendix A (Nos 1 and 2). For the fixed-ended beam, the stiffness is $4EI/L$ and the carry-over factor from A to B is $(-2EI/L)/(4EI/L) = -0.5$. In the pin-ended beam, the stiffness is $3EI/L$ and the carry-over factor is zero.

Distribution factors

It was stated above that the distribution of the internal moments is carried out in proportion to the relative rotational stiffness of adjacent members meeting at a joint. More specifically, the proportion of moment carried by each member meeting at a joint is the total moment at the joint, M, multiplied by the distribution factor for each member. The distribution factor (DF) for member j is defined as the ratio of the rotational stiffness K_j of member j to the sum of the rotational stiffnesses of all members, that is:

$$\text{Distribution factor, member } j = \frac{K_j}{\sum\limits_i K_i} \tag{4.18}$$

where $i = 1, 2, 3, \ldots, m$ and m is the total number of members. Thus, the proportion of the total moment M taken by member j is given by:

$$M_j = \frac{K_j}{\sum\limits_i K_i} M \tag{4.19}$$

Note that the sum of the distribution factors for all members meeting at a joint is always equal to unity.

The techniques involved in moment distribution are explained here for the example illustrated in Fig. 4.11.

Example 4.6 **Continuous beam**

Problem Using the moment distribution method, determine the internal bending moment diagram for the continuous beam of Fig. 4.11 (*EI* constant).

Solution As with the stiffness method, it is sufficient to fix this structure at B, as a fixity at that point isolates the beams AB and BC from one another. Thus, the fixed structure and the associated bending moment diagram are, from Appendix B, Nos 6 and 8, as illustrated in Fig. 4.17. The external fixity at B corresponds to an applied external moment. Corresponding to this applied external moment is the discontinuity in the bending moment diagram at B. It can be seen from Fig. 4.17(c) that the moments in the free-body diagram would generate an

Fig. 4.17 (a) Fixed structure; (b) bending moment diagram for fixed structure; (c) free body diagram at B

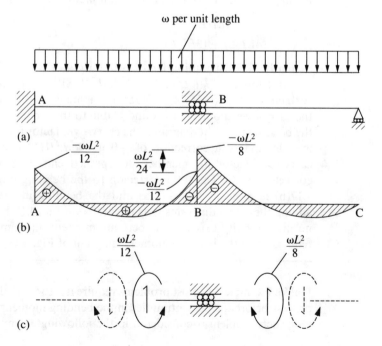

Fig. 4.18 Moment distribution:
(a) correction of moment discontinuity at B; (b) corrected bending moment diagram (equal to sum of (a) and Fig. 4.17(b))

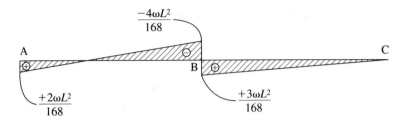

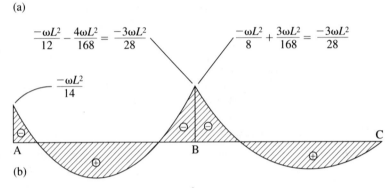

(a)

(b)

external moment reaction of $\omega L^2/24$ at B. To remove the external reaction at this point, the discontinuity in the bending moment diagram must be removed. This is done by applying a correction at B. Essentially, the hog moment to the left of B is increased (add negative number) while the hog moment to the right of B is reduced (add positive number).

For end B of member AB, the rotational stiffness is $4EI/L$, while for end B of member BC the rotational stiffness is $3EI/L$ (see Appendix A, Nos 1 and 2). From equation (4.18), the distribution factor to the left of B is given by:

$$\frac{4EI/L}{(4EI/L) + (3EI/L)} = \frac{4}{7}$$

Similarly, the distribution factor to the right of B can be shown to be $\frac{3}{7}$. Thus, a correction of $-\frac{4}{7}(\omega L^2/24) = -(4\omega L^2/168)$ is made to the left of B and a correction of $+\frac{3}{7}(\omega L^2/24) = 3\omega L^2/168$ is made to the right of B. In addition, the carry-over moments to A and C due to these corrections must be added to the bending moment diagram. The carry-over factor for member AB (fixed end) is -0.5 and so a correction of $-0.5(-4\omega L^2/168)$ is carried over to A. The carry-over factor for member BC (pinned end) is zero and so there is no correction at C. The total correcion to the bending moment diagram of Fig. 4.17(b) is given in Fig. 4.18(a) and the bending moment diagram found by adding these two diagrams is as illustrated in Fig. 4.18(b). After this correction, no discontinuity exists in the bending moment diagram. Thus, no iteration is required and the bending moment diagram of Fig. 4.18(b) is the final solution.

Unlike Example 4.6, most problems require iteration of the moment distribution process before the discontinuities in the bending moment diagram are removed. One such problem is considered in the following example.

Example 4.7   Braced frame (no sidesway)

Problem   Using moment distribution, determine the bending moment diagram for the structure and loading of Fig. 4.19 if the second moment of area of the columns is half that of the beams.

Fig. 4.19   Braced frame

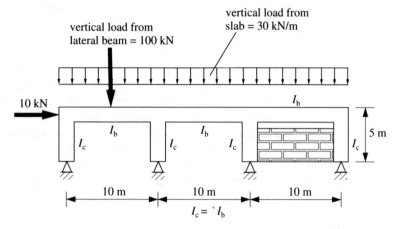

Solution   The structure of Fig. 4.19 is a 'braced' frame, that is a frame in which a structure other than the frame itself is provided to resist horizontal deflection. In this case, a masonry wall is present. Of course, the frame will carry some moment due to the horizontal load at the eaves but, as the wall is far stiffer than the frame, this moment is small and, for a structure such as this, it is reasonable to ignore it.

The fixed frame and the associated bending moment diagram are illustrated in Fig. 4.20. It can be seen that there are discontinuities in the bending moment diagram at A (375 kNm), B (125 kNm) and D (250 kNm).

Fig. 4.20   Fixed structure and associated bending moment diagram

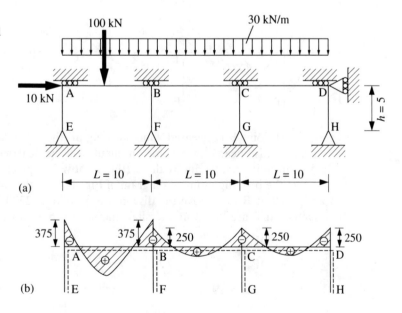

119

The moment distribution process is started by releasing the fixity at A. The stiffness of member AE is $3EI_c/L$ and the stiffness of member AB is $4EI_b/L$ where I_c and I_b are the second moments of area of the columns and the beams, respectively. Thus, the distribution factors at A are given by:

$$DF_{AE} = \frac{(3EI_c/h)}{(4EI_b/L) + (3EI_c/h)} = \frac{(3EI_c/5)}{(4E(2I_c)/10) + (3EI_c/5)} = \frac{3}{7}$$

and:

$$DF_{AB} = 1 - DF_{AE} = \tfrac{4}{7}$$

Therefore, to remove the discontinuity of 375 kNm at A, a correction of $-\tfrac{3}{7}(375)$ kNm is made to the left of (i.e. below) A and a correction of $+\tfrac{4}{7}(375)$ kNm is made to the right of A. In addition, the carry-over moments to E and B due to these corrections must be added to the bending moment diagram. The carry-over factor for member AB (Appendix A, No. 1) is -0.5 and so a 'correction' of $-0.5\tfrac{4}{7}(375)$ kNm is carried over to B. The carry-over factor for member AE (pinned end) is zero and so there is no correction at E. Thus, the first correction to the bending moment diagram of Fig. 4.20(b) is given in Fig. 4.21.

Fig. 4.21 (a) Bending moment diagram associated with release of fixity at A; (b) bending moment diagram after release of fixity at A (equal to sum of (a) and Fig. 4.20(b))

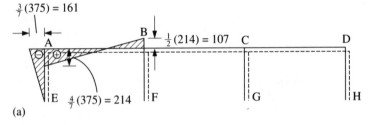

(a)

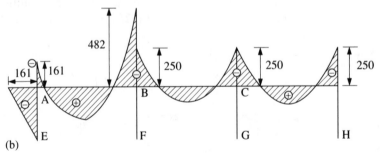

(b)

Next, joint B is released while keeping all the other fixities, including A, in place. This means that some moment will be 'carried over' to A and a discontinuity reintroduced there (this is how the process becomes iterative). From the bending moment diagram of Fig. 4.21(b), the correction which is now required at B to remove the discontinuity there is 232 kNm. The distribution factors at B are found in a similar manner to those for joint A. They are:

$$DF_{BA} = \tfrac{4}{11}$$

$$DF_{BF} = \tfrac{3}{11}$$

$$DF_{BC} = \tfrac{4}{11}$$

Fig. 4.22 Correcting
moments at B

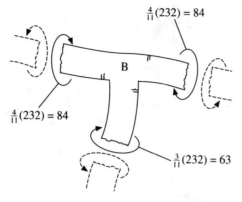

$\frac{4}{11}(232) = 84$

B

$\frac{4}{11}(232) = 84$

$\frac{3}{11}(232) = 63$

Fig. 4.23 (a) Bending
moment diagram
associated with release
of fixity at B;
(b) bending moment
diagram after release of
fixity at B (equal to sum
of (a) and Fig. 4.21(b));
(c) free body diagram at
B after release there

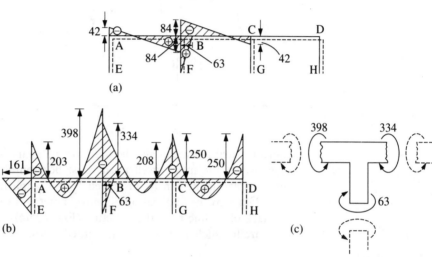

The carry-over factor for the two fixed-ended members, AB and BC, is -0.5
and the carry-over factor for the pin-ended member, BF, is zero. Thus, in order
to remove the discontinuity at B, a sag moment must be introduced just left of
B and a hog moment just right of it. This corresponds to a clockwise rotation
of the joint as illustrated in Fig. 4.22. The total correction to the bending
moment diagram associated with the release of B is shown in Fig. 4.23(a). This
yields the new bending moment diagram illustrated in Fig. 4.23(b). Moment
equilibrium is now satisfied at B (see Fig. 4.23(c)) even though there is a
discontinuity in beam moment there. This discontinuity is due to the moment
taken by the column. Now, however, joints A, C and D are out of equilibrium.

The completion of the first iteration requires the release of the fixities at the
remaining joints, namely C and D. The entire process, that is relaxing the fixities
at each of A, B, C and D, must then be repeated a number of times until moment
equilibrium is established simultaneously at all joints in the frame. The final
solution for this example is given in Fig. 4.24. However, if the process is being
used for an approximate analysis or to check the output from a computer
analysis, it is often unnecessary to achieve complete convergence of the solution
since a result of sufficient accuracy can usually be found very quickly.

Fig. 4.24 Final bending moment diagram after convergence

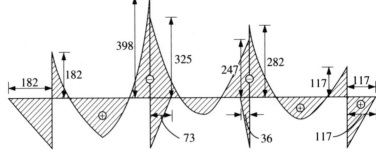

Fig. 4.25 Unbraced frames

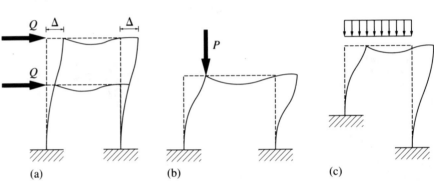

(a)   (b)   (c)

Moment distribution can also be used for the analysis of 'unbraced' or 'sway' frames where the frame itself must resist horizontal displacement. The horizontal displacement can be produced by the application of horizontal loads (Fig. 4.25(a)), by the application of non-symmetrical vertical loads (Fig. 4.25(b)) or by lack of symmetry in the structure (Fig. 4.25(c)). The moment distribution procedure for such structures is illustrated below in Example 4.8.

Example 4.8 Unbraced frame

Problem Using moment distribution, determine the internal bending moments in the frame of Fig. 4.19 for the applied ultimate loads specified, if the wall in the end bay is removed.

Solution The removal of the shear wall from the end bay of the structure of Fig. 4.19 produces a sidesway of the frame under the applied loading. Each of the joints A, B, C and D will be displaced by an amount, Δ, as illustrated in Fig. 4.26(a). This sway frame is equivalent to a frame which is restrained against displacement at D (Fig. 4.26(b)) with a horizontal reaction of R_d, plus a frame subject to an external force, F_d, equal and opposite to R_d, at D (Fig. 4.26(c)). The approach then is to apply moment distribution to the restrained frame of Fig. 4.26(b) to find the internal moments and the external reaction, R_d. The internal moments in the frame of Fig. 4.26(c), for the horizontal force F_d, are then found again using moment distribution. Subsequently, the principle of superposition is used to combine the moments in each frame (Figs 4.26(b) and (c)) to yield the actual internal moments in the sway frame.

  The internal bending moments for the restrained frame of Fig. 4.26(b) can be taken straight from Example 4.7 since the shear wall of that structure is

Fig. 4.26 Sway frame: (a) deflected shape of original frame; (b) frame restricted against sway; (c) frame subjected to sway force only. (a) = (b) + (c)

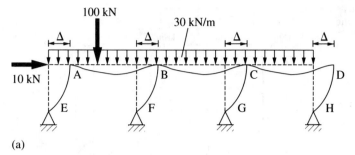

(a)

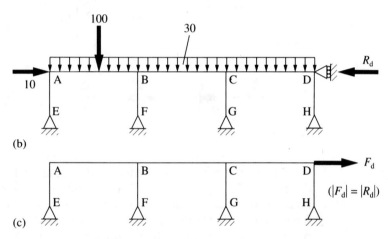

(b)

(c)

equivalent to a pinned restraint at D. Thus, the external reaction, R_d, can be determined from the bending moment diagram of Fig. 4.24. In this case, equilibrium of horizontal forces dictates that the reaction, R_d, is equal to the sum of the applied horizontal load plus the horizontal reactions at the bases of the columns. (As the applied vertical loading is non-symmetrical, it tends to induce a small horizontal sway in the structure.) The column reactions are found by applying the equations of equilibrium to the free-body diagrams of each column, illustrated in Figs 4.27(a)–(d). For instance, by taking moments about point A of Fig. 4.27(a), the reaction at E is given by:

$$5H_e = 182$$

$\Rightarrow \qquad H_e = 36.4 \text{ kN}$

In a similar manner, the other column reactions are found to be:

$$H_f = 14.6 \text{ kN}$$

$$H_g = 36/5 = 7.2 \text{ kN}$$

$$H_h = 23.4 \text{ kN}$$

By horizontal equilibrium of the entire structure:

$$10 + H_e - H_f + H_g - H_h - R_d = 0$$

$\Rightarrow \qquad R_d = 10 + H_e - H_f + H_g - H_h$

$\Rightarrow \qquad R_d = 15.6 \text{ kN}$

123

Fig. 4.27 Free body diagrams to determine reactions: (a) column AE; (b) column BF; (c) column CG; (d) column DH

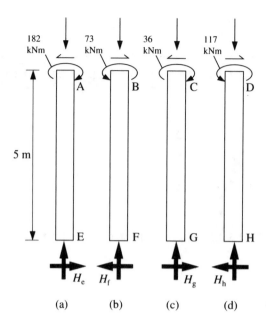

The internal bending moments for the frame of Fig. 4.26(c) must now be determined for $F_d = 15.6$ kN. There is no simple direct method by which moment distribution can be used to find the internal moments due to a sway force such as F_d. However, for any given sway, Δ', the internal moments can be found and from these the force required to induce that sway, F'_d, can be calculated. The internal moments due to F_d are these moments scaled by the ratio F_d/F'_d.

The first step in this process is to fix the joints A, B, C and D against rotation and to apply the force, F'_d, and thus the sway, Δ', as illustrated in Fig. 4.28(a). Each column is now like a cantilever with the fixed end at the top, so the bending moment diagram is, from Appendix A, No. 4, as illustrated in Fig. 4.28(b). An equivalent, but more convenient, form of displaying the bending moment diagram is illustrated in Fig. 4.28(c). Now, to release the moment fixity at A, a correction of $-\frac{3}{7}(100)$ is made to the left of (below) A and a correction of $+\frac{4}{7}(100)$ is made to the right of A. In addition, a correction of $-0.5\frac{4}{7}(100)$ is carried over to B. The carry-over factor for member AE (pinned end) is zero and so there is no correction at E. Thus, the first correction to the bendng moment diagram of Fig. 4.28(c) is as given in Fig. 4.29.

Next, joint B is released while keeping all the other fixities, including A, in place. From the bending moment diagram of Fig. 4.29(b), the free-body diagram for joint B is given in Fig. 4.30. By moment equilibrium at this joint, it can be seen that the correction which is now required at B is 71 kNm. Thus, the correction to the bending moment diagram of Fig. 4.29(b) is as illustrated in Fig. 4.31(a). This yields the new bending moment diagram of Fig. 4.31(b), in which equilibrium is satisfied at B.

As in Example 4.7, the problem is completed by proceeding to balance joints C and D and iterating the entire process until complete convergence is achieved. The final internal moments due to F'_d are illustrated in Fig. 4.32(a). The value

Fig. 4.28 Sway frame fixed against joint rotation: (a) deflected shape; (b) bending moment diagram; (c) bending moment diagram where moments displayed are to be multiplied by $3EI_c\Delta'/(100h^2)$

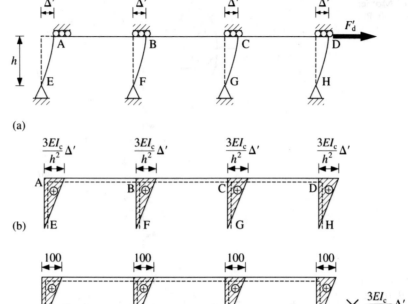

Fig. 4.29 (a) Release of fixity at A; (b) bending moment diagram after release of fixity at A (equal to sum of (a) and Fig. 4.28(c))

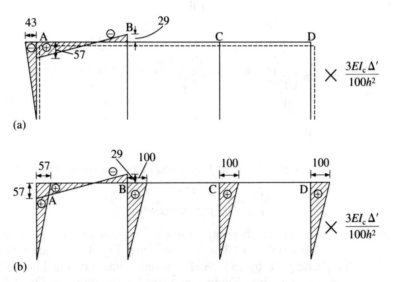

of F'_d is calculated in a similar manner to F_d and is given by:

$$F'_d = 58(3EI_c\Delta'/100h^2)$$

An internal force of this magnitude causes a deflection of Δ' and induces the internal moments given in Fig. 4.32(a). Since the deflection is linear elastic, the force F_d will develop moments in the frame that are proportional to those developed by F'_d. Therefore, the internal moments induced by the force F_d are

125

Fig. 4.30 Moment
equilibrium at B

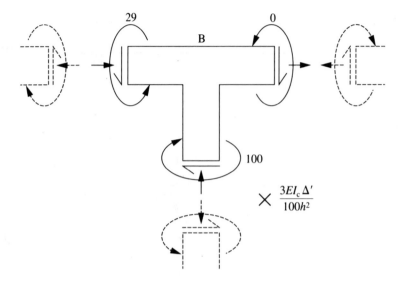

Fig. 4.31 (a) Release of
fixity at B; (b) bending
moment diagram after
release of fixity at B
(equal to sum of (a) and
Fig. 4.29(b))

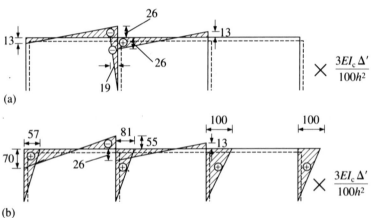

equal to those in Fig. 4.32(a) factored by an amount:

$$\frac{F_d}{F'_d} = \frac{15.6}{58(3EI_c\Delta'/100h^2)}$$

Thus, the bending moment diagram due to F_d is found by removing the
'multiplier' of $3EI_c\Delta'/100h^2$ from Fig. 4.32(a) and scaling the numbers in the
diagram by 15.6/58. The result is illustrated in Fig. 4.32(b).

The final bending moment for the frame of Fig. 4.26(a) is found by direct
superposition of Fig. 4.32(b) and Fig. 4.24, and is shown in Fig. 4.33.

4.4 Non-linear analysis of indeterminate structures

The linear elastic analysis of structures is based on the assumption that there
is a linear relationship between the stress and strain in a member, that is to

Fig. 4.32 Bending moment diagrams due to sway forces: (a) due to F_d'; (b) due to F_d

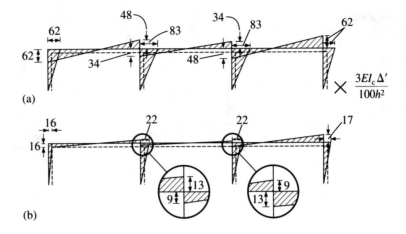

(a)

(b)

Fig. 4.33 Bending moment diagram for sway frame of Fig. 4.26(a) (equal to sum of Fig. 4.24 and Fig. 4.32(b))

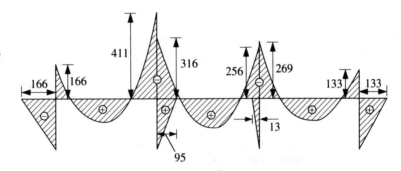

say that stress is given by:

$$\sigma = E\varepsilon \tag{4.20}$$

where ε is strain and E is the elastic modulus (Young's modulus) for the material in question. Recall also that for a section in bending (Fig. 4.34), the linear elastic moment–curvature relationship is given by:

$$\frac{M}{I} = \frac{E}{R} \tag{4.21}$$

where I is the second moment of area of the cross-section and R is the radius of curvature (see Fig. 4.34(b)). Alternatively, the linear elastic relationship can be expressed as:

$$M = EI\kappa \tag{4.22}$$

where $\kappa = 1/R$ is known as the curvature. Thus, for a linear elastic section, the moment–curvature relationship is linear as illustrated in Fig. 4.34(c).

Equation (4.22) and the second moment of area in this equation are applicable to beams of homogeneous cross-section. Reinforced concrete is not homogeneous as it consists of two materials (concrete and steel) which have considerably different values for the elastic modulus. However, it will be shown in Chapter 7 that it is possible to transform a reinforced concrete section into

Fig. 4.34 Linear elastic bending: (a) beam in bending; (b) segment of beam in bending; (c) moment/curvature relationship for linear elastic and homogeneous beam

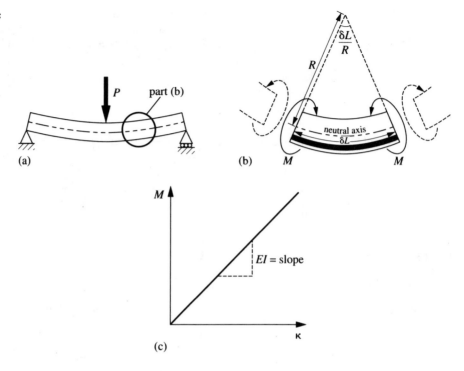

(a)

(b)

(c)

an equivalent homogeneous concrete section and to calculate an equivalent second moment of area. When the internal moment, M, is very small, the concrete is uncracked and the equivalent second moment of area is denoted I_u. Denoting the elastic modulus for concrete as E_c, equation (4.22) becomes:

$$M = E_c I_u \kappa \qquad (4.23)$$

However, at quite low moment, the section cracks, the equivalent second moment of area drops to a much lower value, and equation 4.23 becomes:

$$M = E_c I_c \kappa \qquad (4.24)$$

where I_c is the equivalent second moment of area of the cracked section. This relationship is represented diagrammatically in Fig. 4.35.

Fig. 4.35 Moment/curvature relationship for reinforced concrete

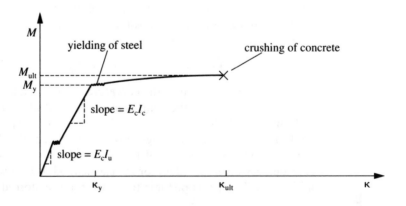

Equation (4.24) remains valid until the material behaviour becomes non-linear. For a properly designed reinforced concrete section, the steel yields before the concrete crushes. This happens at an applied moment of M_y, as illustrated in Fig. 4.35. As steel is a ductile material, the section too is ductile, and beyond the yield point the curvature increases greatly for a relatively small increase in the applied moment. Complete failure of the section occurs when the concrete at the extreme fibre in compression crushes. The curvature at this stage is denoted κ_{ult}.

The moment–curvature relationship of Fig. 4.35 can be idealized by the simplified relationship illustrated in Fig. 4.36. Linear elastic analysis is based on the assumption that $\kappa \leq \kappa_y$ at all sections of all members. Thus, linear elastic analysis techniques, such as the stiffness method and moment distribution, are based on this assumption. In this section, non-linear methods of analysis are considered for which there is no such restriction on κ.

Fig. 4.36 Idealized moment/curvature relationship for reinforced concrete

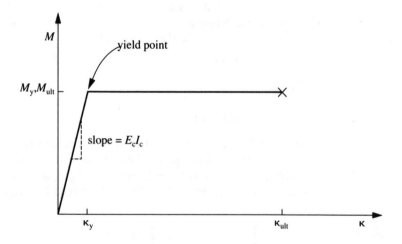

Elastic–plastic analysis

In reinforced concrete structures close to collapse, the curvatures at certain sections will exceed κ_y. In such situations, methods of elastic–plastic analysis can be used to predict accurately the behaviour of the structure. The general procedure is illustrated below with Example 4.9.

Example 4.9 Continuous beam

Problem Using elastic–plastic analysis, determine the critical value for the load factor, λ, to cause failure of the beam of Fig. 4.37. The beam has sufficient reinforcement to resist a maximum bending moment, both in sag and hog, of M_y.

Solution We assume that the loading is proportional, that is that the factor λ increases linearly with time. Thus, the beam starts with zero load and both loads increase simultaneously until the beam fails. Initially, for a load of $\lambda_1 P$, say, where λ_1 is small, all sections of the beam are in the elastic zone. For this situation, linear elastic analysis is perfectly accurate and the bending moment diagram (found

Fig. 4.37 Beam of
Example 4.9

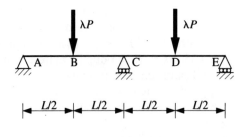

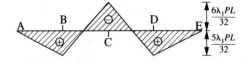

Fig. 4.38 Elastic
bending moment
diagram for beam of
Fig. 4.37

using, say, moment distribution) is as illustrated in Fig. 4.38. It can be seen from the figure that a maximum moment of $6\lambda_1 PL/32$ occurs at the internal support. If we define λ_y as the load factor at which the first yield occurs, then point C becomes plastic (i.e. yields) when:

$$6\lambda_y PL/32 = M_y$$

from which:

$$\lambda_y = 32M_y/6PL$$

For point C, we are then at the yield point illustrated in Fig. 4.36. For any further increase in load, point C can continue to resist the moment M_y but can provide no further resistance to rotation. Thus, for any additional increase, λ_a, in the load factor, point C is effectively hinged (see Fig. 4.39) and is termed a

Fig. 4.39 Beam after
plastic hinge has
formed: (a) original load
where $\lambda > \lambda_y$;
(b) portion of load for
which beam is elastic;
(c) additional load after
yield at C ($\lambda_a = \lambda - \lambda_y$).
(a) = (b) + (c)

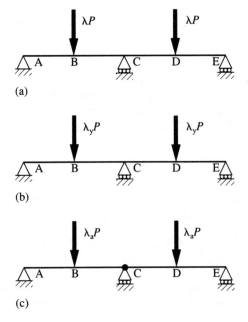

Fig. 4.40 Components of bending moment diagram for $\lambda > \lambda_y$ corresponding to loads of Fig. 4.39: (a) total bending moment diagram equal to sum of (b) and (c) ($\lambda = \lambda_a + \lambda_y$); (b) elastic bending moment diagram (λ_y); (c) bending moment diagram due to additional load beyond the elastic (λ_a)

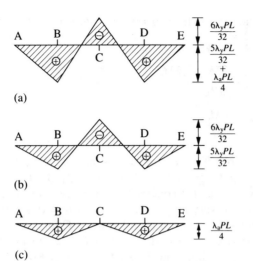

(a)

(b)

(c)

Fig. 4.41 Collapse mechanism, $\lambda = \lambda_m$

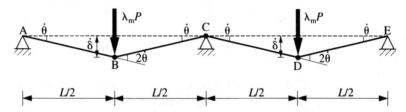

'plastic' hinge. The structure of Fig. 4.39(c) is statically determinate and can readily be analysed to give the bending moment diagram of Fig. 4.40(c). Thus, for a load factor of $\lambda = \lambda_y + \lambda_a$, the total bending moment diagram is found by adding the bending moments associated with the loads of Figs 4.39(b) and (c). This is illustrated in Fig. 4.40.

If the beam is ductile at C, that is $\kappa_{ult} \gg \kappa_y$, the load factor can be increased beyond λ_y until the moment at points B and D reaches the yield moment. At this stage, each span forms a mechanism and collapses as illustrated in Fig. 4.41. From Fig. 4.40(a), the moments at B and D are given by:

$$M_B = M_D = \frac{5\lambda_y PL}{32} + \frac{\lambda_a PL}{4}$$

At collapse, $M_B = M_D = M_y$, from which:

$$\lambda_a = \frac{4}{PL}\left(M_y - \frac{5\lambda_y PL}{32}\right)$$

But:

$$\lambda_y = \frac{32M_y}{6PL}$$

$$\Rightarrow \quad \lambda_a = \frac{4}{PL}\left[M_y - \frac{5PL}{32}\left(\frac{32M_y}{6PL}\right)\right]$$

$$\Rightarrow \quad \lambda_a = \frac{2M_y}{3PL}$$

The load factor at which the mechanism occurs is termed λ_m and is given by:

$$\lambda_m = \lambda_y + \lambda_a$$

$$\Rightarrow \qquad \lambda_m = \frac{32M_y}{6PL} + \frac{2M_y}{3PL}$$

$$\Rightarrow \qquad \lambda_m = \frac{6M_y}{PL} \qquad\qquad (4.25)$$

All of the above is based on the assumption that, at all times up to collapse, the curvature at point C is less than κ_{ult}. In fact, this assumption is not necessarily true, as for concrete beams (especially deeper ones) the difference, $\kappa_{ult} - \kappa_y$, is not very large. The implication of this is that it is possible for the beam to collapse owing to total failure at C for some λ such that $\lambda_y < \lambda < \lambda_m$. From Appendix A, No. 6, the rotation at the end of a simply supported beam due to a central point load, F, is $FL^2/16E_cI_c$. At C in Fig. 4.39(c), plastic hinge rotation is occurring on both sides of the point with the result that the rotation after initial yield is:

$$\theta = 2\left(\frac{\lambda_a PL^2}{16E_cI_c}\right) = \frac{\lambda_a PL^2}{8E_cI_c}$$

If the plastic hinge forms over a length, L_h, of beam near C, the rotation is related to the radius of curvature by:

$$\theta = L_h/R$$

$$\Rightarrow \qquad \theta = L_h\kappa$$

Hence, the capacity to rotate after formation of the plastic hinge is:

$$\theta_{ult} = L_h(\kappa_{ult} - \kappa_y)$$

Thus, rotation at C reaches its ultimate capacity when:

$$\frac{\lambda_a PL^2}{8E_cI_c} = L_h(\kappa_{ult} - \kappa_y)$$

$$\Rightarrow \qquad \lambda_a = \frac{8E_cI_c}{PL^2}L_h(\kappa_{ult} - \kappa_y)$$

The load factor at which this beam fails through inadequate rotational capacity is:

$$\lambda_\theta = \lambda_y + \lambda_a$$

$$\Rightarrow \qquad \lambda_\theta = \frac{32M_y}{6PL} + \frac{8E_cI_c}{PL^2}L_h(\kappa_{ult} - \kappa_y) \qquad\qquad (4.26)$$

The value for λ at which this beam fails is the lesser of λ_θ and λ_m (equations (4.25) and (4.26)).

The concept of plastic moment redistribution (not to be confused with moment distribution), which is described in Chapter 5, is an approximation to elastic–plastic analysis.

Plastic analysis

Plastic analysis is totally different in its approach from linear elastic or elastic–plastic analysis. The value of λ_m is determined from a consideration of the conditions at collapse with no regard to the deformations that occur prior to this. For the structure of Example 4.9, the conditions **during** collapse are as illustrated in Fig. 4.41. Thus, during collapse, the external load at B is moving at a rate of $\dot{\delta}$ (deflection per unit time) while the pin at A is rotating at a rate of $\dot{\theta}$ where $\dot{\delta} = (L/2)\dot{\theta}$. By simple geometry, the rate of rotation of the hinge at C is $2\dot{\theta}$ ($\dot{\theta}$ in each span) and hence the rates of rotation at B and D are also $2\dot{\theta}$. Note that elastic deformations are also present between the plastic hinges but these have no influence on the calculations. It has been proven that, during collapse of the frame by this mechanism, the internal rate of work done at the plastic hinges equals the external rate of work done by the loads, $\lambda_m P$. The rate of work done at a plastic hinge during collapse is equal to the rate of rotation of the hinge multiplied by the ultimate moment capacity of the member, M_{ult} (taken to be equal to the yield moment). Thus, the total internal rate of work done, $\dot{W}_i$, by the three hinges during collapse is:

$$\dot{W}_i = 3M_{ult}(2\dot{\theta}) = 6M_{ult}\dot{\theta}$$

The rate of work done by the load during collapse is its magnitude multiplied by its rate of deflection. Thus, the total external rate of work done, $\dot{W}_e$, during collapse is:

$$\dot{W}_e = 2\lambda_m P\dot{\delta} = 2\lambda_m P(L/2)\dot{\theta}$$

Equating the internal with the external rate of work done gives:

$$6M_{ult}\dot{\theta} = 2\lambda_m P(L/2)\dot{\theta}$$

$$\Rightarrow \qquad \lambda_m = \frac{6M_{ult}}{PL}$$

As can be seen from the simplified moment–curvature diagram of Fig. 4.36, the yield moment, M_y, is assumed to be the same as the ultimate moment capacity, M_{ult}. Thus, the result is the same as the result of elastic–plastic analysis (equation (4.25)).

It should be noted that there is often more than a single possible collapse mechanism for a given structure. In this case, each mechanism must be considered in turn and the lesser of the load factors is taken as λ_m.

Plastic analysis is very easy to perform but the results give no information on how much plastic hinge rotation occurs at the plastic hinges before the structure collapses. This renders it unsuitable for concrete beams and, in particular, concrete beams for which ductility is relatively limited. However, concrete slabs, not being deep, are far more ductile than concrete beams and a method of plastic analysis known as yield line theory (see Chapter 5) is used in their analysis.

4.5 Finding shear force, axial force and deflecton

A complete analysis of a structure involves not only the determination of internal moments but also other quantities such as shear force, axial force and deflection. However, as will be seen from the following examples, it is a relatively simple matter to determine these quantities once the distribution of bending moment has been determined.

Shear force

Shear is an important phenomenon in reinforced and prestressed concrete. For determinate structures, shear can be found using the principles of equilibrium. For indeterminate structures, the stiffness method can be used to determine the shear distribution directly. Alternatively, the distribution of moment can be found using either the stiffness method or moment distribution, and the shear distribution found from this.

The first step in finding the shear force at any section in any member of a structure is to determine the reactions. This can be done by considering the equilibrium of free-body diagrams. Once the reactions are known, the shear force diagram can readily be drawn.

Example 4.10 **Shear distribution in continuous beam**

Problem The linear elastic bending moment diagram for the beam of Fig. 4.42(a) has been found using moment distribution and is given in Fig. 4.42(b). Determine the shear force diagram for this beam.

Fig. 4.42 Shear distribution in continuous beam: (a) geometry and loading; (b) bending moment diagram

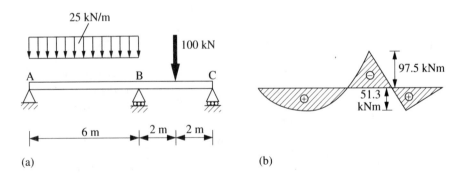

(a)

(b)

Solution The free-body diagrams for member AB, joint B and member BC are illustrated in Figs 4.43(a), (b) and (c), respectively. The reactions R_1, R_2 and R_3 can be found by taking moments about the point B in both Figs 4.43(a) and (c). By equilibrium, the clockwise moments must equal the anti-clockwise moments.

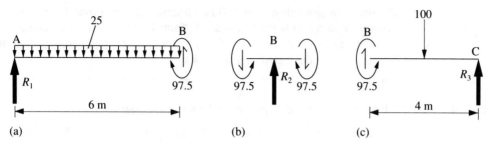

Fig. 4.43 Free body diagrams for beam of Fig. 4.42: (a) member AB; (b) joint B; (c) member BC

That is for Fig. 4.43(a):

$$6R_1 + 97.5 = 25 \times 6 \times 3$$

$$\Rightarrow \quad R_1 = 58.8 \text{ kN}$$

and for Fig. 4.43(c):

$$4R_3 + 97.5 = 100 \times 2$$

$$\Rightarrow \quad R_3 = 25.6 \text{ kN}$$

By equilibrium of vertical forces in the entire structure:

$$R_1 + R_2 + R_3 = 25 \times 6 + 100$$

$$\Rightarrow \quad R_2 = 165.6 \text{ kN}$$

Once the reactions are known, the shear force diagram can be plotted in the same way as for determinate structures (see sections 4.1 and 4.2), that is by starting at one end of the beam and 'following the arrows'. Thus, start at point A, say, by going up by R_1, as shown in Fig. 4.44(a). The uniformly distributed loading of 25 kN/m acts downwards and so the diagram next goes down linearly from R_1 to a minimum of $R_1 - (25 \times 6)$ just to the left of point B. At point B, the diagram again goes up, by an amount R_2. Shear then remains constant in the section of span BC where there is no applied loading. At mid-span, however,

Fig. 4.44 Shear force diagrams for Example 4.10: (a) construction of diagram; (b) completed shear force diagram

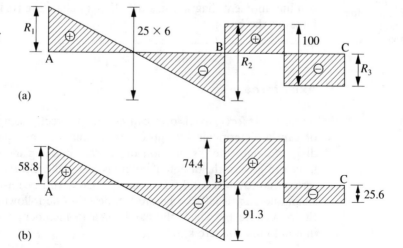

the diagram goes down by 100 kN because of the point load. The diagram remains level again from mid-span to point C. The rise in the diagram at point C is equal to the reaction R_3. The complete shear force diagram thus constructed is illustrated in Fig. 4.44(b).

Example 4.11   Shear distribution and non-linear analysis

Problem   For the continuous beam of Fig. 4.42(a), the internal bending moments have been found using plastic moment redistribution (an approximation to elastic–plastic analysis) and are given in Fig. 4.45(a) (note that the bending moments from this non-linear analysis are different from those from the linear elastic analysis illustrated in Fig. 4.42(b)). Using this new bending moment diagram, determine the distribution of shear force for the beam.

Solution   Although the internal moments in the beam have changed somewhat, equilibrium still holds (equilibrium **always** holds). Hence, the reactions can be determined from the free-body diagrams of individual members in the manner described in Example 4.10. In this case, the support reactions are:

$$R_1 = 62 \text{ kN}$$

$$R_2 = 157.5 \text{ kN}$$

$$R_3 = 30.5 \text{ kN}$$

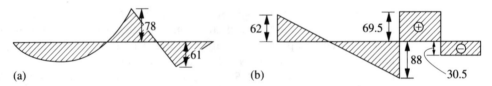

(a)   (b)

Fig. 4.45 Bending moment and shear force diagrams for Example 4.11: (a) bending moment diagram after plastic moment redistribution; (b) shear force diagram

The corresponding shear force diagram is illustrated in Fig. 4.45(b). By comparison of Figs 4.44(b) and 4.45(b), it can be seen that the change in the bending moment diagram has the effect of changing (redistributing) the shear forces in the beam.

Axial force

As for shear force, axial force can be found directly using the stiffness method or can be determined by applying the equations of equilibrium to free-body diagrams once the distribution of internal bending moment is known. Axial force can govern the design of columns and walls. On the other hand, the axial forces due to horizontal wind loads which act in beams and slabs tend to be small and can usually be ignored in design. The following example illustrates the procedure by which the axial force in members can be found once the internal moments are known.

Example 4.12   Axial forces in a braced frame

Problem   For the braced frame of Fig. 4.46(a), plastic moment redistribution was used to determine the internal bending moments illustrated in Fig. 4.46(b). Use the redistributed bending moment diagram to find the axial force in each of the columns. All horizontal forces are transferred by a floor slab attached to ABC to shear walls elsewhere in the building.

Fig. 4.46 Frame example: (a) geometry and loading; (b) redistributed bending moment diagram

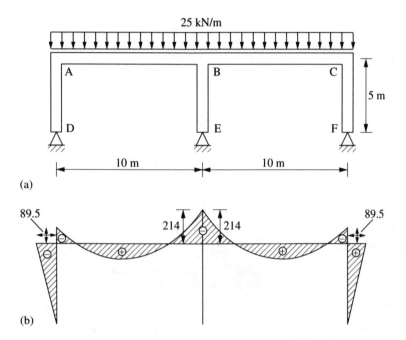

Solution   The free-body diagrams for member AB, member AD and the joint between the two members are illustrated in Fig. 4.47. From the free-body diagram for the joint, it can be seen that the shear force, V_{ab}, at the left end of member AB is transferred to member AD as an axial (compressive) force, A_{ad}, and that $V_{ab} = A_{ad}$. The magnitude of this force cannot be determined directly from a consideration of the equilibrium of member AD. However, V_{ab} can be found by taking moments about the point B in the free-body diagram for member AB. By equilibrium:

$$10V_{ab} + 214 = 25 \times 10 \times 5 + 89.5$$

$$\Rightarrow \quad V_{ab} = 112.5 \text{ kN}$$

Thus, the axial force, A_{ad}, in member AD is 112.5 kN. By symmetry, the axial force, A_{cf}, in member CF is also 112.5 kN.

By equilibrium of the entire frame, the axial force in the centre column, A_{be}, is given by:

$$A_{be} = 2 \times 25 \times 10 - A_{ad} - A_{cf}$$

$$\Rightarrow \quad A_{be} = 275 \text{ kN}$$

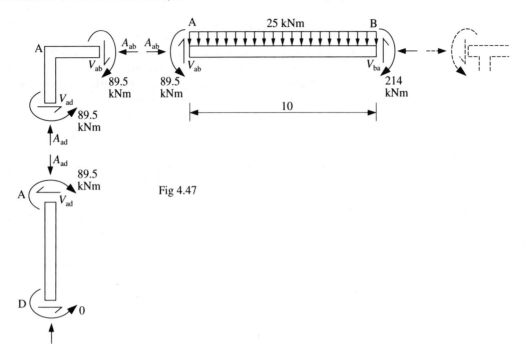

Fig. 4.47 Free body diagrams for AD, A and AB

By taking moments about D in the free-body diagram for member AD, it can be shown that the shear force, V_{ad}, is 17.9 kN. Thus, by equilibrium of joint A, the axial force in member AB is 17.9 kN.

Deflection

It is necessary to limit serviceability limit-state deflections in many types of concrete structures as excessive deflection can cause adverse effects such as the cracking of partitions. Therefore, much time is devoted in structural analysis courses to the calculation of deflection in linear elastic structures. However, all of these methods are of limited usefulness for a number of reasons. Not least of these are the problems of selecting the appropriate second moment of area for reinforced concrete and making allowance for time-dependent deflection in both reinforced and prestressed concrete. Prestressed concrete beams do not crack (except for partial prestressing) and it is usually sufficiently accurate to calculate the second moment of area ignoring the presence of a reinforcement. Reinforced concrete beams generally do crack and so it is necessary to calculate the equivalent second moment of area of the cracked section taking account of the reinforcement present. However, calculation of the cracked second moment of area requires a knowledge of the quantities of steel reinforcement present in the beam and this is not known at the early stages of design. It is also necessary to know which parts of the beam are in hog and which are in sag as the cracked second moments of area will, in general, be different for each case. Finally, the beam will not usually be cracked everywhere, the degree of cracking being dependent on the internal bending moment, so it may be excessively conservative to use a cracked second moment of area for the calculation of deflection. The time-dependent deflection due to creep is normally accounted for by reducing the modulus of elasticity. Such calculations are often hugely inaccurate as creep is extremely difficult to predict.

Fortunately, the Eurocode for concrete, EC2, specifies that it is not necessary to calculate the deflections in beams and slabs if certain span/depth ratios (listed in Table 6.5) are not exceeded. In circumstances where these span/depth ratios are exceeded, an assumption of elastic behaviour is perfectly reasonable for the calculation of deflections.

Once values of the modulus of elasticity (E) and the second moment of area (I) have been selected, the calculation of deflection can be done in many ways. The stiffness method, for instance, can be used to determine deflections directly. However, some computer programs only determine deflections and rotations at the nodes, that is the points at which two or more members meet. From this data, the moment–area theorems can be used to determine the deflection at any other point in the structure. The moment–area theorems are as follows:

Theorem 1: The difference in slope between two points on a linear elastic member is equal to the area under the M/EI diagram between these two points. Thus, for two points, A and B, the difference in slope is given by (see Fig. 4.48):

$$\theta_A - \theta_B = \int_A^B \frac{M}{EI}\,dx \tag{4.27}$$

Fig. 4.48 Moment area theorems: (a) segment of beam in bending; (b) bending moment diagram divided by EI

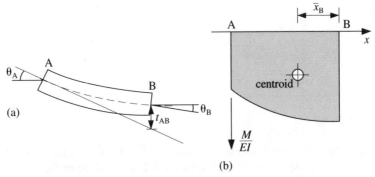

Theorem 2: The vertical deviation, t_{AB}, of the tangent to a point A on a linear elastic member from a point B (see Fig. 4.48) is equal to the 'moment' of the M/EI diagram between the two points about point B, that is:

$$t_{AB} = \bar{x}_B \int_A^B \frac{M}{EI}\,dx \tag{4.28}$$

where $\bar{x}_B$ is the distance from point B to the centroid of the portion of the M/EI diagram between the two points.

The application of these two theorems is illustrated below in Example 4.13.

Example 4.13 Deflections in a sway frame

Problem The bending moment diagram for the frame of Fig. 4.49(a) is illustrated in Fig. 4.49(b). The deformation of the frame under the applied load is shown in Fig. 4.49(c). Results from a computer program indicate that the frame undergoes a rotation of 0.007 radians at point B. If the second moment of area of the beam, I_b, in both hog and sag is constant throughout its length and is 1.3×10^{-3} m^4, determine the maximum vertical deflection in member BC. Assume that the modulus of elasticity, E, equals 27×10^6 kN/m^2.

Fig. 4.49 Portal frame of Example 4.13: (a) geometry and loading; (b) bending moment diagram; (c) deflected shape

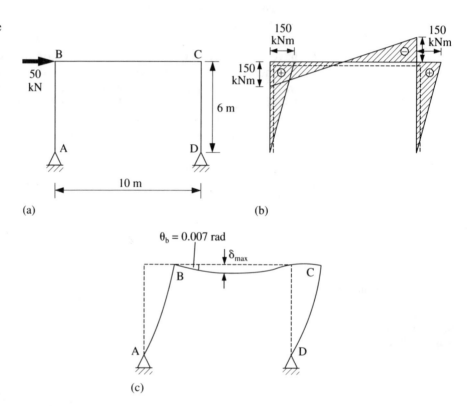

Solution From the deflected shape of the frame, it can be seen that the maximum vertical deflection occurs in the left half of member BC at a point E, say. Assume that the location of point E is a distance y from B where $y < 5$ m. From Fig. 4.50(a), the area under the bending moment diagram, A_M, between the points B and E is found to be:

$$A_M = \left(\frac{150 + M_y}{2}\right)y$$

but, by linear interpolation of the bending moment diagram:

$$M_y = 150 - 300y/10$$
$$= 150 - 30y$$

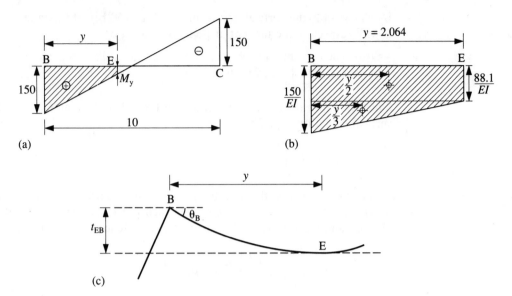

Fig. 4.50 Member BC in frame of Fig. 4.49: (a) bending moment diagram for BC; (b) segment of M/EI diagram for BE; (c) deflected shape of BE

Therefore:

$$A_M = \left(\frac{150 + 150 - 30y}{2}\right)y$$

$$\Rightarrow \quad A_M = 150y - 15y^2$$

By Theorem 1, the difference in slope between the two points B and E is given by equation (4.27), that is:

$$\theta_B - \theta_E = \int_B^E \frac{M}{EI}\,\mathrm{d}x = \frac{A_M}{EI} \tag{4.29}$$

The slope, θ_E, at the point of maximum deflection is zero. Thus equation (4.29) becomes:

$$0.007 - 0 = \frac{150y - 15y^2}{EI}$$

Substitution for $I = 1.3 \times 10^{-3}\,\mathrm{m}^4$ and $E = 27 \times 10^6\,\mathrm{kN/m}^2$ gives:

$$245.7 = 150y - 15y^2$$

$$\Rightarrow \quad y^2 - 10y + 16.38 = 0$$

This quadratic equation is solved to give the distance from point B at which maximum deflection occurs, that is:

$$y = 2.064\ \mathrm{m}$$

The corresponding bending moment, M_y, is:

$$M_y = 150 - 30(2.064) = 88.1\ \mathrm{kNm\ (sag)}$$

141

By Theorem 2, the vertical deviation, t_{EB} (Fig. 4.50(b)), of the tangent at point E with respect to the tangent at point B is given by:

$$t_{EB} = \bar{x}_B \int_B^E \frac{M}{EI} \, dx$$

Since the tangent to E is horizontal, the total deflection at E is equal to the vertical deviation, t_{EB}. Thus:

$$\delta_{max} = \bar{x}_B \int_B^E \frac{M}{EI} \, dx$$

Rather than calculate the area under the M/EI diagram and the value for x_B separately, it is convenient to consider each portion of area under the M/EI diagram and the distance of its centroid from B. Hence, referring to Fig. 4.50(b):

$$\delta_{max} = \left(\frac{88.1}{EI}\right)y\left(\frac{y}{2}\right) + \frac{1}{2}\left(\frac{61.9}{EI}\right)y\left(\frac{y}{3}\right)$$

Substituting for $y = 2.064$ m and for E and I gives:

$$\delta_{max} = 0.0066 \text{ m}$$
$$= 6.6 \text{ mm}$$

Problems

Sections 4.1 and 4.2

4.1   For the beam illustrated in Appendix B, No. 6, verify that equilibrium is satisfied and draw the shear force diagram for the beam.

4.2   Verify the reactions and the bending moment diagram and construct the shear force diagram for the beam of Appendix B, No. 3.

Section 4.3

4.3   Use the stiffness method to find the bending moment diagram for the structure illustrated in Fig. 4.51.

Fig. 4.51
Structure for
Problem 4.3

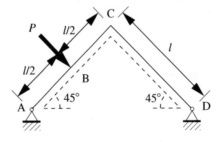

4.4   Use moment distribution to find the bending moment diagram for the structure of Problem 4.3.

4.5 Find the bending moment diagram for the beam whose geometry and loading are as illustrated in Fig. 4.52.

Fig. 4.52 Beam of Problem 4.5

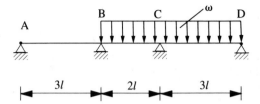

4.6 The structure illustrated in Fig. 4.53 resists horizontal forces by frame action. Find the bending moment diagram due to the horizontal loading given that the second moment of area of the beam, I_b, is twice that of the columns, I_c.

Fig. 4.53 Structure for Problem 4.6

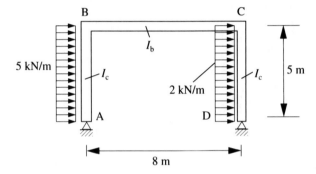

Section 4.4

4.7 For the structure of Problem 4.4, the yield moment for both sag and hog is $Pl/5$. Use elastic–plastic analysis to determine the yield load factor and the mechanism load factor.

4.8 For the structure of Problem 4.7, use plastic analysis to determine the mechanism load factor.

4.9 For the beam of Problem 4.5, the yield moment in sag is $0.8ql^2$, while that in hog is $0.6ql^2$. Determine the mechanism load factor.

Section 4.5

4.10 Find the shear force and axial force diagrams for the structure of Problem 4.7:

(a) for linear elastic bending;
(b) for the mechanism collapse condition.

5

Applications of Structural Analysis to Concrete Structures

5.1 Introduction

Owing to their monolithic construction, the majority of *in situ* concrete structures are made up of continuous statically indeterminate members. Precise methods of analysis of such three-dimensional structures can effectively only be carried out using a powerful computer and considerable effort on the part of the engineer. Therefore, in practice, most concrete structures are broken up into smaller, more manageable, two-dimensional sub-structures. These sub-structures are derived in such a way that they are easier to analyse (by hand or by computer) even though they may still be indeterminate. At the same time, these sub-structures model the behaviour of the actual structure within acceptable margins of accuracy. This chapter presents some of the more popular approaches to the modelling and analysis of concrete structures.

The more popular methods of analysis of two-dimensional indeterminate sub-structures, such as the stiffness method or the finite element method, are most conveniently executed on computer. It is widely accepted that these methods are, in most cases, too laborious to be carried out by hand. However, even when using computer programs, it is important to be able to check the results. For this reason, less exact or 'approximate' methods of analysis are required which are simple enough to be carried out by hand. These methods of analysis also give the designer insight into the behaviour of a structure. Finally, approximate methods of analysis can be used where an exact solution is not required or where time does not permit a more thorough analysis. For these reasons, this chapter concentrates on hand methods, rather than computer methods, suitable for the approximate analysis of common concrete sub-structures.

Apart from the yield line method for slabs, the methods of analysis presented in this chapter assume a linear elastic behaviour of the structural material. To allow for the actual inelastic behaviour of reinforced concrete under ultimate loads, the results of the elastic analyses are adjusted using a method known as plastic moment redistribution (not to be confused with the moment distribution method of analysis).

5.2   Continuous beam analysis

Continuous beam analysis is perhaps the most common analysis problem in concrete buildings. It can be applied to the approximate analysis of strips through one-way spanning slabs in addition to beams. Where a number of load cases have to be considered for continuous beams, computer programs (using the flexibility or stiffness method) are probably the most convenient method of analysis. However, to check the results from a computer, or to carry out a quick analysis, formulae such as those given in Appendix D can be applied to the member. Alternatively, a solution can be found using moment distribution, as illustrated in the following example.

Example 5.1   Continuous beam analysis

Problem   For the floor system of Fig. 5.1, find upper limits for (a) the maximum moment in a 1 m strip of slab at section X–X or Y–Y and (b) the vertical load in column B due to a factored uniformly distributed loading of 10 kN/m^2 acting throughout the slab.

Solution   (a) Maximum moment in slab

The slab at section X–X of Fig. 5.1 is spanning one way between beams AB, DE and GH (the slab will also span north–south to some degree but it is

Fig. 5.1   Geometry of floor for Example 5.1: (a) plan view; (b) section X–X; (c) section Y–Y

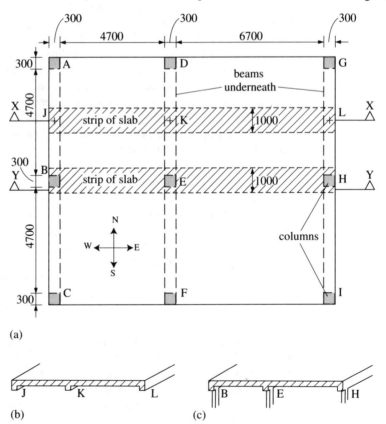

common practice in such situations to assume it to span between beams). Each of the beams acts as a support to the slab and, in turn, transfers the reactions from the slab to the columns. While the torsional stiffness of the supporting beams provides some resistance to rotation at points J, K and L, this is not generally very significant and it is common to assume that the strip of slab is simply supported by the beams as illustrated in Fig. 5.2. The moment distribution method can now be used to determine the bending moment diagram in this model due to the applied loading on a 1 m wide strip.

Fig. 5.2 Strip of slab at section X–X

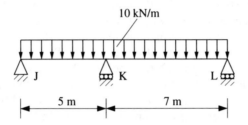

The fixed structure and the associated bending moment diagram are as illustrated in Figs 5.3(a) and (b). The process of moment distribution is started by releasing point K. The distribution factors at this point are equal to:

$$DF_{KJ} = \frac{3EI/L_{JK}}{(3EI/L_{JK}) + (3EI/L_{KL})} = \frac{1/5}{1/5 + 1/7} = \frac{7}{12}$$

$$DF_{KL} = 1 - DF_{KJ} = 1 - \frac{7}{12} = \frac{5}{12}$$

Therefore, to release the discontinuity of 30 kNm at point K, a correction of $-\frac{7}{12}(30)$ is made to the left of K and a correction of $+\frac{5}{12}(30)$ is made to the right of K. The associated carry-over factors are zero and the bending moment diagram associated with this correction is as illustrated in Fig. 5.3(c). The

Fig. 5.3 Analysis of continuous beam of Fig. 5.2 by moment distribution: (a) fixed structure; (b) bending moment diagram for fixed structure; (c) bending moment diagram associated with release of K; (d) bending moment diagram after release of K (equal to sum of (b) and (c))

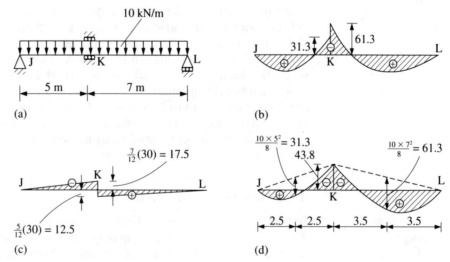

bending moment diagram after the release of point K is the sum of those illustrated in Figs 5.3(b) and (c) and is illustrated in Fig. 5.3(d). As equilibrium is now satisfied at all joints, no further distribution is required and the bending moment diagram of Fig. 5.3(d) is the final one.

From Fig. 5.3(d), it can be seen that the maximum moment in the strip of slab occurs at point K and is equal to 43.8 kNm. Since the ends of the slab are not really simply supported, there will in fact be a small hogging moment along the edge and, as a result, a nominal amount of top reinforcement should be provided along the edges of the slab.

For the strip of slab at section X–X, mid-way between the supporting columns, the assumed model of Fig. 5.2 is sufficiently accurate. However, for strips nearer the supporting columns, such as the strip at section Y–Y (see Fig. 5.1), the resistance to rotation at points B, E and H is more significant. Therefore, a more appropriate model for the strip of section Y–Y would be one which would allow for the rotational stiffness of the columns at these points. If the columns were very stiff, a suitable model would be that illustrated in Fig. 5.4(a) and the bending moment diagram would be as illustrated in Fig. 5.4(b) (see Appendix B, No. 8). The actual bending moment diagram at section Y–Y will, in fact, be something between those illustrated in Figs 5.3(d) and 5.4(b). For preliminary purposes, the envelope of moments from these figures could be used for design. For a more exact result, a frame analysis is necessary, as described in section 5.5.

Fig. 5.4 (a) Beam with infinitely stiff columns; (b) associated bending moment diagram

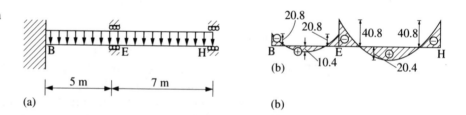

It can be seen that the central support and in-span moments in Fig. 5.4(b) are approximately the same as or smaller than those for the model of Fig. 5.2 and so the previous results for the simply supported model can be used for the design of all 1 m strips through the slab. An exception to this is at the outer supports where, as can be seen at B and H in Fig. 5.4(b), there exists a substantial hogging moment. For this reason, it is necessary to design the slab for a hog moment (up to 20.8 or 40.8 kNm/m, as appropriate) along its edges in the vicinity of all external columns. The precise distance away from each column in the north–south direction within which top reinforcement is provided is a matter for engineering judgement.

(b) Vertical load in column B

To find the maximum vertical load in column B, it is first necessary to find the reactions from each 1 m wide strip of slab in the east–west direction. Using the results from above, the reactions for strips at sections X–X and Y–Y can be

Fig. 5.5 Free body
diagram for part of
section X–X

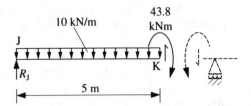

found by simple equilibrium. Consider, for example, the strip of slab at section
X–X. The free-body diagram for the strip of slab between J and K, derived
from the bending moment diagram of Fig. 5.3(d), is illustrated in Fig. 5.5. By
taking moments about point K we have:

$$5R_J + 43.8 = 10 \times 5 \times \tfrac{5}{2}$$

$$\Rightarrow \qquad R_J = 16.4 \text{ kN}$$

If the columns at section Y–Y are assumed to be infinitely stiff as in the model
of Fig. 5.4(a), half the reaction is taken at each support and:

$$R_B = \frac{10 \times 5}{2} = 25 \text{ kN}$$

Thus, the reaction transferred to beam ABC from any strip parallel to section
X–X is something between 16.4 kN and 25 kN. For the calculation of the
maximum force in column B, it is conservative (perhaps too much so) to assume
the greater of these values for all strips. The resulting loading on beam ABC
is a uniformly distributed load of 25 kN/m. The beam can now be analysed to
determine the reaction at B.

If the columns were assumed to have infinite rigidity, each span (AB and
BC) would be fixed at either end and the reaction at B would equal half the
load from each span, that is:

$$2\left(\frac{25 \times 5}{2}\right) = 125 \text{ kN}$$

If, on the other hand, the column stiffnesses were ignored, the beam would
be simply supported at A, B and C and the reaction at B (from Appendix C,
No. 4) would be:

$$R_B = \frac{5\omega L}{4} = \frac{5 \times 25 \times 5}{4} = 156 \text{ kN}$$

Hence, a conservative estimate for the maximum axial compression in column
B is 156 kN.

5.3 Plastic moment redistribution

Reinforced and prestressed concrete members with bending (i.e. beams and
slabs) are designed to have a certain ductility under ultimate loads. This

ductility ensures that the member is capable of undergoing a certain amount of rotation after yielding of the tension steel reinforcement and before crushing of the concrete in compression. The idealized moment–curvature relationship assumed for a member in bending can be seen in the graph of Fig. 4.36. The portion of this graph between κ_y and κ_{ult} represents the ductility. It will be shown in Chapter 7 that a limitation on the neutral axis depth is the mechanism by which ductility is guaranteed by EC2. As was seen in section 4.4, when members in bending have this ductility, they have the potential to continue to resist load beyond the time of initial yield. This feature was demonstrated in Example 4.9. If a member in bending was to be designed using the linear elastic assumption only, the maximum load factor would be the value at which the applied moment first equals M_{ult}. For the member of Example 4.9, this occurs at:

$$\lambda_y = \frac{32M_{ult}}{6PL} = \frac{5.33M_{ult}}{PL}$$

while the mechanism load factor for this beam has been shown to be:

$$\lambda_m = \frac{6M_{ult}}{PL}$$

Clearly, when a member has sufficient ductility to achieve the mechanism load factor, considerable savings can be achieved by recognizing this extra load-carrying capacity. Elastic–plastic analysis is impractical for everyday design because it requires a knowledge of the moment capacity of all members in advance of the analysis. A much simpler approach which circumvents this requirement is that of plastic moment redistribution. This is an approximate method by which the elastic bending moment diagram is adjusted to account for the ductile behaviour of reinforced and prestressed members in bending. EC2 allows the original elastic moment in continuous members to be reduced by an amount of up to 30 per cent. The precise amount of redistribution allowed is dependent on the grade of the concrete and on the ductility characteristics of the reinforcement as well as the neutral axis depth. Concrete is graded by the strength of test cylinders or cubes. Specifically, the following limits are imposed on the ratio of the redistributed moment to the moment before redistribution, δ. For concrete grades with cylinder compressive test strength of less than or equal to 35 N/mm²:

$$\delta \geq 0.44 + 1.25\frac{x}{d} \tag{5.1}$$

For concrete with strengths greater than 35 N/mm²:

$$\delta \geq 0.56 + 1.25\frac{x}{d} \tag{5.2}$$

where x is the distance from the extreme compressive fibre in the section to the neutral axis, and d is the effective depth to the tension reinforcement. Where high-ductility steel reinforcement is used, δ must be greater than or equal to 0.7 (i.e. maximum of 30 per cent redistribution). Where normal ductility steel is used, δ must be greater than or equal to 0.85.

The saving which results from reducing the elastic moment at a given section in a member using plastic moment redistribution is invariably offset by the fact that equilibrium requires the bending moment at other points in the beam to be increased. For this reason, the optimum amount of moment redistribution which should be carried out depends greatly on the geometry and loading of the member. The application of the technique is illustrated below in Example 5.2.

Example 5.2 Plastic moment redistribution

Problem The bending moment diagram for the two-span beam of Fig. 5.6(a), derived using an elastic method of analysis, is illustrated in Fig. 5.6(b). Two arrangements of reinforcement are available for design: two 20 mm diameter bars giving a moment capacity of 254 kNm, or two 25 mm diameter bars giving a moment capacity of 306 kNm. As the beam cross-section is rectangular, either pair of bars can be placed near the top of the section to resist hog or near the bottom to resist sag. Using plastic moment redistribution, determine the most suitable arrangement of reinforcement if the factored ULS load, P, is 150 kN and L is 10 m. It may be assumed that up to 30 per cent redistribution is allowed.

Fig. 5.6 Two-span beam: (a) geometry and loading; (b) elastic bending moment diagram; (c) reinforcement to resist elastic moments

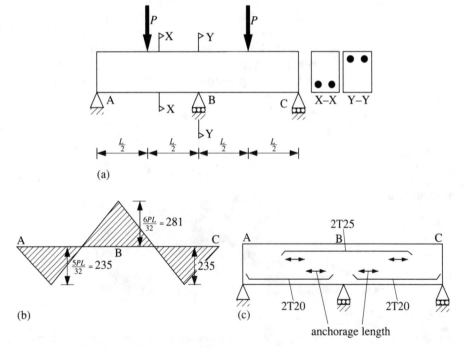

Solution For a factored ULS load of 150 kN, the original elastic bending moment over the central support is equal to 281 kNm (hog) and the maximum elastic in-span moment is equal to 235 kNm (sag). To resist these design elastic moments, two 20 mm diameter bars (2T20) of combined moment capacity 254 kNm are required in the interior of each span and two 25 mm diameter bars (2T25) of combined moment capacity 306 kNm are required over the central support (see Fig. 5.6(c)).

151

In Chapter 4, it was shown that a bending moment diagram can be constructed by superimposing the bending moment diagrams for simply supported spans on the bending moment diagram associated with the support moments. Thus, the elastic bending moment diagram of Fig. 5.6(b) is the sum of the bending moment diagrams illustrated in Figs 5.7(a) and (b). When a

Fig. 5.7 Components of bending moment diagram: (a) elastic bending moment diagram if spans are simply supported: (b) elastic bending moment diagram associated with support moment; (c) bending moment diagram associated with reduced support moment

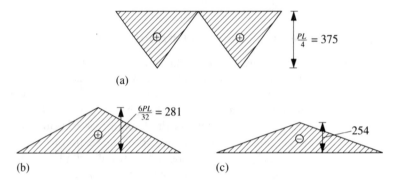

plastic hinge forms at a support it prevents any further increase in moment at that point. Now, if the moment capacity at B were to be limited to 254 kNm (two 20 mm diameter bars), a hinge would form there for an applied load less than 150 kN and, for any further increase in load, spans AB and BC would behave as if simply supported. The new bending moment diagram is constructed simply by superimposing the diagram associated with $M_B = -254$ kNm (Fig. 5.7(c)) on the diagram of Fig. 5.7(a). The result of this 10 per cent redistribution at B is illustrated in Fig. 5.8 (the elastic bending moment is shown dotted for

Fig. 5.8 Bending moment diagram before and after plastic redistribution

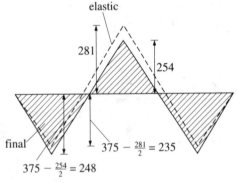

comparison). It can be seen that, even though the in-span moment has increased as a result of the redistribution, it can still be resisted using two 20 mm bars (capacity = 254 kNm). Hence, for a redistribution of 10 per cent, a significant saving in reinforcement has been made. It is clear that any further increase in the amount of moment redistribution would result in greater steel requirements at the interior of each span (i.e. redistribution moment would exceed 254 kNm).

5.4   The implications of lower-bound methods

Lower-bound methods are methods of design where a distribution of bending moment is assumed which satisfies the requirements of equilibrium but not necessarily compatibility. Thus, it may not be the actual distribution of internal moment in the structure that would be found by linear elastic analysis. It has been proven by a theorem known as the static theorem (Heyman 1971) that a structure of infinite ductility is safe under its applied loads if a distribution of internal bending moment can be found that satisfies the equilibrium requirements and nowhere exceeds the ultimate moment capacity. The following examples illustrate the application of lower-bound methods.

Example 5.3   Two-way spanning slab

The square reinforced concrete slab illustrated in Fig. 5.9(a) has been provided with reinforcement in both directions. If the moment capacities in both directions are sufficiently high that no plastic hinge rotation takes place, then the bending moment diagrams will equal those found by a linear elastic (finite element) analysis (see Fig. 5.9(b)). Thus, if the moment capacity of the slab exceeds 18.4 kNm in both directions, the slab will not yield under the applied loading.

Now consider the same slab reinforced in one direction only, say with bars parallel to the Y-axis. The bending moment diagram for a beam spanning in one direction only and carrying a uniformly distributed load of 12 kN/m is illustrated in Fig. 5.9(c). Thus, if a series of slabs each 1 m wide are laid spanning from AB to CD and if each has a moment capacity of 54 kNm, the structure will remain safe even though no reinforcement parallel to the X-axis is present. It would seem logical, therefore, to assume that if a band of reinforcement with a moment capacity of 54 kNm per metre width is provided parallel to the Y-axis in the solid slab of Fig. 5.9(a), then the slab will not collapse.

The kind of logic used in Example 5.3 is referred to as the lower-bound approach and it is reasonable if and only if the structure has sufficient ductility to allow the assumed bending moment distribution to occur. In the case of the slab reinforced in one direction, there would be considerable cracking parallel to the reinforcement (i.e. yielding in unreinforced direction) before the load is taken by the bands of reinforcement. A certain minimum quantity of reinforcement is provided in all concrete structures (more in Chapter 7). It is generally felt that if the assumed distribution of bending moment is reasonably close to the linear elastic distribution, then slabs will usually have sufficient ductility to allow design by lower-bound methods.

Example 5.4   Portal frame

For the portal frame of Fig. 5.10(a), preliminary sizing rules for the members result in the following preliminary dimensions (rules for preliminary sizing are

Fig. 5.9 Two-way spanning slab: (a) plan view; (b) distribution of moment per unit width from linear elastic analysis; (c) distribution of moment in one-way spanning beam

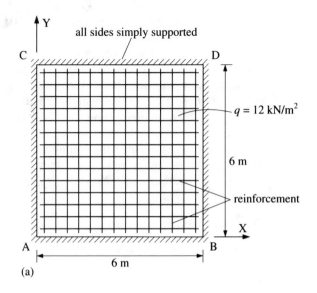

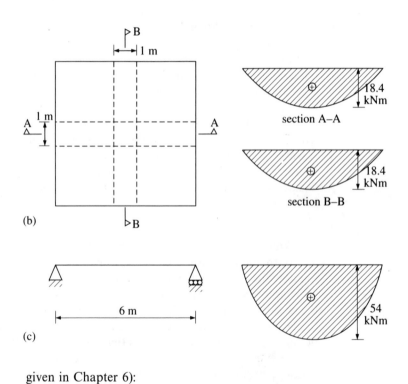

given in Chapter 6):

> beam depth, h = 450 mm
>
> beam breadth, b = 250 mm
>
> column depth and breadth = 250 mm

It is common practice to calculate the second moments of area of members

Fig. 5.10 Portal frame: (a) geometry and loading; (b) bending moment diagram for $I_{beam} = 1.9 \times 10^{-3}$; (c) bending moment diagram for $I_{beam} = 5.72 \times 10^{-3}$

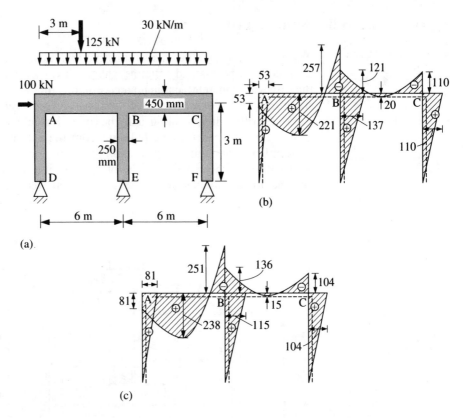

neglecting the reinforcement and cracking of the members. Thus:

$$I_{beam} = bh^3/12 = 1.9 \times 10^{-3} \text{ m}^4$$
$$I_{col} = h^4/12 = 0.325 \times 10^{-3} \text{ m}^4$$

A computer analysis (stiffness method) of the frame under the applied loads, using these second moments of area, gives the bending moment diagram of Fig. 5.10(b). In checking the moment capacity of each member against these applied moments, it is found that the preliminary size of the beam is inadequate for the applied moment of 257 kNm at B. Therefore, the beam section throughout ABC is increased to 650 mm which increases I_{beam} to 5.72×10^{-3} m^4. It has been shown previously in Example 4.5 that the relative values for the second moments of area in members affect the distribution of moment. However, the distribution of Fig. 5.10(b) does satisfy equilibrium even if it is now incorrect. Therefore, by the static theorem, it is not necessary to reanalyse the structure if it has sufficient ductility in the members to facilitate redistribution of the moments. For this example, the increase in beam stiffness is so large that there may not be adequate ductility.

Note: Linear elastic analysis for the same geometry and loads but using the new $I_{beam} = 5.72 \times 10^{-3}$ m^4 results in the bending moment diagram of Fig. 5.10(c). It can be seen that, at some points (point A, for example), the moment has increased significantly.

155

5.5 Analysis of frames

Apart from flat slab (beamless) construction, most concrete buildings contain a structure of beams and columns which, when rigidly connected (as is usually the case with concrete), make up a continuous frame. Figure 5.11 illustrates the structural framework of one such multi-storey concrete building. The framework of this building would usually be concealed behind cladding panels which protect the occupants of the building from the external environment. As can be seen from the figure, lateral stability of this frame in one direction (E–W) is provided by shear walls. In the other direction, no shear walls are present and the horizontal stability of the frame is achieved through frame action.

Fig. 5.11 Three-dimensional skeletal frame

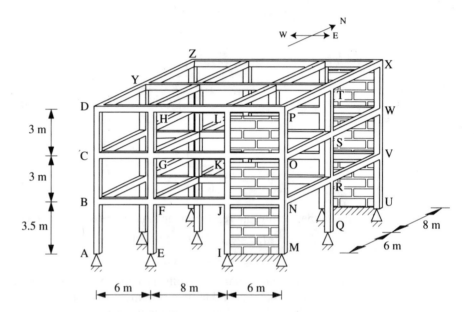

The analysis of a complete three-dimensional frame, such as that in Fig. 5.11, can be carried out by hand or by computer using any appropriate method such as, for example, the stiffness method. However, the mathematical complexity of the solution process generally makes it unfeasible to analyse a complete three-dimensional structure by hand. Even when analysing by computer, the solution may become unduly complex. One particular aspect of analysis which makes it as yet impractical to design a complete three-dimensional structure is the need to consider all possible arrangements of load. In theory, every possible combination of permanent, variable and wind loading must be considered to determine the critical load effects in each member. The greater the number of members in the frame, the greater the number of possible combinations of applied load. For this reason, certain assumptions and simplifications are commonly made before the structure is analysed.

To overcome the complexity of considering the full multi-storey skeletal structure and to facilitate a solution by hand, a common simplification is to represent the three-dimensional frame with smaller, two-dimensional sub-frames. This substantially reduces the total number of load cases which must be considered for each sub-frame and simplifies the process of describing the structural model to the computer. The precise method of simplification depends on whether or not the original frame is braced against horizontal loads. A frame which is braced against horizontal loads using substantial bracing members such as cores and/or shear walls is termed a **non-sway frame**. Owing to the presence of such stiff bracing members, there is little or no lateral deflection in a non-sway frame. For this reason, such a frame is designed to resist only the applied vertical loads (i.e. the bracing members are designed to carry the horizontal loads). A frame which does not have shear walls, cores or other bracing members transmits the horizontal loads to its foundations by frame action. This type of frame undergoes significant horizontal deflection under applied horizontal loads and hence is known as a **sway frame**. Sway frames must be designed to resist both vertical and horizontal loads.

Analysis of non-sway frames

Consider the frame of Fig. 5.11 which is braced against lateral loads in an E–W direction by masonry shear walls and contains *in situ* concrete floor slabs at each floor level. The first simplification which can be made is to assume that, in the E–W direction, the frame can be represented by three two-dimensional non-sway frames, as illustrated in Fig. 5.12. Note that the horizontal load applied to the central frame is transmitted by membrane action in the floor slabs to the shear walls. Note also that the vertical loadings for the two outer plane frames are the same and hence only one needs to be analysed. The central plane frame carries a greater vertical load since it supports a greater floor area. The masonry shear walls are designed to resist horizontal loads and are detailed so as not to resist any vertical loads.

Fig. 5.12 Two-dimensional frame

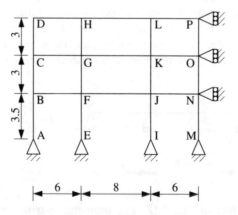

The plane frame of Fig. 5.12 can readily be analysed by computer for each possible arrangement of load. However, two alternative methods are available for further simplifying the plane frame to facilitate a hand solution. The first of these methods is to divide the plane frame into a set of sub-frames, each of

Fig. 5.13 Sub-frames
for the frame of Fig.
5.12: (a) top; (b) middle;
(c) bottom

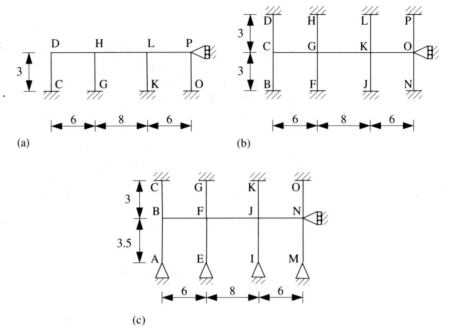

(a) (b)

(c)

which is analysed separately. Each sub-frame is made up of the beams at one
level together with the columns connected to these beams. For example, the
plane frame of Fig. 5.12 can be divided into the three sub-frames of Fig. 5.13.
The columns meeting the beams are assumed to be fixed at their ends as
illustrated (except when the assumption of a pin-ended column is more suitable
as would often happen at foundation level). These sub-frames can readily be
analysed by hand using, say, the moment distribution method to give the
moments, shears, etc., in both the beams and the columns.

A less exact, and not necessarily more conservative, simplification than the
sub-frame is to assume that the beams are continuous over the supporting
columns and that the columns provide no restraint to the rotation of the
beams. Therefore, the beams in each plane frame can be analysed as continuous
beams on simple supports. Figure 5.14 shows one such continuous beam for

Fig. 5.14 Continuous
beam model

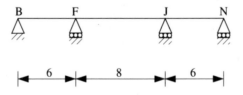

the plane frame of Fig. 5.12. The moment distribution method may then be
used to determine the moments, shears and reactions in each beam. The
moments in the columns can be determined by analysing models such as that
of Fig. 5.15 on the assumption that the ratio of I to L for the beams is at
one-half of their actual values.

Fig. 5.15 Sub-frames
for column analysis

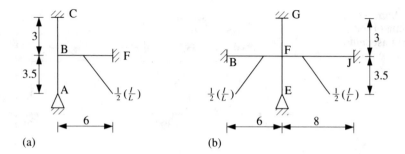

(a)   (b)

The analysis procedure for a typical braced frame, indicating the arrangements of loading which must be considered for a typical model, is illustrated
by the following example.

Example 5.5   Analysis of non-sway frame

Problem   For the structure of Fig. 5.11, determine the maximum sagging moment in
member FJ. The characteristic floor loads are:

$$\text{permanent gravity (incl. self-weight)} = 6.0 \text{ kN/m}^2$$
$$\text{imposed} = 4.0 \text{ kN/m}^2$$
$$\text{wind (any direction)} = 1.0 \text{ kN/m}^2$$

The flexural ridigity, *EI*, is constant for all members throughout the frame.
The floor consists of precast prestressed concrete slab strips which span
N–S (only). The distribution of horizontal wind load through the slab to the
shear walls is achieved through a 75 mm thick structural screed on top of the
precast units.

Solution   The maximum bending moment in member FJ is found by analysing the plane
frame of Fig. 5.12. Owing to the presence of the shear wall in the end bays, the
frame is effectively braced against wind from an easterly or westerly direction,
as illustrated in the figure. Thus, to factilitate a hand solution, the frame
can be simplified into the three sub-frames of Fig. 5.13 of which only that of
Fig. 5.13(c) is of interest for member FJ.

The specified permanent load acting at each level of the frame of Fig. 5.12
includes the weight of the portion of slab carried by the frame and the self-weight
of the beams. Figure 5.16 illustrates two sections, E–W and N–S, through a
typical slab in the frame of Fig. 5.11. The weight of the slab spanning N–S (Fig.
5.16(a)) is carried equally by the supporting beams. Thus, the total characteristic
permanent load acting on the beams in the outer frame is:

$$g_k = (0.5)(6 \text{ m})(6 \text{ kN/m}^2)$$
$$= 18.0 \text{ kN per metre length}$$

In a similar manner, the characteristic variable gravity floor load carried by
the frame is given by:

$$q_{ik} = (0.5)(6 \text{ m})(4 \text{ kN/m}^2) = 12.0 \text{ kN/m}$$

159

Fig. 5.16 Sections through slab: (a) N–S (view from East); (b) E–W (view from North)

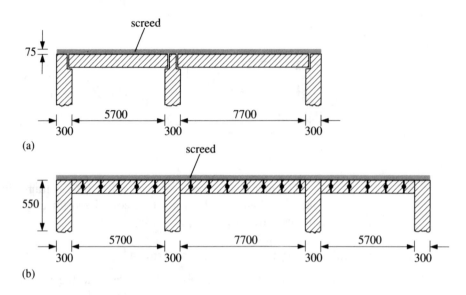

(a)

(b)

Load can have either a favourable or an unfavourable effect on a member and it is necessary to determine which, before the appropriate load factors can be selected. For the sub-frame of Fig. 5.13(c), the maximum distributed load for design at ULS is:

$$1.35(18.0) + 1.5(12.0) = 42.3 \text{ kN/m}$$

and the minimum distributed load for design at ULS is:

$$1.0(18.0) = 18.0 \text{ kN/m}$$

The loading arrangement which gives the maximum sagging moment in member FJ is maximum distributed load on member FJ and minimum distributed load elsewhere, as illustrated in Fig. 5.17. Using the moment distribution method, the sub-frame of Fig. 5.17 can readily be analysed by hand. The resulting bending moment diagram is illustrated in Fig. 5.18. From this figure, it can be seen that the maximum sagging moment in member FJ is approximately equal to 129 kNm.

Fig. 5.17 Loading for maximum sag in member FJ

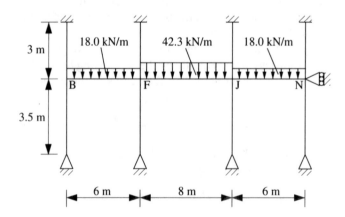

Fig. 5.18 Bending
moment diagram for
loading of Fig. 5.17

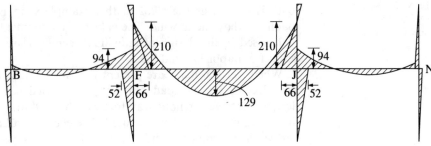

Fig. 5.19 Continuous
beam model:
(a) geometry and
loading; (b) bending
moment diagram

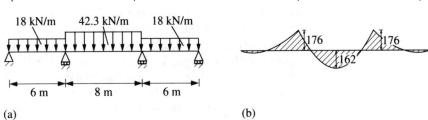

(a)

(b)

Alternatively, the maximum sagging moment in member FJ can be estimated
using the continuous beam approximation of Fig. 5.19(a). The bending moment
diagram for this model is illustrated in Fig. 5.19(b). Comparing the diagrams
of Figs 5.18 and 5.19(b), it can be seen that, for this example, there is a significant
variation in results between the two models with the simpler model of Fig.
5.19(a) giving more conservative results for the maximum sagging moment in
member FJ.

Analysis of sway frames

Consider the frame of Fig. 5.11 which is not braced against lateral loads in the
N–S direction. As for non-sway frames, the first simplication which can be made
is to assume that, in the N–S direction, the frame can be represented by four
two-dimensional sway frames, one of which is illustrated in Fig. 5.20. Each
of the four plane frames is then analysed for all realistic combinations of

Fig. 5.20 Sway frame

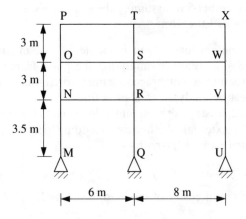

load. However, as the slabs in this example span N–S, the only significant loading is that due to wind. The values of the moments, shears and axial forces to be used in the design of individual members are the critical values which result from applied wind load.

Where vertical loads are present in a sway frame, an analysis to determine the effect of vertical load only may be carried out in the same manner as described above for non-sway frames. Thus sub-frame models, such as those illustrated in Fig. 5.13, can be used. The effects of all applied loads, horizontal and vertical, acting together are then found by superposing the effects of the vertical loads with those of the horizontal loads.

If the effects of the horizontal loads are to be calculated by hand, the plane frames may be simplified by assuming that points of contraflexure occur at the centre of all members, as illustrated in Fig. 5.21. As the moment at points of contraflexure is zero, a hinge can be assumed at the centre of each member, as illustrated in Fig. 5.21(b). This simplification then allows the frame to be analysed using either the **portal method** or the **cantilever method**. The following example illustrates the analysis of a sway frame using the portal method.

Fig. 5.21 (a) Deflected shape; (b) points of contraflexure

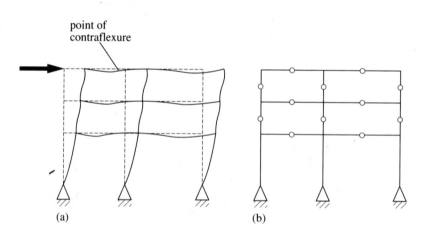

point of contraflexure

(a)   (b)

Example 5.6 Analysis of sway frame

Problem With reference to the data from Example 5.5, estimate the maximum hogging moments in member SW, assuming the southerly wind results in a characteristic load intensity of $1.0 \, kN/m^2$.

Solution Using the same approach as in Example 5.5, the maximum bending moment in member SW is found by analysing the plane frame of Fig. 5.20. However, since the structure is not braced against horizontal loads, the plane frame of Fig. 5.20 must be analysed as a sway frame.

Figure 5.22 illustrates the assumed load paths for the horizontal wind loads acting on the external south-facing cladding panels. The characteristic force acting on each panel is given by:

$$Q_{wk} = q_{wk} A_p$$

where q_{wk} is the intensity of the load, given as $1.0 \, kN/m^2$, and A_p is the total

Fig. 5.22 Assumed
load paths

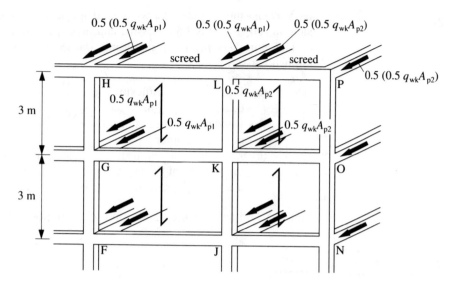

area of the panel. Each panel is assumed to act as a simply supported, one-way spanning vertical slab, transferring the load to the beams at the top and bottom edges of the panel. In the external panels, half of the load is then transferred by the structural screed to the outer frames, as illustrated in Fig. 5.22. Thus, the characteristic force acting at the top of the external frames is:

$$0.5(0.5q_{wk}A_{p2}) = 0.25q_{wk}A_{p2}$$

that is, the tributary area is one-quarter the area of the complete external panel. For point P, the applied force is thus:

$$0.25q_{wk}A_{p2} = 0.25(1)(3 \times 6) = 4.5 \text{ kN}$$

Similarly, for points O and N, the tributary areas are $0.5(3 \times 6) \text{ m}^2$ and $0.5(3.25 \times 6) \text{ m}^2$ respectively and the forces are:

point O: $1 \times 0.5(3 \times 6)$ $= 9 \text{ kN}$

point N: $1 \times 0.5(3.25 \times 6) = 9.75 \text{ kN}$

Thus, the total loading on the frame is as illustrated in Fig. 5.23. For the

Fig. 5.23 Frame
loading

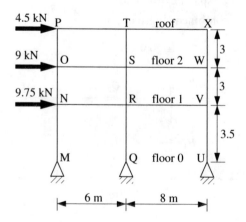

required moment effect, the wind load is the principal variable load, so the factored ULS load is 1.5 times the characteristic value.

The portal method for analysing frames with lateral loads makes the following assumptions:

(a) Points of contraflexure occur at the centre of all columns and beams (except pin-ended columns).
(b) Following from assumption (a), a hinge is assumed at the centre of each member.
(c) At each level, the frame is divided into single portal frames and the portion of the lateral load at that level which is carried by each frame is proportional to its span length.

Figure 5.24 illustrates the application of assumptions (a) and (b) to the frame

Fig. 5.24 Frame showing ULS loading and assumed hinges

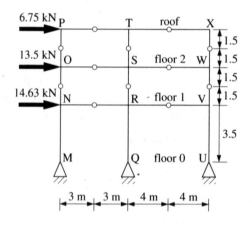

of Fig. 5.23. In this figure also, the wind loads have been factored to their ultimate (ULS) values. At roof level, the frame can be divided into two separate portal frames, as illustrated in Fig. 5.25. By assumption (c), $Q_1 + Q_2 = 6.75$ kN where the magnitudes of Q_1 and Q_2 are given by:

$$Q_1 = \frac{\text{span length of portal}}{\text{combined length of portals}} \times 6.75$$

$$= \frac{6}{6+8}(6.75) = 2.89 \text{ kN}$$

Fig. 5.25 Roof level frames: (a) left portion; (b) right portion

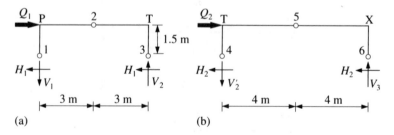

Similarly:

$$Q_2 = \frac{8}{6+8}(6.75) = 3.86 \text{ kN}$$

By symmetry of Figs 5.25(a) and (b), we have:

$$H_1 = \tfrac{1}{2}Q_1 = \tfrac{1}{2}(2.89) = 1.45 \text{ kN}$$

and:

$$H_2 = \tfrac{1}{2}Q_2 = \tfrac{1}{2}(3.86) = 1.93 \text{ kN}$$

Taking moments about hinge 2 of Fig. 5.25(a) gives:

$$3V_1 = 1.5H_1$$
$$\Rightarrow \quad V_1 = 0.73 \text{ kN}$$

Taking moments about hinge 5 of Fig. 5.25(b) gives:

$$4V_3 = 1.5H_2$$
$$\Rightarrow \quad V_3 = 0.73 \text{ kN}$$

Since V_1 and V_3 are of equal magnitude but opposite in direction, so too are V_2 and V_2'. Hence, it can be seen that the portal method assumes that the only vertical forces are in the two outer columns. From these results, the free-body diagrams for the two portal frames of Fig. 5.25 can be redrawn as illustrated in Fig. 5.26. The bending moment diagram of Fig. 5.27 can then be deduced from Fig. 5.26.

Fig. 5.26 Roof level frames with reactions: (a) left portion; (b) right portion; (c) complete frame

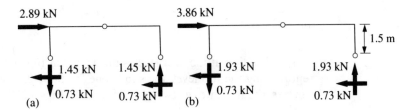

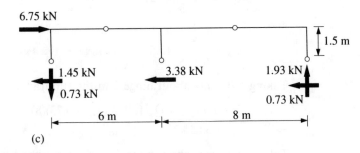

Fig. 5.27 Bending moment diagram for roof level frame

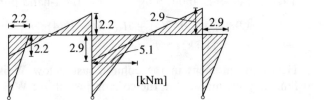

Fig. 5.28   Floor 2
frames: (a) complete
frame; (b) left portion;
(c) right portion

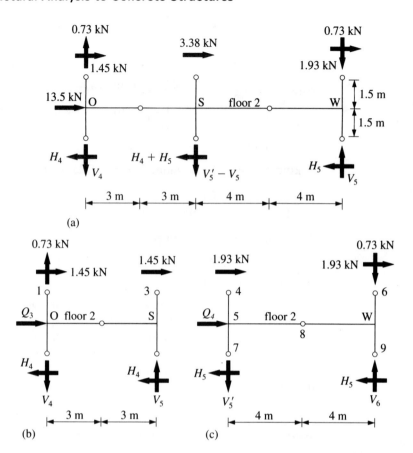

The situation at the next level is illustrated in Fig. 5.28. As for the sub-frame representing the top level, the sub-frame for floor 2, illustrated in Fig. 5.28(a), can be separated into the two portal frames of Figs 5.28(b) and (c). As before:

$$Q_3 = \frac{6}{6+8} (13.5) = 5.79 \text{ kN}$$

and:

$$Q_4 = \frac{8}{6+8} (13.5) = 7.71 \text{ kN}$$

Taking moments about hinge 7 in the sub-frame of Fig. 5.28(c):

$$2 \times 1.93(3) + Q_4(1.5) = (V_6 - 0.73)(8)$$
$$\Rightarrow \quad V_6 = 3.63 \text{ kN}$$

Taking moments about hinge 8 in the right-hand portion of that sub-frame:

$$1.93(1.5) + 0.73(4) + H_5(1.5) = V_6(4)$$
$$\Rightarrow \quad H_5 = 5.80 \text{ kN}$$

Hence the moment in the column just below W is ($H_5 \times 1.5 =$) 8.7 kNm. Similarly the moment in the column just above W is (1.93 × 1.5 =) 2.9 kNm.

Fig. 5.29 Bending moment diagram for sub-frame of Fig. 5.28(c)

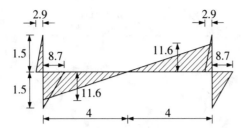

By equilibrium of moments at W, the moment in the beam at this point is (2.9 + 8.7 =) 11.6 kNm. Hence the bending moment diagram for this sub-frame is as illustrated in Fig. 5.29 and the maximum moment due to wind in member SW is 11.6 kNm.

5.6 Analysis of slabs

Slabs are the structural systems which actually create the living/working floor space in a structure by spanning the voids between framing members. Most permanent and variable gravity loads are applied directly to the floor slabs which then transfer the load primarily by bending action to the supporting beams and/or columns. It was shown in section 2.2 that there are many different forms of construction for slabs. The precise system which is used depends on a number of factors including the required span distances and the magnitude of the applied vertical loads.

While beams extend only in one direction at a time, slabs extend in two directions, thus filling a complete two-dimensional surface. Hence, slabs can be thought of as the two-dimensional equivalent of beams since bending occurs about two perpencidular axes, as illustrated in Fig. 5.30. Using this analogy, the moment–curvature relationship for a beam is replaced by a two-dimensional equivalent in the case of a slab.

The distributions of stress in a small segment of slab are illustrated in Fig. 5.30(c). Direct stress on the face perpendicular to the X-axis, known as the X-face, is denoted σ_x. As in a beam, the distribution of direct stress is the result of an internal bending moment. On the X-face, this moment per unit breadth is denoted m_x, as illustrated in Fig. 5.30(d). Note that, for slabs, m_x is not moment per unit breadth **about** the X-axis but moment per unit breadth for which **reinforcement parallel** to the X-axis would be required. Similarly, there is a direct moment per unit breadth on the Y-face which is associated with the distribution of the direct stress, σ_y.

In addition to these direct moments, there are two twisting moments per unit breadth of equal magnitude. They are m_{xy}, on the X-face, and m_{yx}, on the Y-face, and it is them which cause the shear stresses τ_{xy} and τ_{yx}, respectively. These twisting moments tend to cause warping in the slab as illustrated in Fig. 5.31. Finally, there are vertical shear forces in slabs, V_x and V_y, on the X- and Y-faces, respectively. These shear forces result in the distribution of stresses τ_{xz} and τ_{yz}, respectively.

167

Fig. 5.30 Slab with moments and stresses: (a) slab; (b) beam; (c) stresses in portion of slab; (d) moments and shear forces on portion of slab

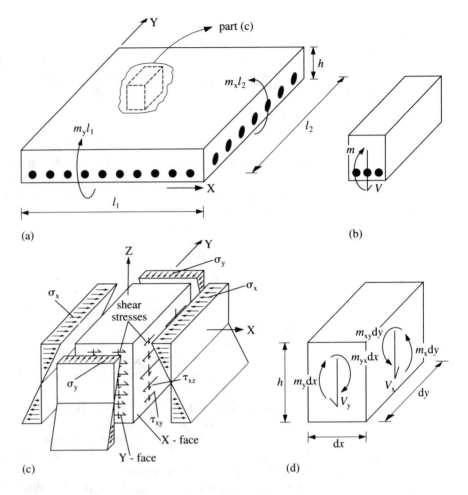

(a)

(b)

(c)

(d)

From a design viewpoint, slabs are hardly more complex than beams. Reinforcement parallel to the X-axis (ordinary reinforcement or prestress) is provided to resist m_x and reinforcement parallel to the Y-axis is provided to resist m_y. The concrete is assumed to be capable of resisting compressive stresses σ_x and σ_y simultaneously without any reduction in strength. Similarly, V_x and V_y are resisted in the same way as for beams. The twisting moments, m_{xy} and m_{yx}, however, are unique to slabs. Depending on their sign, these moments can act with or against m_x and m_y as can be seen from Fig. 5.30(d) (m_{yx} with/against m_x and m_{xy} with/against m_y). Formulae for the combination of these moments for design purposes are given in Appendix E. Fortunately, for most slabs, the twisting moments are relatively small and can frequently be neglected. It is only when slabs are skewed or where the corners of rectangular slabs are held down that there may be areas where the twisting moments are of significant magnitude locally.

Although all slabs extend in two directions, some forms of slab construction result in the applied loads being transferred predominantly in one direction. This may arise either if the slab is supported only along two parallel edges or

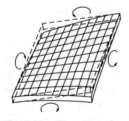

Fig. 5.31 Warping of slab

Fig. 5.32 One-way
spanning slab: (a) slab;
(b) strips of slab treated
as beams; (c) section
X–X

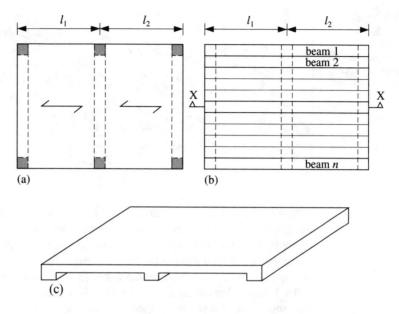

if its span in one direction is much longer than its span in the other direction. In the latter case, the majority of the load is transferred across the shorter span since this has a greater stiffness. In slabs of this type, known as one-way spanning slabs, the moment about one axis is negligible compared with the moment about the other axis and the slab can be thought of as a series of adjoining beams, as illustrated in Fig. 5.32. Consequently, the analysis of one-way spanning slabs is considerably more straightforward than the analysis of two-way spanning slabs.

EC2 allows a slab to be analysed as one-way spanning if either:

(a) it possesses two unsupported edges which are (sensibly) parallel; or
(b) it is the central part of a (sensibly) rectangular slab supported on four edges with a ratio of the longer to shorter span greater than two.

Linear elastic methods

As described in Chapter 4, linear elastic methods of analysis are based on the assumption of linear elastic behaviour. Linear elastic analysis of one-way spanning slabs is carried out in the same way as the linear elastic analysis of beams. In effect, this means taking, say, 1 metre wide strips of slab and carrying out a beam analysis for a typical strip. An example of a one-way spanning continuous slab analysis was given in Example 5.1.

Linear elastic analysis of two-way spanning slabs, both flat (beamless) slabs and slabs supported by beams, is usually carried out on computer using one of two common methods, both applications of the stiffness method: grillage analysis and finite element analysis. In a grillage analysis, the slab is idealized as a mesh of beams, as illustrated in Fig. 5.33, which, being skeletal, can readily be analysed by the stiffness method. Each beam in the grillage mesh represents

169

Fig. 5.33 Grillage
analysis of slab:
(a) actual slab;
(b) grillage of beams
(from Hambly (1991)

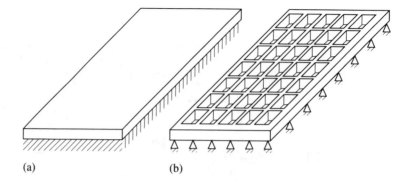

(a)   (b)

a strip of the slab and the stiffnesses of the beams in the mesh can be calculated
on this basis. Grillage analysis is reasonably accurate for most problems except
in areas of high concentrated loads such as near a support. However, there are
a number of special guidelines recommended for the proper application of the
grillage method. Interested readers should refer to Hambly (1991).

In a finite element analysis, the slab is idealized as a collection of discrete
slab segments or elements joined only at nodes, as illustrated in Fig. 5.34. The

Fig. 5.34 Finite
element analysis of slab:
(a) actual slab; (b) finite
element mesh; (c) detail
between two nodes

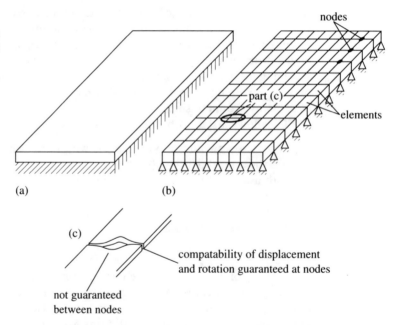

analysis of such a model can be viewed as a special case of the stiffness method.
For accurate results, a reasonably large number of elements should be
used especially in regions where moments are changing rapidly, such as over
supports. Unlike the grillage method, guidelines for the application of the
finite element method are relatively straightforward. Elements should not be
excessively long and narrow as this tends to result in an accumulation of errors
due to the rounding off of large numbers, especially in less accurate computers.

Fig. 5.35 Length to breadth ratios for finite elements: (a) good; (b) poor

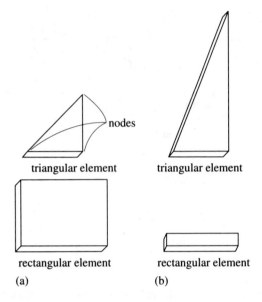

triangular element   triangular element

nodes

rectangular element   rectangular element

(a)   (b)

As a general guide, results should be reasonable provided the length/breadth ratio of elements is less than or equal to two, as illustrated in Fig. 5.35. It is also important to remember that elements are only joined at the nodes (which are usually at the corners in less sophisticated elements). Thus, for good modelling, it is necessary to avoid 'T-junctions' such as illustrated at point A in Fig. 5.36(b). At such a point, elements 2 and 3 are joined to each other but element 1, not having a node at A, is not joined to either of the other two elements. For this reason, it is possible for element 1 to have a different deflection than elements 2 and 3 at this point.

Fig. 5.36 Connections between elements: (a) good; (b) poor

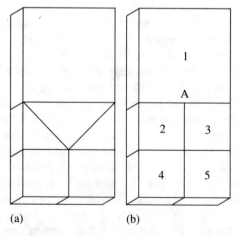

(a)   (b)

Example 5.7   Finite element method

Problem   For the slab of Fig. 5.37(a), use a linear elastic analysis to find the distribution of internal bending moment due to an applied uniformly distributed vertical loading of 15 kN/m².

Fig. 5.37 Slab of Example 5.7: (a) actual slab; (b) finite element mesh

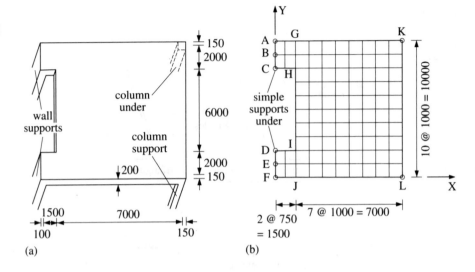

(a)

(b)

Solution It is quite inappropriate, for the purposes of everyday design, to expend an excessive amount of effort to obtain a highly accurate linear elastic analysis of a minor building slab. This is particularly true considering the implications of the assumptions such as the degree of rotational fixity at the supports. Nevertheless, available software now makes the generation of a finite element mesh very easy and the solution time is negligible on typical microcomputers.

For the mesh of elements illustrated in Fig. 5.37(b), the 1 m element width in the Y-direction has been chosen partly for convenience and partly to give a minimum of two elements across the panel ACHG. Similarly, in the X-direction, two elements are provided across this panel and the spacing is chosen so that the corner points, H and I, coincide with the nodes between elements. In a more accurate analysis, additional elements would be provided in the region of points H and I (because the moment in this region is rapidly changing) but this seems hardly appropriate for everyday design.

The resistance of the supporting columns and walls to rotation has been ignored and only simple (translational) supports have been provided at these points. This means that provision for the hog moment that will arise in this region will need to be made using engineering judgement. (The rotational stiffness of the columns can, in fact, be modelled by representing each column as a simple support plus springs which resist rotation. If the column is fixed at its other end, the spring stiffnesses are, from Appendix A, No. 1, $4EI_x/h$ and $4EI_y/h$ for the two coordinate directions.) Analysis of this model gives the distributions of moment per unit breadth illustrated in Fig. 5.38. Reinforcement parallel to the X-axis should be provided to resist m_x and reinforcement parallel to the Y-axis should be provided to resist m_y. Some additional reinforcement in both directions is required near the corners to resist m_{xy}. The calculation of this is given in Appendix E.

The grillage method is currently widely used in design offices for the analysis of larger and more complex slabs (such as bridge decks). The finite element

172

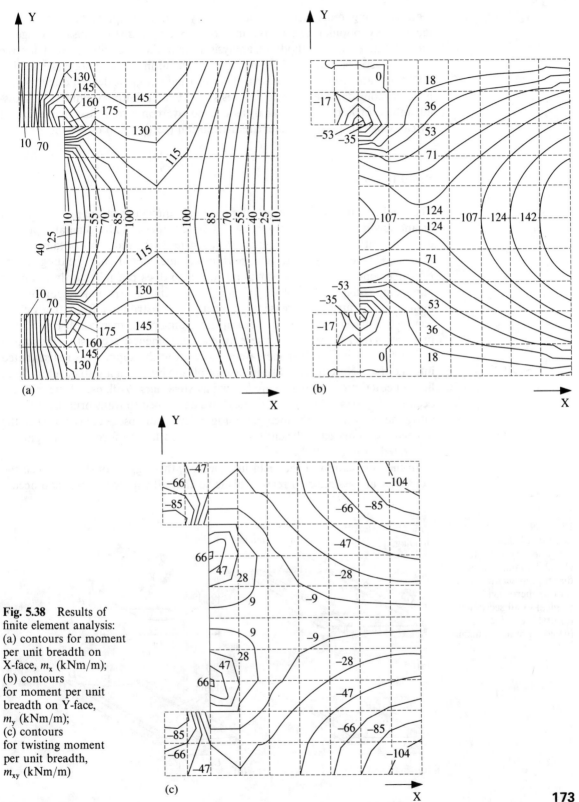

Fig. 5.38 Results of finite element analysis: (a) contours for moment per unit breadth on X-face, m_x (kNm/m); (b) contours for moment per unit breadth on Y-face, m_y (kNm/m); (c) contours for twisting moment per unit breadth, m_{xy} (kNm/m)

method, being more accurate and usually easier to apply, has become an increasingly popular alternative in recent years. A distinct disadvantage of these linear elastic methods of analysis is that they do not account for the elastic–plastic behaviour of the members at the ultimate limit state (see section 4.4). For slabs (more than beams) there is usually a great deal of ductility and hence a considerable capacity to resist load after initial yielding. This disadvantage of linear elastic analysis can be overcome to a reasonable extent by carrying out plastic moment redistribution (section 5.3) of the results.

Yield line method

The yield line method is the two-dimensional equivalent of the plastic method of analysis described in section 4.4. As stated above, slabs generally possess great ductility and hence a capacity to redistribute the moments obtained from a linear elastic analysis. It is important to understand that yielding of the reinforcement at any particular point in a slab does not necessarily imply an impending collapse of the slab. When a plastic hinge occurs in a region of high moment, any further increase in moment is redistributed from the yielded section to adjacent sections. For example, yielding of the member of Fig. 5.39(a) first occurs over the supports where the elastic moment is greatest, as illustrated in Fig. 5.39(c). Any further increase in moment at the supports is redistributed into the span. Note that, unlike in beams, plastic hinges in a slab develop along lines of equal moment which are known as **yield lines**. Collapse of the member occurs only when sufficient yield lines have developed to transform the member into a mechanism. For the member of Fig. 5.39(a), collapse occurs only when the moment is increased sufficiently to cause a yield line to form at mid-span, as illustrated in Fig. 5.39(d).

In the application of the yield line method, the regions of slab between the yield lines are assumed to remain plane so that all rotation of the slab occurs

Fig. 5.39 Yield line analysis of slab: (a) geometry; (b) linear elastic bending moment diagram at section X–X on one metre strip (typical of all sections); (c) first yield lines; (d) collapse mechanism

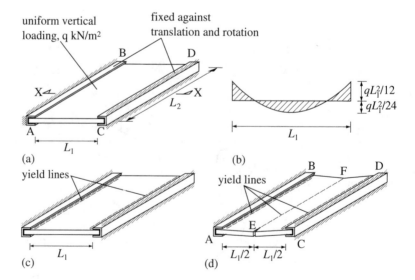

Fig. 5.40 Typical yield line patterns in slabs

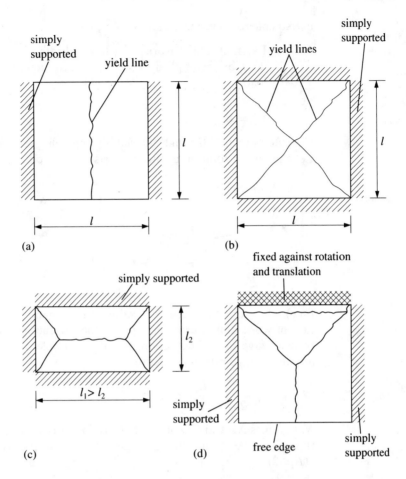

(a)

(b)

(c)

(d)

along the yield lines. Since the regions of slab between the yield lines remain plane, the yield lines must always be straight. Figure 5.40 illustrates the yield line pattern in a number of slabs subject to uniform load acting perpendicular to the plane of the slabs.

The yield line method of analysis for slabs can be used to determine the magnitude of the ultimate loads which cause full plastic (mechanism) failure of the member. The following two examples illustrate the application of the method.

Example 5.8

Yield line method applied to one-way spanning slab

Problem

For the slab of Fig. 5.39(a) with known reinforcement, the moment capacities per unit breadth in sag and hog, m_x and m'_x respectively, have been calculated as 50 kNm per metre breadth. Find the intensity of the uniformly distributed load, q, which causes collapse of the slab if $L_1 = 8$ m and $L_2 = 6$ m. Assume an infinite capacity for plastic hinge rotation.

Solution

For the slab of Fig. 5.39(a) to collapse, the applied moment must be sufficiently large to develop yield lines at three locations in the member, as illustrated in Fig. 5.39(d). The kinematic approach is based on the conditions in the slab

175

during collapse. During collapse:

$$\left.\begin{array}{l}\text{External rate of work}\\\text{done by the applied}\\\text{loads}\end{array}\right\} = \left\{\begin{array}{l}\text{Internal rate of work}\\\text{dissipated along the}\\\text{yield lines}\end{array}\right.$$

that is:

$$\dot{W}_e = \dot{W}_i \tag{5.3}$$

Consider a section through the slab during collapse as illustrated in Fig. 5.41. Let the rate of rotation at the supports during collapse be $\dot{\theta}$ (radians per unit

Fig. 5.41 Section X–X through slab of Fig. 5.39(a) during collapse

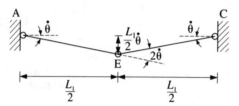

time). Therefore, the rate of deflection at mid-span is equal to $L_1\dot{\theta}/2$. Now, the rate of work done during collapse is the product of the external forces and their average rate of deflection. For this example, the external load is qL_1L_2 and its average rate of deflection is:

$$\frac{1}{2}\left(\frac{L_1\dot{\theta}}{2}\right) = \frac{L_1\dot{\theta}}{4}$$

Note: Lines AB and CD of Fig. 5.39(d) do not move at all while line EF moves through $L_1\dot{\theta}/2$. Hence, the average deflection of panels ABEF and CDEF is $\frac{1}{2}(L_1\dot{\theta}/2)$.

Therefore, the external rate of work done by the applied loads is:

$$\dot{W}_e = qL_1L_2\frac{L_1\dot{\theta}}{4} = q(L_1)^2L_2\frac{\dot{\theta}}{4}$$

The internal rate of work dissipated along the yield lines is the product of their ultimate moment capacities and their rates of rotation. Now, the capacity to resist hog moment per metre width is m'_x. Therefore, the total ultimate moment capacity for hog, M'_x, is given by:

$$M'_x = m'_x L_2$$

The rate of rotation along the yield lines at the edges is $\dot{\theta}$. Hence, the internal rate of work dissipated along each fixed support is:

$$M'_x\dot{\theta} = m'_x L_2\dot{\theta}$$

Similarly, along the mid-span yield line the ultimate moment capacity for sag is M_x and the rate of rotation is $2\dot{\theta}$. Hence, the internal rate of work dissipated at mid-span is:

$$M_x(2\dot{\theta}) = 2m_x L_2\dot{\theta}$$

The total rate of work dissipated, $\dot{W_i}$, is the sum of the rates of work dissipated along each yield line, that is:

$$\dot{W_i} = 2(m'_x L_2 \dot{\theta}) + 2m_x L_2 \dot{\theta}$$

By equation (5.3), we then have:

$$\dot{W_e} = \dot{W_i}$$

$$\Rightarrow \quad q(L_1)^2 L_2 \frac{\dot{\theta}}{4} = 2(m'_x L_2 \dot{\theta}) + 2m_x L_2 \dot{\theta}$$

$$\Rightarrow \quad q = \frac{4(2m_x + 2m'_x)}{(L_1)^2} = \frac{4(100 + 100)}{(8)^2}$$

$$\Rightarrow \quad q = 12.5 \text{ kN/m}^2$$

Example 5.9 Yield line method applied to two-way spanning slab

Problem The slab of Fig. 5.42(a) is isotropic, that is, it has the same moment capacity per unit breadth in all directions, m. Given that m is 60 kNm per unit width, and that the slab is simply supported, find the intensity of the uniformly distributed load, q, which causes collapse of the slab. Assume infinite capacity for plastic hinge rotation and take $L = 7$ m.

Fig. 5.42 Slab of Example 5.9: (a) geometry; (b) yield pattern

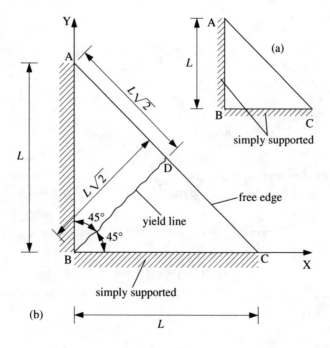

Solution   The only way in which the slab of Fig. 5.42(a) can fail is with a yield line along the middle as illustrated in Fig. 5.42(b) (symmetry requires D to be mid-way between A and C). Since the yield line is not parallel to either of the coordinate axes the solution of this example proves to be more complex than the previous example.

Panel ABD must rotate about the Y-axis as it is supported at A and B. Let the rate of rotation of this panel equal $\dot{\theta}_y$. Then, on a section through D, say S–S (Figs 5.43(a) and (b)), the rate of deflection at D is $L\dot{\theta}_y/2$. Similarly, panel BCD will rotate about the X-axis. Its rate of rotation, $\dot{\theta}_x$, can be related to $\dot{\theta}_y$. From Fig. 5.43(c), the rate of deflection of D is $L\dot{\theta}_x/2$. Equating the rates of deflection at D gives:

$$\frac{L\dot{\theta}_y}{2} = \frac{L\dot{\theta}_x}{2} \Rightarrow \dot{\theta}_y = \dot{\theta}_x = \dot{\theta}\text{(say)}$$

Now, the average rate of deflection of panel ABD is the rate of deflection of the centroid of the triangle which occurs at one-third of the distance from AB to D, that is:

$$\frac{1}{3}\frac{L\dot{\theta}}{2} = \frac{L\dot{\theta}}{6}$$

The average rate of deflection of panel BCD is the same. Hence, the total external rate of work done is:

$$\dot{W}_e = 2\left(\tfrac{1}{2}qL\frac{L}{2}\frac{L\dot{\theta}}{6}\right) = \frac{qL^3\dot{\theta}}{12}$$

Fig. 5.43 Sections through slab of Example 5.9: (a) slab; (b) section S–S; (c) section T–T; (d) section U–U; (e) section V–V

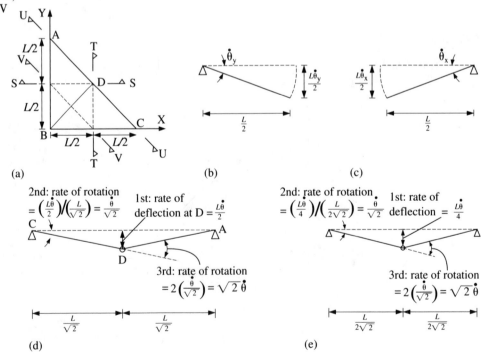

(a)

(b)

(c)

(d)

(e)

The internal rate of work done, $\dot{W}_i$, is found by determining the rate of rotation of the yield line. To calculate this rate of rotation, it is necessary to take a cross-section perpendicular to the yield line, such as U–U (Figs 5.43(a) and (d)). Along section U–U, the rate of rotation at D is found to be $\sqrt{2}\dot{\theta}$ from Fig. 5.43(d). For a section V–V, half-way between B and D (Figs 5.43(a) and (e)), although the length and the rate of deflection at the yield line are halved, the rate of rotation at the yield line remains the same as that at U–U. Hence, the rate of rotation is constant throughout the length of the yield line and the internal rate of work done is:

$$\dot{W}_i = m\left(\frac{L}{\sqrt{2}}\right)\sqrt{2}\dot{\theta} = mL\dot{\theta}$$

By equation (5.3), we have:

$$\dot{W}_e = \dot{W}_i \Rightarrow \frac{qL^3\dot{\theta}}{12} = mL\dot{\theta}$$

$$\Rightarrow \quad q = \frac{12m}{L^2} = \frac{12(60)}{49} = 14.7 \text{ kN/m}^2$$

A major disadvantage of the yield line method is that it gives no information on the response of slabs under service loads (for instance, it gives no information on elastic deflections). In addition, the method provides no check on the extent of the plastic hinge rotations required to achieve the critical yield pattern. It is presumably for this reason in particular that the yield line method is not approved in the British Standard for concrete design, BS 8110, unless the bending moment diagram is similar to that obtained by linear elastic analysis. Of course, similarity can only be checked if an elastic analysis has been performed, in which case these results with plastic moment redistribution can be used directly. EC2 takes a more lenient view, allowing full use of the yield line method provided certain ductility requirements are satisfied.

Lower-bound approaches and Hillerborg strip method

Consider the rectangular, simply supported slab of Fig. 5.44(a) which is subjected to a uniform load of intensity q kN/m^2. Suppose we were to assume that all load on this slab spans in one direction across the shorter span, that is in an E–W direction (the arrows in the figure indicate the assumed direction of span). Under this assumption, the maximum bending moment in this member is $ql_1^2/8$ per metre width (from Appendix B) as illustrated in Fig. 5.44(b). As pointed out in section 5.4, if we provide reinforcement with a moment capacity of no less than $ql_1^2/8$ in the E–W direction and there is infinite ductility, then the member will not collapse.

Such lower-bound methods are widely used in practice and are a safe method of analysis for slabs under ultimate loads provided the slab is sufficiently ductile. However, the results of a lower-bound method do not guarantee a satisfactory

Fig. 5.44 Rectangular slab: (a) plan view; (b) section X–X and bending moment diagram

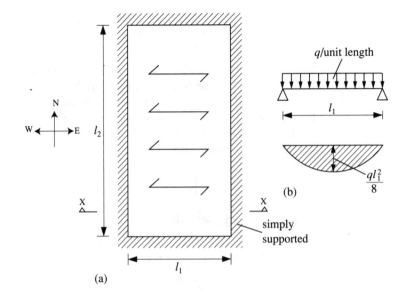

(a)

(b)

design at SLS as cracking may occur if the assumed load path is not reasonable. Furthermore, the slab may not necessarily be safe if it has insufficient ductility to redistribute the moments from the linear elastic values to the final values assumed. Slabs will generally have sufficient ductility to do this provided the moments assumed do not greatly differ from those of a linear elastic analysis.

One particular lower-bound approach which can be very useful for the analysis of slabs is the Hillerborg strip method. Its main advantage is that it is easy to implement and can be done without the aid of a computer. Like the example described above, the Hillerborg strip method is based on the assumption that load tends to be carried to the nearest support. However, the strip method extends this approach by assuming that load is **always** carried to the nearest support.

To illustrate the Hillerborg strip method, consider the slab of Fig. 5.44(a). The slab is first divided into regions, shown in Fig. 5.45(a), which indicate to which supports the load in that portion of the slab is transferred. The lines defining the regions are plotted at an angle of 45° from each corner. The slab is then divided into a number of strips spanning E–W and N–S as illustrated in Fig. 5.45(b). Then, load on segment S (Fig. 5.45(b)), say, is deemed to be spanning E–W by strip B–B. Load on segment T, on the other hand, is deemed to be spanning N–S by strip C–C. The bending moment diagrams for the various strips can readily be determined. For strip A–A, the bending moment diagram (from Appendix B) is as illustrated in Fig. 5.45(c). For strip B–B, the vertical reactions at each end must equal half the applied load of $2qy$. The moment is then found from a consideration of the free-body diagram taken from the left support to a distance y from it. The resulting diagram is illustrated in Fig. 5.45(d) and the bending moment diagram for strip C–C is illustrated in Fig. 5.45(e). In this way, the bending moments at any section of the slab can be estimated.

180

Fig. 5.45 (a) Plan geometry showing regions; (b) strips through slab; (c) strip A–A and bending moment diagram; (d) strip B–B and bending moment diagram; (e) strip C–C and bending moment diagram

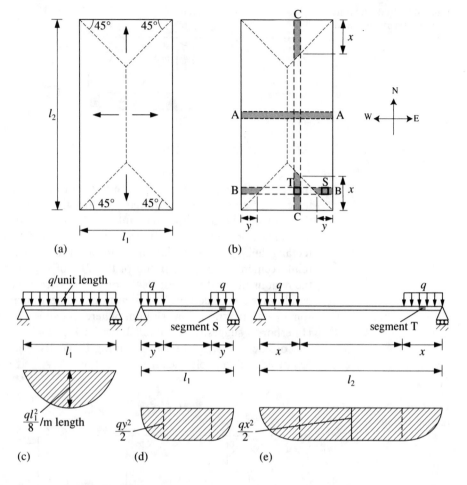

Example 5.10   Lower-bound approach

Problem   Using a lower-bound approach, specify the required moment capacities for the slab shown in Fig. 5.46 if the uniformly distributed ultimate load is 20 kN/m².

Fig. 5.46 Slab of Example 5.10

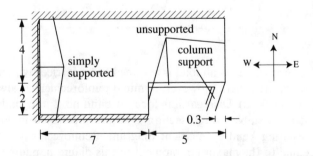

Solution   At the western side of the slab, the regions of slab which are supported by each edge are illustrated using arrows in Fig. 5.47. At the eastern end, two strong bands are provided to transfer the reaction from the slab to the column and

Fig. 5.47 Slab showing regions and strong bands

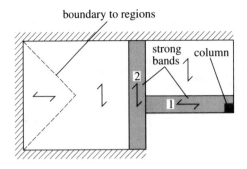

boundary to regions

the supported edges. In general, the moment differs between strips of slab and hence the reinforcement required to resist them also differs. For example, in the slab of Fig. 5.45, the maximum moments in strips A–A and B–B were different. For convenience, therefore, it is common to divide the slab into rectangular regions in which the strips are provided with the same quantity of reinforcement. The western end of the slab of Fig. 5.46 has been divided into the regions illustrated in Fig. 5.48, for which there are two differently reinforced bands in each direction. The selection of the dimensions of these bands is somewhat arbitrary but should, where possible, be reasonably similar to Hillerborg's suggested 45° line (shown in Fig. 5.47).

Consider the strips of slab spanning E–W. By Fig. 5.48, two strips need to be considered, namely the strips at sections A–A and B–B. Analysis of section

Fig. 5.48 Slab showing revised regions

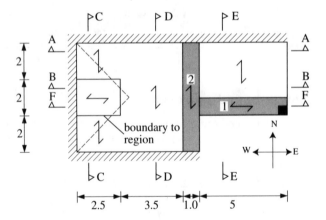

A–A is trivial as no load is assumed to be carried by it and so no reinforcement is required there (however, nominal reinforcement is always provided throughout all slabs). Disregarding for now band no. 1, the loads carried by a 1 m wide strip of slab through section B–B are illustrated in Fig. 5.49 along with the resulting bending moment diagram. Reinforcement with a moment capacity equal to the maximum moment in this diagram, namely 50.2 kNm/m, must be provided running E–W, as illustrated in Fig. 5.50.

The two strips of slab running N–S which need to be analysed are those at sections C–C and D–D of Fig. 5.48. The loads and resulting bending moment

Fig. 5.49 Section B–B:
(a) loading; (b) bending
moment diagram

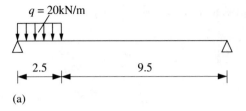

(a)

(b)

Fig. 5.50
Reinforcement
requirements as
calculated using lower
bound approach

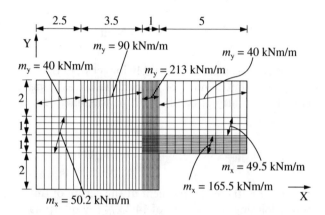

diagrams for 1 m wide strips through each section are illustrated in Fig. 5.51. The corresponding requirements for N–S reinforcement are illustrated in the diagram of Fig. 5.50.

In addition to the reinforcement requirements calculated above, reinforcement is required for strips at section E–E and in the strong bands, 1 and 2. These strong bands are assumed to have a width of 1 m. First, consider the 1 m wide strip at section E–E (Fig. 5.48) which is illustrated in Fig. 5.52(a). The strong band (no. 1) running E–W effectively acts as a support to this strip.

Fig. 5.51 Loading and
bending moment
diagrams: (a) section
C–C; (b) section D–D

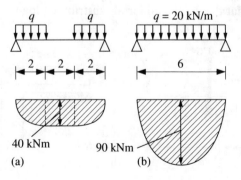

(a) (b)

183

Fig. 5.52 Section E–E:
(a) geometry and
loading; (b) bending
moment diagram

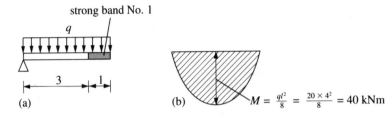

$$M = \frac{ql^2}{8} = \frac{20 \times 4^2}{8} = 40 \text{ kNm}$$

While the span to the centre of the support (i.e. the band) is 3.5 m, it has been found using finite element analysis to be more accurate to assume that the strip spans the full 4 m to the end of the band. The corresponding bending moment diagram for the strip is illustrated in Fig. 5.52(b). From this diagram, it can be seen that the reinforcement running N–S in this region must have a moment capacity of 40 kNm. The reaction on band 1 from these strips is given by:

$$q_r = \frac{qL}{2} = \frac{20 \times 4}{2} = 40 \text{ kN/m}$$

As the E–W span is considerably longer than the N–S span, it has been decided to assume that band 1 only spans as far as band 2. Assuming a span (from column to centre of band) of 5.5 m, the maximum bending moment in band 1 becomes:

$$M_{max} = \frac{q_r L^2}{8} = \frac{40(5.5)^2}{8} = 151 \text{ kNm}$$

In addition, there is a moment at the centre of this assumed span of 5.5 m due to the loading of Fig. 5.49 of 14.2 kN (by interpolation in Fig. 5.49(b)). Hence, reinforcement with a moment capacity of 165.5 kN m must be provided in this band.

The reaction from band 1 acts as an applied load on band 2. Excluding the load where bands 1 and 2 intersect, this reaction is given by:

$$R = \frac{q_r L}{2} = \frac{40 \times 5}{2} = 100 \text{ kN}$$

Thus, the total load acting on band 2 is as illustrated in Fig. 5.53(a). The reaction due to this load can be found by superimposing the reactions due to the point load and that due to the uniformly distributed load. Hence, from Appendix B,

Fig. 5.53 Band 2:
(a) geometry and
loading; (b) bending
moment diagram

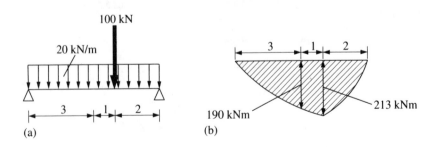

Nos 1 and 3, the reaction at the left support is:

$$\frac{20 \times 6}{2} + \frac{100 \times 2}{6} = 93.3 \text{ kN}$$

The corresponding bending moment diagram is illustrated in Fig. 5.53(b). From this, it can be seen that the reinforcement in band 2 must have a moment capacity not less than 213 kNm.

All of the reinforcement requirements are illustrated in the diagram of Fig. 5.50. The linear elastic bending moment diagram for this slab has been found using finite element analysis and, for comparison, the reinforcement requirements for the corresponding parts of the slab have been calculated and are illustrated in Fig. 5.54.

Fig. 5.54
Reinforcement
requirements as
calculated using finite
element analysis
(without plastic moment
redistribution)

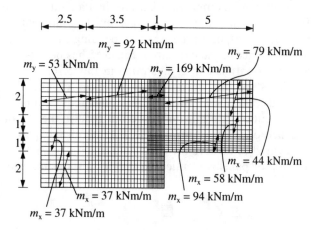

Lower-bound analysis of flat slabs

In flat slab construction, the slab is supported directly on columns, as illustrated in Fig. 5.55, rather than on beams. The strip of slab, ABC, in the figure could reasonably be modelled as a continuous beam supported by the columns at A and C. The strip, DEF, could also be modelled as a continuous beam as it is supported at D and F by strips of slab running N–S, that is D provides some support to DEF by virtue of its proximity to the columns at A and G. However, the support at D and F is less effective than that provided directly by a

Fig. 5.55 Flat slab
construction

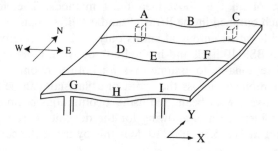

Fig. 5.56 Bending moment diagram for m_x for flat slab of Fig. 5.55

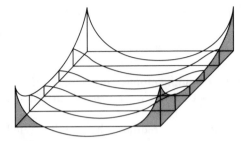

column – D and F are in effect spring supports capable of significant deflection when load is applied to them.

The analysis of a continuous beam on spring supports gives a significantly different bending moment diagram than a continuous beam on rigid supports. In general, when supports are 'springy', hog moments reduce and sag moments become correspondly larger. The bending moment diagram for the moment on the X-face in the flat slab of Fig. 5.55 has been found by linear elastic (finite element) analysis and is illustrated in Fig. 5.56.

Flat slabs can be analysed by linear elastic, yield line or lower-bound methods. Of these, linear elastic methods, coupled with plastic moment redistribution, are most effective but require the use of a computer. The yield line method is less satisfactory as a greater deal of plastic moment redistribution is required to achieve the yield pattern (Fig. 5.57). In fact, the yield pattern for an interior panel is identical to that for a slab supported on all sides by beams,

Fig. 5.57 Yield pattern for interior panel in flat slab construction

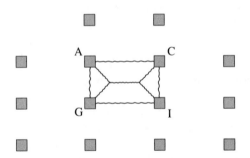

even though the elastic bending moment diagrams for two types of slab are significantly different. Thus, in many countries, lower-bound methods have traditionally been favoured over the yield line method. Eurocode EC2 does not provide any specific recommendations for the analysis of flat slabs but allows the use of all the above-mentioned methods. Therefore, the lower-bound approach outlined in the British Standard, BS8110, is perfectly acceptable. The method is rather empirical and is only dealt with briefly here. Full details are given in BS8110 itself and in the book by Allen (1988).

It is reasonable to assume that, for both horizontal and vertical loads, the average moment across the width of a flat slab can be found by treating the slab as a very wide beam and using a continuous beam or frame analysis. The only problem then is allowing for the deviations from that mean which are apparent in Fig. 5.56. BS8110 does this by using the concept of column strips

Fig. 5.58 Column and middle strips in flat slab: (a) N–S section; (b) plan

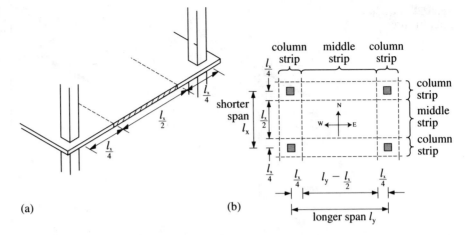

(a)

(b)

and middle strips of slab. These are strips within the slab which are reinforced to resist different portions of the total moment (see Fig. 5.58). The column strip, although not as wide, is resting directly on the columns and, as a result, it takes more of the total moment than the middle strip. However, the effect is less pronounced for the sag moment as the effect of the spring support on the middle strip is to reduce its hog and increase its sag. BS8110 recommends that the column strip be assigned 75 per cent of the total hog moment and 55 per cent of the sag moment. The middle strip is designed to resist the remaining 25 per cent and 45 per cent of the hog and sag moments, respectively. This process is carried out for both m_x and m_y. The reinforcement running E–W can be viewed as transferring the load towards the supports in this direction while the reinforcement running N–S transfers the load towards the supports in the N–S direction. Both are necessary in order to transfer the load to the column supports.

5.7 Analysis of shear wall systems

Shear wall systems are commonly incorporated into a multi-storey building to provide lateral stability. Shear walls, which are made up of concrete or masonry walls, act as vertical cantilevers in resisting horizontal forces (see Fig. 5.59(a)). Their great cross-sectional depth, d, gives them an extremely high second moment of area ($I = bd^3/12$) and hence a great stiffness against horizontal deflection. As discussed in Chapter 1, when the depth, d, exceeds the height, h, the mechanism by which shear walls resist load changes from flexure to membrane action. This, however, does not reduce the effectiveness of shear walls in any way.

Another most effective way of resisting horizontal forces is the vertical cantilever of box section known as a core such as illustrated in Fig. 5.59(b). Cores are commonly incorporated into multi-storey buildings where they can serve the additional function of providing fire resistance around elevators and stairwells (holes for doorways, if sufficiently small, do not greatly reduce the

Fig. 5.59 (a) Shear wall; (b) core

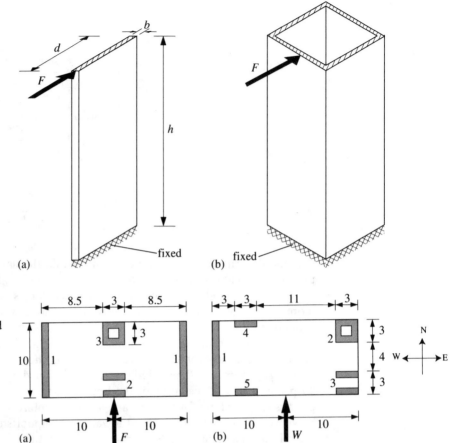

Fig. 5.60 Shear wall/core systems for resisting horizontal load (plan views): (a) symmetrical shear wall system for N–S load; (b) non-symmetrical system

effectiveness of the core in resisting horizontal loads). Cores are more efficient than shear walls at resisting horizontal loads because they have a much higher second moment of area with a significant portion of the cross-sectional area concentrated at the ends of the member. Non-symmetrical shear wall assemblies, such as that illustrated in Fig. 5.60(b), can have a twisting effect on the building and consequently are more complex to analyse than symmetrical shear wall assemblies. For this reason, shear walls and cores placed symmetrically within a building are generally favoured whenever possible.

The horizontal forces acting on a building are usually transferred by the floor slabs to the shear wall members. For this reason, it is normal to assume that the forces acting on shear walls do so at floor levels. In general, the force at each level is then carried to the foundations by each shear wall in proportion to its relative stiffness. The following two examples illustrate a method for determining the proportion of force carried by each member in both symmetrical and non-symmetrical shear wall assemblies.

Example 5.11 Symmetrical shear wall assembly

Problem Figure 5.60(a) illustrates the plan of a typical multi-storey building of height, h. The structure is braced against wind load in a N–S direction by the symmetrical arrangement of shear walls illustrated in the figure. Determine the portion of the wind load, F, carried by each shear wall and hence derive an expression for the moment due to horizontal load at the base of the external walls. Assume all walls have a thickness of 200 mm.

Solution   Each of the four shear walls and the core of Fig. 5.60(a) acts as a cantilever in transferring the wind load to the foundation. Although all the cantilevers are of the same height, they have differing stiffnesses since each has a different second moment of area. Thus, if the individual cantilevers were to be subjected to an applied horizontal force, P, each would deflect by a different amount. However, in most shear wall assemblies, the cantilevers are connected to the floor slabs at each level. Since the floor slabs are much stiffer against horizontal force than the shear walls, it can be assumed that each cantilever is forced to deflect by the same amount, as illustrated in Fig. 5.61. Ignoring the

Fig. 5.61 Horizontal deflection of shear wall/core system

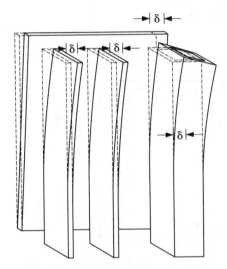

resistance of the floor slab to rotation, the horizontal force, P, required to cause a deflection, δ, in a cantilever is, from Appendix A, No. 4:

$$P = \frac{3EI\delta}{h^3} \tag{5.4}$$

where h is the height of the cantilever and I is the second moment of area of the cantilever section about the relevant axis. Let F_1 be the force carried by each of the two outer shear walls, F_2 be the force carried by each of the two internal shear walls and F_3 be the force carried by the core. Then the total force required is:

$$F = 2F_1 + 2F_2 + F_3$$

Each shear wall deflects by the same amount, δ say, at the level of the floor slab. Thus, as equation (5.4) holds for each, we have:

$$F_1 = \frac{3EI_{y1}\delta}{h^3} \quad F_2 = \frac{3EI_{y2}\delta}{h^3} \quad \text{and} \quad F_3 = \frac{3EI_{y3}\delta}{h^3}$$

189

where I_{y1}, I_{y2} and I_{y3} are the appropriate second moments of area. Since all the shear wall members are the same height, the force resisted by each member can be seen to be proportional to its second moment of area. Referring to Fig. 5.60(a), the second moments of area are:

$$I_{y1} = \frac{bd^3}{12} = \frac{0.2(10)^3}{12} = 16.67 \text{ m}^4$$

Similarly:

$$I_{y2} = \frac{3(0.2)^3}{12} = 0.002 \text{ m}^4$$

With reference to Fig. 5.62, the second moment of area for the core is:

$$I_{y3} = \left(\frac{3(3)^3}{12} - \frac{2.6(2.6)^3}{12} \right)$$

$$= 2.94 \text{ m}^4$$

Fig. 5.62 Cross-section through core

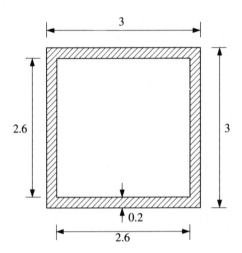

Hence:

$$\sum I_y = 2I_{y1} + 2I_{y2} + I_{y3}$$

$$= 2(16.67) + 2(0.002) + 2.94$$

$$= 36.28 \text{ m}^4$$

The portion of the total wind load, F, taken to member i, is then given by:

$$F_i = F \left(\frac{I_{yi}}{\sum I_y} \right) \tag{5.5}$$

Hence, the portion taken by each external shear wall is:

$$F_1 = \frac{F(16.67)}{36.28} = 0.46F$$

In a similar manner, the portion of the load carried by the internal shear walls and the core is:

$$F_2 = \frac{F(0.002)}{36.28} = 0.00006F$$

$$F_3 = \frac{F(2.94)}{36.28} = 0.081F$$

Thus, it can be seen that the two external shear walls carry the majority of the load (92 per cent) and the contribution of the internal shear walls is negligible. In practice, it is common to ignore the resistance of members, such as the internal shear walls of Fig. 5.60(a), which are bending about their weak axes. The moment at the base of each external wall is simply:

$$F_1 h = 0.46Fh$$

Example 5.12   **Non-symmetrical shear wall assembly**

Problem   The structure illustrated in Fig. 5.60(b) is braced against wind load in the N–S direction by a non-symmetrical arrangement of shear walls. Determine the portion of the wind load, W, to be carried by the west shear wall and the moment at the base of wall 4. Assume all walls have a thickness of 200 mm.

Solution   As mentioned previously, when shear walls/cores are non-symmetrical, we must allow for the resulting twist in the building as a whole. In this example, twisting results from the fact that the external west shear wall is far less flexible than the core (the stiffness of the other shear walls for N–S bending is negligible as can be seen from the previous example). This twisting is then resisted by the E–W stiffness of all the shear walls and the core.

The analysis procedure for a non-symmetrical shear wall system is similar to that for a group of bolts in a structural steel connection. The rotation is assumed to act about a stationary point known as the shear centre, S. If a force is applied in line with the shear centre, as illustrated in Fig. 5.63, no twisting will occur in the building. Thus, the first step is to determine the location of the shear centre. To do this we must calculate the stiffness of each core/wall in each direction (E–W and N–S). Since all members are the same height, the stiffness of each is proportional to its second moment of area. These are given

Fig. 5.63 Deflection due to force at shear centre

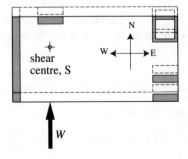

191

Fig. 5.64 Springs representing wall/core stiffnesses

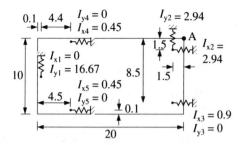

in Fig. 5.64 where each shear wall/core is represented as springs parallel to the coordinate directions. The shear centre is then the centre of resistance of the springs.

The first step in finding the shear centre is to calculate the total stiffness in each direction, that is:

$$\sum I_x = 0.45 \times 2 + 2.94 + 0.9 = 4.74 \text{ m}^4$$

and:

$$\sum I_y = 16.67 + 2.94 = 19.61 \text{ m}^4$$

Now, to find the centre of resistance, S (see Fig. 5.65(a)), take 'moments' about any point, A, say. Hence:

$$1.5I_{x2} + 8.5I_{x3} + 0.1I_{x4} + 9.9I_{x5} = \bar{d}_y \sum I_x$$

$$\Rightarrow \quad \bar{d}_y = 3.494 \text{ m}$$

and:

$$19.9I_{y1} + 1.5I_{y2} = \bar{d}_x \sum I_y$$

$$\Rightarrow \quad \bar{d}_x = 17.141 \text{ m}$$

Fig. 5.65 (a) Applied load; (b) equivalent loading at shear centre

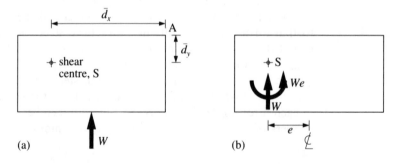

The central wind load, W, is equivalent to a force and a torque at the shear centre as illustrated in Fig. 5.65(b). The magnitude of the torque is given by:

$$T = We \tag{5.6}$$

where e is the eccentricity of the force as illustrated in the figure. In this example:

$$T = W(17.141 - 10) = 7.141W$$

This torsion generates a force on member i in the X-direction according to the expression:

$$F_{xi} = \frac{T(\bar{d}_y - d_{yi})I_{xi}}{\sum_j [I_{xj}(\bar{d}_y - d_{yj})^2 + I_{yj}(\bar{d}_x - d_{xj})^2]} \tag{5.7}$$

where $(\bar{d}_x - d_{xi})$ and $(\bar{d}_y - d_{yi})$ are the distances of member i from the shear centre in the E–W and N–S directions respectively (note that if member i is to the left of the shear centre, $(\bar{d}_x - d_{xi})$ will be negative).

In the Y-direction, there is a corresponding torsional force but there is also a direct axial force which is proportional to the member stiffness. Thus, for member i:

$$F_{yi} = \frac{WI_{yi}}{\sum I_{yj}} + \frac{T(\bar{d}_x - d_{xi})I_{yi}}{\sum_j [I_{xj}(\bar{d}_y - d_{yj})^2 + I_{yj}(\bar{d}_x - d_{xj})^2]} \tag{5.8}$$

Thus, the total portion of the wind load, W, carried by the external west shear wall in the N–S direction is:

$$F_{y1} = \frac{16.67W}{19.61} - \frac{7.141W(19.9 - 17.141)(16.67)}{904.0}$$

$$= 0.487W$$

The portion of W carried by wall 4 in the E–W direction is:

$$F_{x4} = \frac{7.141W(0.45)(3.494 - 0.1)}{904.0}$$

$$= 0.0121W$$

and the moment at its base is:

$$M_4 = 0.0121Wh$$

Problems

Section 5.2

5.1 One span of a continuous beam is illustrated in Fig. 5.66(a). It is stated in Appendix D that the bending moment diagram for such a span can be constructed by superimposing the bending moment diagram associated with the support moments (Fig. 5.66(b)) on the bending moment diagram due to the applied loading on the simply supported span (Fig. 5.66(c)). Verify that this is indeed true for the beam illustrated in Fig. 5.66.

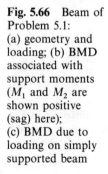

Fig. 5.66 Beam of Problem 5.1: (a) geometry and loading; (b) BMD associated with support moments (M_1 and M_2 are shown positive (sag) here); (c) BMD due to loading on simply supported beam

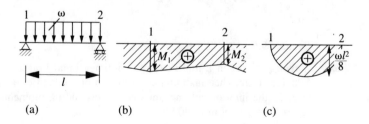

(a)　　　　　　(b)　　　　　　(c)

5.2   For the three-span beam illustrated in Fig. 5.67, find the factored ultimate bending moment envelope which gives the maximum hogging and sagging moments at all points. The unfactored permanent loading is 15 kN/m on span AB and 10 kN/m on spans BC and CD. The unfactored variable gravity loading is 20 kN/m on all spans.

Fig. 5.67   Beam of Problem 5.2

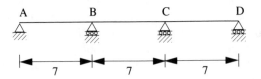

5.3   Determine the reaction at B for the beam whose geometry and loading are illustrated in Fig. 5.68.

Fig. 5.68   Beam of Problem 5.3

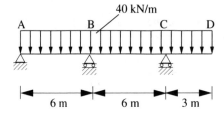

Section 5.3

5.4   The structure of Problem 4.7 is to be analysed by linear elastic analysis with up to 25 per cent plastic moment redistribution. Determine an appropriate design bending moment diagram and compare the results with those of Problem 4.7.

Section 5.4

5.5   In Example 5.4, the initial bending moment diagram is as illustrated in Fig. 5.10(b), but a change in the beam depth results in this being revised to the bending moment diagram illustrated in Fig. 5.10(c). Moment capacities have been provided based on Fig. 5.10(b), i.e. without reanalysis, and are illustrated in Fig. 5.69. Determine if this design is safe given that the frame is sufficiently ductile to allow 20 per cent plastic moment redistribution at all points.

Fig. 5.69   Problem 5.5: (a) capacities to resist positive moments; (b) capacities to resist negative moments

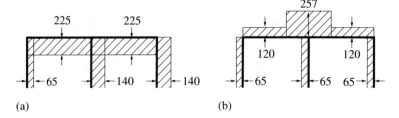

Section 5.5

5.6   A typical storey in a multi-storey two-dimensional braced frame is illustrated in Fig. 5.70. Determine the maximum contribution of the load on a typical storey to the axial force in the interior column and the corresponding moment. The total ULS loading varies from 15 kN/m to 40 kN/m.

Fig. 5.70 Frame
of Problem 5.6

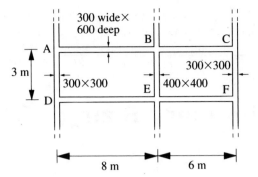

Section 5.6

5.7 For the slab illustrated in Fig. 5.71, the ratio of the larger to the smaller span dimension
is 2:1 and hence EC2 allows the central part to be designed as if the slab were one-way
spanning. Use the finite element method to analyse this slab for an applied vertical
loading of 8 kN/m² given that the slab is 200 mm thick.

(a) Present a contour plot of the moments per unit breadth in each direction. Compare
the results to those which are obtained assuming one-way spanning behaviour.
(b) Determine the twisting moment per unit breadth in the corners of the slab.
(c) There are in fact 300 × 300 mm columns at the four corners of the slab which extend
3 m to the slabs above and below. Revise the computer model to allow for the
rotational stiffness provided by these columns and determine the effects on m_x and
m_y. The modulus of elasticity of the concrete in the columns can be assumed to be
32×10^6 kN/m².

Fig. 5.71 Plan
view of slab of
Problems 5.7 and
5.8

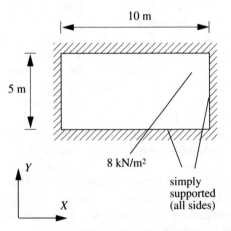

5.8 If the slab of Problem 5.7 is provided with moment capacities of 15 kNm/m in all
directions, calculate the mechanism load factor using the yield line method.

6

Preliminary Sizing of Members

6.1 Introduction

The analysis and design of concrete structures is essentially a trial and error process. The magnitude of permanent gravity loads, for instance, is dependent on the member sizes and hence an initial estimate of member sizes is required before the structure can be analysed. If, however, the final dimensions required to resist, say, bending moment in a member differ from the initial estimate of the member size, then the design process must be repeated with revised member dimensions until these initial assumptions are satisfied. The more accurate the initial size estimates, the less iterations will be involved as the solution will converge more rapidly to the precise requirements.

The approximate methods and rules of thumb for the preliminary sizing of concrete members described in this chapter provide the means to make a reasonable first estimate of member sizes for the analysis and design of a viable structural scheme. The advantage of these methods is that they can be carried out on a slip of paper or a hand calculator. In addition, these methods can be used to estimate the material requirements of different schemes, thus allowing a quick assessment of the viability of each scheme.

The variables involved in the sizing of concrete members are the geometry (span and depth of beams and slabs, effective height of columns, breadth of ribs and beams), material grades (concrete and steel strengths) and areas of reinforcement. The factors which will affect the choice of each variable include strength, durability and fire-resistance requirements. It is traditional to decide on the material grades at an early stage of the design and to employ the same grades for members of the same type. Thus, material grades are not normally a variable at the member design stage.

Within the overall design process, the preliminary sizing process itself can also be iterative. For example, the required depth for beams, say, is determined relatively simply but the minimum breadth can depend on the breadth required to fit the reinforcement. This causes problems as it is necessary first to know the breadth before the required reinforcement can be calculated. Thus, it is usual to adopt trial geometric dimensions and to calculate the required area of reinforcement for these dimensions. If the area of reinforcement is unacceptably high for practical or performance reasons, the geometry is revised and the process is repeated.

It is important to note that the material in this chapter is to be used for the 'approximate' preliminary design of concrete structures and is not sufficient for a final design. There should always be a proper design and analysis to check the adequacy of each structural member. For the rapid appraisal of alternative structural schemes, these approximate methods can be used in conjunction with the approximate values for permanent and variable gravity loads given in sections 3.4 and 3.5.

6.2 Material grades

Concrete

A concrete can be designed to give a wide variety of strengths, depending on the relative proportions of its constituents. For this reason, it is usual to specify a particular **grade** of concrete when designing reinforced and prestressed concrete structures. Since compressive strength is the most important property of concrete, the grade is measured by the characteristic compressive strength at 28 days, that is the strength below which not more than 5 per cent of specimens are expected to fail when tested 28 days after casting. The characteristic strength is determined by means of standardized tests carried out on either cylindrical or cubed specimens. The characteristic cylinder strength, f_{ck}, is slightly smaller than the characteristic cube strength, f_{cu}, for a particular design mix (the ratio of f_{ck}/f_{cu} is approximately equal to 0.8 in most cases; see Table 6.1). EC2 design rules are based on cylinder strengths while BS8110 rules are

Table 6.1
Characteristic strengths of cylinders and cubes (from EC2)

Characteristic cylinder strength, f_{ck} (N/mm^2)	Characteristic cube strength, f_{cu} (N/mm^2)
12	15
16	20
20	25
25	30
30	37
35	45
40	50
45	55
50	60

based on cube strengths. Thus, for a design in accordance with EC2, a specimen having grade 30 concrete has a characteristic **cylinder** strength of 30 N/mm^2 (the corresponding cube strength is 37 N/mm^2). For a design in accordance with BS8110, a specimen having a concrete grade of 30 has a characteristic **cube** strength of 30 N/mm^2 (the corresponding cylinder strength is 25 N/mm^2). To avoid ambiguity, grades that give both cylinder and cube strengths (in that order) are commonly used. Hence, EC2 will refer to C30/37 or C25/30 for the above concretes. For most applications, concrete is only specified in multiples of five from a lower grade of about C25/30 to an upper grade of about C50/60. For preliminary design, a popular choice would be concrete of characteristic cylinder strength of 25 N/mm² (i.e. C25/30) for all beams and slabs and 30÷35 N/mm² (i.e. C30/37 or C35/45) for all columns.

Reinforcement

The grade of steel reinforcement is measured by its characteristic yield strength, f_y. Unlike concrete, however, steel reinforcement is only widely available in the UK in two grades: mild steel ($f_y = 250$ N/mm²) and high-yield steel ($f_y = 460$ N/mm²). Both types of reinforcement can commonly be obtained only in the following standard diameters: 8, 10, 12, 16, 20, 25, 32 and 40 mm. Mild steel is generally not favoured for use as longitudinal tensile reinforcement because of its weaker strength. However, mild steel is more ductile than high-yield steel and so it can be bent through tighter radii, making it particularly useful for link reinforcement.

Note: At the time of going to press it seems likely that the two grades referred to above will be withdrawn, and that two new grades, $f_y = 400$ N/mm² and $f_y = 500$ N/mm², will be introduced at some time in the future. EC2 also has a separate classification for ductility suggesting that steel of either strength may be available with low- or high-ductility ratings.

6.3 Preliminary sizing of beams

Fire resistance and durability

The size of structural beams is influenced by the cover needed to ensure adequate resistance to fire and corrosion. The concrete cover is the **distance between the outermost surface of reinforcement (usually the stirrups) and the nearest concrete surface**, as shown in Fig. 6.1. Minimum values of cover required for corrosion resistance are governed by the environmental conditions to which the beams are exposed. EC2 defines the five exposure classes of Table 6.2 corresponding to different environmental conditions. The minimum cover requirements for all reinforced concrete members (including beams) corresponding to these five exposure classes are given in Table 6.3. In addition, it is specified that the cover to main reinforcement must exceed the bar diameter. An allowance for construction tolerance, of the order of 5–10 mm must be added to those values and the total specified on the working drawings.

Steel reinforcement loses its strength when subjected to high temperatures. Concrete is a good insulator, with the result that it protects the steel and extends the time taken for the temperature in the steel to reach a dangerous level. Hence, greater cover means greater time for occupants to escape from a burning building before it collapses. Covers for specific fire resistance times are given in that part of EC2 relating to structural fire design. Minimum beam and web breadth are also given for specified fire resistance times. For the **preliminary**

Fig. 6.1   Concrete cover

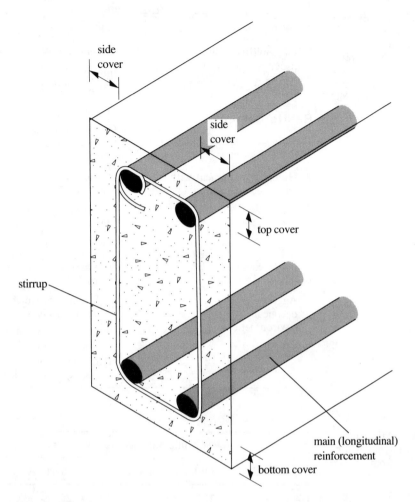

sizing of continuous beams in mild and moderate environments (e.g. exposure classes 1 and 2 of Table 6.2), the values of minimum beam breadth and **cover to main reinforcement** can be taken from Table 6.4. These values are sufficient to satisfy the requirements of both fire resistance and durability. For severe exposures and/or simply supported beams, larger values of cover should be used. Although the values in the table are those recommended in the ISE manual for a preliminary design in accordance with BS 8110, they are equally applicable for preliminary design to other codes of practice such as EC2.

Effective depth of beams

The strength of beams in flexure is governed principally by the effective depth, that is, the depth from the extreme compression fibre of the beam to the centroid of the tension steel. A large effective depth results in a relatively small required quantity of reinforcement, as shown in Fig. 6.2(a). However, deeper beams at each floor level increase the permanent gravity load to the columns and result in a higher building overall which requires extra cladding. Reduced structural

199

Table 6.2 Exposure classes related to environmental conditions. Exposure class 5 may occur alone or in combination with other classes (from EC2)

Exposure class		Examples of environmental conditions
1. Dry environment		Interior of buildings for habitation or offices
2. Humid environment	a Without frost	Interior of buildings where humidity is high (e.g. laundries) Exterior members Members in non-aggressive soil and/or water
	b With frost	Exterior members exposed to frost Interior members exposed to frost in humid areas
3. Humid environment with frost and de-icing salts		Interior and exterior members exposed to frost and de-icing agents
4. Sea-water environment	a Without frost	Members submerged in sea water Members in saturated salt air (coastal areas)
	b With frost	Members partially submerged in sea water or in splash zone and exposed to frost Members in saturated salt air and exposed to frost
5. Aggressive chemical environment	a	Slightly aggressive chemical environment (gas, liquid or solid)
	b	Moderately aggressive chemical environment (gas, liquid or solid)
	c	Highly aggressive chemical environment (gas, liquid or solid)

Table 6.3 Minimum cover requirements for durability (from EC2)

Exposure class	Minimum cover to reinforcement (mm)	
	Ordinary reinforcement	Prestressing steel
1	15	25
2a	20	30
2b	25	35
3	40	50
4a	40	50
4b	40	50
5a	25	35
5b	30	40
5c	40	50

Table 6.4 Minimum preliminary design values of overall breadth and cover to *main* reinforcement for continuous beams (from the ISE manual)

Fire rating (h)	Minimum overall breadth (mm)	Minimum cover to *main* reinforcement (mm)
1	200	45
2	200	50
4	240	70

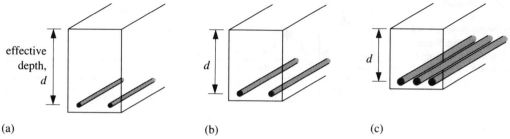

Fig. 6.2 Effective depth of beams: (a) large effective depth – building higher than economical; (b) ideal effective depth – moderate quantity of reinforcement required; (c) small effective depth – excessive quantity of reinforcement required

depth on each floor can reduce heating costs and could allow an extra floor in a multi-storey building when planning controls restrict the overall height. Reducing the effective depth also has its price as it results in a need for more reinforcement to resist the bending moments due to the applied loads. For extremely small effective depths, the required quantity of reinforcement often becomes impractical in the breadth available (Fig. 6.2(c)).

Engineers in different design offices have different rules of thumb by which they decide on the best effective depth for a given situation. These rules are generally represented in the form of span/depth ratios with modifications for influencing factors. **Ideal** values of span/effective depth ratios, recommended in the ISE manual for the preliminary sizing of reinforced concrete beams, are given in Table 6.5. Also given in this table are the **maximum** span/effective depth ratios recommended by EC2 for deflection control. The values in Table 6.5 are not definitive – it is for every designer to use his/her own discretion. Many designers adopt values between those recommended in the ISE manual and the maximum values given depending on the restraints on structural depth. In the case of prestressed concrete beams, it is usual to adopt span/effective depth ratios about 25 per cent greater than those in Table 6.5.

Table 6.5 Span/effective depth ratios for the preliminary design of reinforced concrete beams. For effective spans in excess of 7 m, multiply by 7/(effective span in metres). For T- and L-shaped beams, deduct 20 per cent from the values

	Ideal (ISE manual)	Maximum (EC2)
Cantilever	6	7
Simply supported	12	18
Continuous	15	25
End spans of continuous beams	Average of simply supported and continuous values	23

Example 6.1 Effective depth of beams

Problem Suggest preliminary depths for the beam spans illustrated in Fig. 6.3 if there are no constraints on structural depth and the fire rating is one hour.

Solution For ease of construction, the same depth will be used in all spans. The depth will be governed by conditions in span AB or span BC. In span BC, adopt the

Fig. 6.3 Calculation of effective depth: (a) sectional elevation of building showing beam and slab construction; (b) section X–X

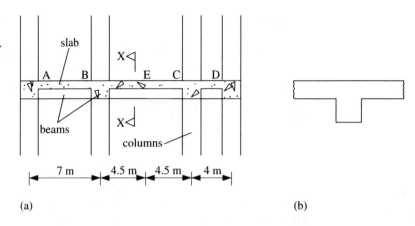

(a)

(b)

recommended ideal span/effective depth ratio of 15:1. Hence:

$$(\text{effective depth})_{BC} = 9000/15 = 600 \text{ mm}$$

While the span is in excess of 7 m (see note to Table 6.5), the beam is continuous over B and C and the **effective** span, that is, the distance between points of zero moment, is about $0.7(9) = 6.3$ m. Hence no adjustment is required.

Span AB is the end span of a continuous beam. Hence, adopt the average of the ideal span/effective depth ratios for simply supported and continuous beams:

$$(\text{span/depth ratio})_{AB} = (12 + 15)/2 = 13.5$$

$$(\text{effective depth})_{AB} = 7000/13.5 = 519 \text{ mm}$$

Clearly, span BC governs. To allow for the fact that the beam is of T-section, increase the effective depth by 20 per cent:

$$(\text{effective depth})_{\text{all spans}} = 600 \times 1.2 = 720 \text{ mm}$$

The total beam depth is the effective depth plus the sum of the cover to reinforcement, the thickness of link and half the diameter of the tension reinforcement bars, as illustrated in Fig. 6.4. Assuming a tension reinforcement diameter of 25 mm, and exposure class 2a, the cover for durability becomes the greater of 25 mm and 20 mm. Assuming a link diameter of 10 mm, the cover

Fig. 6.4 Total depth of beam

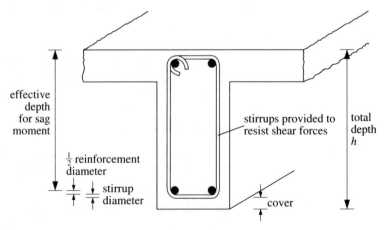

to the main steel then becomes $25 + 10 = 35$ mm. However, for preliminary design purposes, the ISE manual, whose recommendations are given in Table 6.4, specifies a cover to main steel of 45 mm for a one hour fire resistance. While many designers would view this value as being excessively conservative, it is convenient for preliminary purposes, and will be adopted here. Thus, the total beam depth, h, becomes:

$$h = 720 + 12.5 + 45 = 777.5 \text{ mm}$$

which is rounded off to 800 mm (this corresponds to a new effective depth of 742 mm).

Breadth of beams

The breadth of rectangular concrete beams and webs in flanged beams (i.e. T- and L-beams) has a much lesser effect on the resistance of a beam to bending moment than does effective depth. Breadth is frequently governed by the practical consideration of simply fitting all the reinforcement into the section while avoiding congestion. The bar arrangement must allow sufficient concrete on all sides of the bar to transfer the axial forces to and from the bar. This anchorage of the reinforcement in the concrete is commonly referred to as the 'bond'. In addition, the bars must be arranged so that the fresh concrete can be placed and compacted properly. The spacing, therefore, must be great enough to allow the passage of aggregate between the bars. Also, access must be provided for a vibrator hose all the way to the bottom of the beam.

Reinforcement can be arranged in many ways to satisfy minimum spacing requirements, some of which are illustrated in Fig. 6.5. The minimum spacings of bars specified by EC2 are given in Table 6.6. The spacings specified by BS8110 are also included for comparison.

Table 6.6 Minimum spacing of reinforcement bars where h_{agg} is the maximum aggregate size (usually about 20 mm) and ϕ_{max} is the maximum diameter of adjacent bars

	BS8110	EC2
Horizontal spacing	Greater of $h_{agg} + 5$ mm or ϕ_{max}	Greater of ϕ_{max} or 20 mm. If $h_{agg} > 32$ mm, then the spacing must not be less than $h_{agg} + 5$ mm
Vertical spacing	$\frac{2}{3}h_{agg}$	Same as minimum horizontal spacing

For the beam of Fig. 6.5(a), the minimum spacing requirements will not govern the beam breadth. However, for the beam of Fig. 6.5(b), the minimum practical breadth for a design in accordance with EC2 is:

$$2(\text{cover to main reinforcement}) + 7\phi_{max} = 2(45) + 7(25) = 265 \text{ mm}$$

which could be a governing criterion.

In heavily reinforced beams, the bars can be 'bundled' to reduce the breadth required (Fig. 6.5(c)) or placed in more than one layer (Fig. 6.5(d)). To calculate the minimum spacing in the case of bundled bars, the bundle is treated as a single bar of equivalent cross-sectional area. Bars placed in more than one layer should be located vertically above each other to facilitate effective vibration.

Fig. 6.5 Alternative methods of placing reinforcement: (a) one bar at each corner; (b) conventional method; (c) in bundles; (d) in two layers

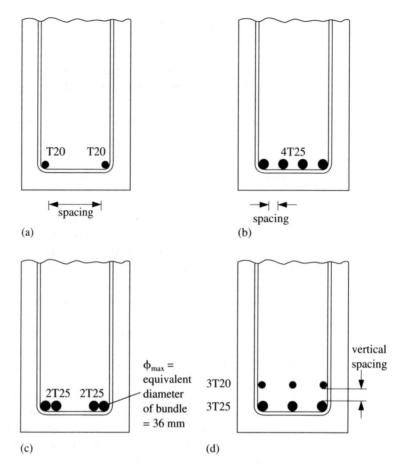

In long beams, continuity of reinforcement is often achieved by 'lapping'. With this technique, the bars are overlapped a specific distance, known as the 'lap length', so that the force can be transferred from one bar to the other. If the bars are placed alongside each other, as shown in Fig. 6.6(a), the breadth requirement of the beam will be increased. Alternatively, one bar may be placed below the other, as shown in Fig. 6.6(b), in which case there is no increase in the breadth requirement. However, this type of arrangement can cause complications if there is more than one layer of reinforcement.

Breadth is a major influencing factor on the shear strength of beams. A method has been proposed in the ISE manual for the estimation of the beam breadths required to satisfy shear strength requirements. The preliminary breadth is determined by limiting the shear stress in beams to 2.0 N/mm² for concrete of minimum characteristic cylinder strength of 25 N/mm². The required breadth, b, for rectangular beams and the required web width, b_w (Fig. 6.7), for flanged beams is given by:

$$b_w = \frac{V}{2.0d} \quad [\text{mm}] \tag{6.1}$$

Fig. 6.6 Lapping of reinforcement: (a) bars placed alongside each other; (b) bars placed one above the other

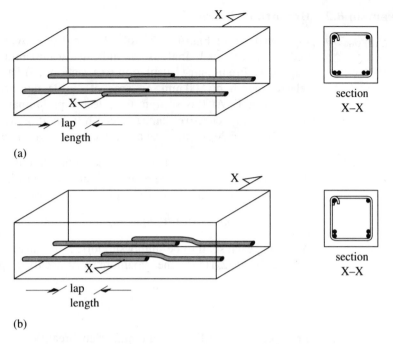

(a)

(b)

Fig. 6.7 Flanged beams: (a) T-beam; (b) L-beam

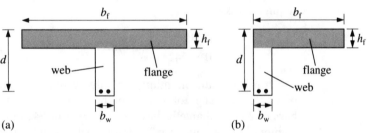

(a)   (b)

where V is the maximum shear force (in newtons) in the member, considered as simply supported, and d is the effective depth (in millimetres). For concrete of cylinder strength less than 25 N/mm², the width should be increased by an amount $25/f_{ck}$, where f_{ck} is the characteristic cylinder strength (in newtons per square millimetre).

Flanged beams occur where the beams are cast monolithically with the slab. The portion of the slab near the beams which acts with the beam in compression is known as the flange of the beam. The effective breadth of the flange depends on the web dimensions, the support conditions and the span lengths. The following recommendations for the calculation of effective flange breadths are given by EC2 and BS8110:

$$\text{T-beam:} \quad b_f = b_w + \tfrac{1}{5}L_0 \tag{6.2}$$

$$\text{L-beam:} \quad b_f = b_w + \tfrac{1}{10}L_0 \tag{6.3}$$

where L_0 is the distance between points of zero moment in the span. In continuous beams of length L, $L_0 = 0.85L$ for external spans, $0.7L$ for internal spans and $2.0L$ for cantilevered spans.

Example 6.2   Breadth of beams

Problem   For the beam of Fig. 6.3, the total factored ultimate loading, ω, is 43 kN/m and an approximate analysis has indicated a maximum moment of 220 kNm (sag) at E. Preliminary calculation (see section 6.5) has indicated that, for the selected depth of 800 mm, an area of tensile reinforcement of 950 mm^2 is required at E. At this section: (a) select a preliminary web breadth and (b) determine the effective flange breadth assuming a characteristic cylinder strength, f_{ck}, of 30 N/mm^2 and a reinforcement tensile strength, f_y, of 460 N/mm^2.

Solution   (a) The effective depth, d, is the total beam depth minus the cover plus the link diameter (or the cover to the main steel) and half the diameter of the reinforcing bars. Using the values from Example 6.1, the effective depth becomes:

$$d = 800 - (45 + 12.5) = 742 \text{ mm}$$

Assuming each span to be simply supported, the shear force at the ends of each member is $\omega L/2$. Hence, the maximum shear occurs in span BC and is given by:

$$V_{max} = \frac{\omega L}{2} = \frac{43 \times 9}{2} = 194 \text{ kN}$$

From equation (6.1), the minimum web breadth which satisfies shear stress requirements is:

$$b_w = \frac{195\,000}{2.0 \times 742} = 132 \text{ mm}$$

which is less than the minimum value given in Table 6.4 for a fire resistance of one hour. The total area of reinforcement required is 950 mm^2. Three bars of 20 mm diameter have a total area of 942 mm^2. Hence, three bars, two 20 mm diameter, and one 25 mm diameter, will be used. It is necessary to check that the web breadth of 200 mm satisfies the minimum spacing requirements for these bars.

From Table 6.6, the minimum bar spacing is 25 mm (assuming EC2 values and that $h_{agg} \leq 32$ mm). The cover to the main reinforcement is assumed to be 45 mm, in accordance with Table 6.4. Thus, the minimum web breadth required for all bars on one level is given by:

$$b_w = 45 + 5(25) + 45 = 215 \text{ mm}$$

which is more stringent than the breadth required for shear and fire. Thus, the preliminary web breadth is taken as 225 mm.

Note: It is often convenient for shuttering purposes to make the beam and column breadths equal. Accordingly, the column breadth may influence the preliminary beam breadth selected.

(b) The distance between points of zero moment in span BC is:

$$L_0 = 0.7(9) = 6.3 \text{ m}$$

Hence, the effective flange breadth is, from equation (6.2):

$$b_f = b_w + \tfrac{1}{5}L_0 = 225 + \tfrac{1}{5}(6300)$$

$$= 1485 \text{ mm}$$

6.4 Preliminary sizing of slabs

Fire resistance and durability

As for beams, the dimensions of slabs may be governed by fire-resistance requirements or the cover needed to ensure adequate resistance to corrosion. The minimum cover requirements for corrosion resistance in the five exposure classes of Table 6.2 are given in Table 6.3. The preliminary fire-resistance requirements in Table 6.7 are those recommended for preliminary design in the ISE manual for slabs spanning continuously over supports in mild and moderate environmental conditions. For more severe conditions and/or simply supported slabs, these minimum sizes and covers should be increased.

Table 6.7 Minimum dimensions and cover to *main* reinforcement for the preliminary design of continuous plain (solid) and ribbed slabs. All values are in millimetres (from the ISE manual)

Fire rating (h)	Plain slabs		Ribbed slabs		
	Minimum depth	Minimum cover	Minimum depth	Minimum rib width	Minimum cover
1	100	35	90	90	35
2	125	35	115	110	35
4	170	45	150	150	55

Effective depth of slabs

As for beams, the effective depth has a major influence on the capacity of a slab to resist bending moment. With the exception of flat slabs and ribbed slabs, the depth of slabs does not control the total depth of structures. However, a small increase in slab depth adds greatly to the self-weight of the structure which significantly increases the bending moments in the slab itself, and adds load to all beams, columns and foundations which support the slab.

In order to minimize the depth of slabs while at the same time ensuring adequate moment and shear capacity, the ISE manual recommends rules for the preliminary sizing of slabs. These rules, for the minimum span/effective depth ratios of slabs, are given in Table 6.8.

The effective depth requirements for flat slabs can also be governed by 'punching shear', the tendency of a supporting column to 'punch up' through the slab (of course it is the slab which pushes down around the column). The approximate method proposed in the ISE manual for ensuring adequate

Table 6.8 Span/effective depth ratios for the preliminary design of slabs. The ratios for two-way slabs are calculated for a square panel. For a 2 × 1 panel, the value for a one-way panel should be used. For panels with a length/breadth ratio between one and two, the value is obtained by interpolation between the one-way and two-way values. Flat slab design is to be based on the longer span dimensions. For exterior panels, 85 per cent of the ratios given should be used. For ribbed slabs, 85 per cent of the ratios given should be used (from the ISE manual)

| Variable loading (kN/m²) | One-way spanning | | | Two-way spanning | | Flat slab |
	Simple supports	Continuous	Cantilever	Simple supports	Continuous	Without drops
5	27	31	11	30	40	36
10	24	28	10	28	39	33

resistance to punching shear in flat slabs is as follows. Check that:

$$\frac{1250w(\text{area supported by column})}{(\text{column perimeter} + 9h)d} \leq 0.6 \text{ N/mm}^2 \tag{6.4}$$

and that:

$$\frac{1250w(\text{area supported by column})}{(\text{column perimeter})d} \leq 0.9\sqrt{f_{ck}} \text{ and } 5 \text{ N/mm}^2 \tag{6.5}$$

where w is the ultimate load per unit area in kilonewtons per square metre, d is the effective depth of the slab at the column in millimetres and h is the total depth of the slab at the column in millimetres. The area supported by the column is in square metres and the column perimeter is in millimetres.

Example 6.3 Effective depth of one-way spanning slabs

Problem A multi-storey structure is to be constructed using a continuous one-way spanning floor system. The layout for the proposed system is illustrated in Fig. 6.8. (a) If a typical floor is to carry a characteristic variable gravity load of 5 kN/m² and the required fire resistance is 1 hour, determine a preliminary depth for the slabs. (b) Comment on the proposed scheme.

Solution (a) As for the beams in Example 6.1, the same depth is used in all slabs for ease of construction. From Table 6.7, a fire resistance of 1 hour requires a minimum depth of 100 mm. At a typical floor level there are two interior panels and two exterior panels, one of which is a cantilever. The recommended span/effective depth ratio for each panel, taken from Table 6.8, is given by:

$$(\text{span/eff. depth})_{\text{ext. panel}} = 0.85(31) = 26.4$$

$$(\text{span/eff. depth})_{\text{int. panel}} = 31$$

$$(\text{span/eff. depth})_{\text{cantilever}} = 11$$

Thus, the required effective depth for each panel is:

Fig. 6.8 Plan view of floor layout

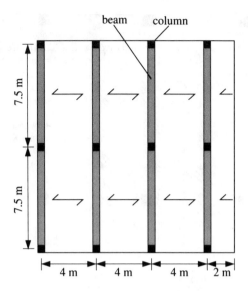

beam   column

7.5 m

7.5 m

4 m   4 m   4 m   2 m

$$(\text{eff. depth})_{\text{ext. panel}} = 4000/26.4 = 152 \text{ mm}$$

$$(\text{eff. depth})_{\text{int. panel}} = 4000/31 = 129 \text{ mm}$$

$$(\text{eff. depth})_{\text{cantilever}} = 2000/11 = 182 \text{ mm}$$

Therefore, an effective depth, d, of 182 mm, as governed by the cantilever panel, is adopted throughout. The total slab depth is the sum of the effective depth, the cover and half the diameter of the tension reinforcement (unlike for beams, links for shear resistance are not commonly provided in slabs). Table 6.3 specifies a minimum cover for durability purposes of 15 mm (exposure class 1). However, for a fire resistance of 1 hour, Table 6.7 gives a preliminary minimum cover of 35 mm. Thus, assuming a reinforcement diameter of 12 mm, the total depth, h, becomes:

$$h = 182 + 35 + \tfrac{1}{2}(12) = 223 \text{ mm}$$

which is rounded up to 225 mm.

(b) The proposed layout is uneconomic owing to the excessively long cantilever. By reducing the length of the cantilever to, say, 1.5 m and increasing the two interior spans to 4.25 m, the required effective depth is reduced to 152 mm and a total depth of $h = 200$ mm becomes feasible.

Example 6.4   Effective depth of flat slabs

Problem   A multi-storey residential building is to be constructed using a flat slab floor system. The layout of the proposed floor system is illustrated in Fig. 6.9 (all columns are 400 mm by 400 mm). Each floor is required to sustain a total characteristic variable gravity load of 2 kN/m^2 plus the weight of non-permanent partitions. Suggest a preliminary depth for the slabs if the required fire resistance is two hours. Assume a concrete cylinder strength, f_{ck}, of 30 N/mm^2.

Fig. 6.9 Flat slab floor
of Example 6.4

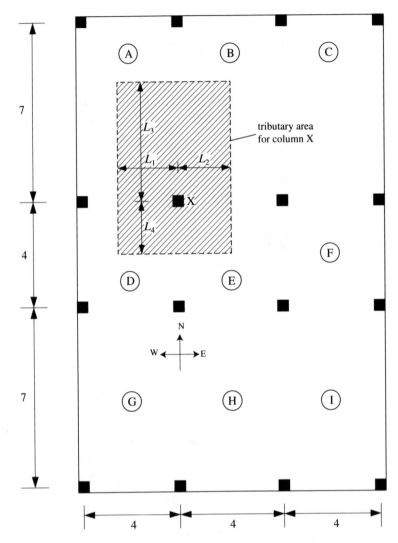

Solution   As for the previous examples, the same depth is used in all slabs. The depth may be governed by conditions in any of the nine panels. However, for convenience, the panels can be classified into three groups (see Table 6.9) depending on their type and geometry. The required depth will be governed by one of these three groups of panels.

Table 6.9   Panel groups
for Example 6.4

Group	Type	Longer span dimension (m)	Panels
I	Exterior	7	A, B, C, G, H, I
II	Exterior	4	D, F
III	Interior	4	E

As the characteristic variable load (of $2 \, kN/m^2$ plus $1 \, kN/m^2$ for partitions) is less than $5 \, kN/m^2$, the lesser of the span/effective depth ratios given in Table 6.8 is adopted for all groups, that is:

span/eff. depth = 36

However, since groups I and II are exterior panels, 85 per cent of this value is used for them, that is:

$(\text{span/eff. depth})_{\text{groups I, II}} = 0.85(36) = 30.6$

Thus, the required effective depth for each group is given by:

$(\text{eff. depth})_{\text{group I}} = 7000/30.6 = 229 \, mm$

$(\text{eff. depth})_{\text{group II}} = 4000/30.6 = 131 \, mm$

$(\text{eff. depth})_{\text{group III}} = 4000/36 = 111 \, mm$

Therefore, an effective depth, d, of 229 mm, as governed by group I panels, is adopted throughout. The total slab depth is equal to the effective depth plus the cover to the reinforcement plus half the diameter of the tension reinforcement. From Table 6.7, the preliminary minimum cover for a 2 hour fire resistance is 35 mm. Assuming a tension reinforcement diameter of 16 mm, the total slab depth, h, becomes:

$h = 229 + \frac{1}{2}(16) + 35 = 272 \, mm$

which is rounded off to 275 mm (note that this is greater than the minimum depth given in Table 6.7 and so is satisfactory). As a result of the rounding off, the effective depth is now $(275 - 35 - 16/2) = 232 \, mm$.

It should be noted that, in this example, the total depth has been calculated assuming that the outermost mat of reinforcement spans in the critical direction. However, in other examples where the effective depth is governed by a square panel, the total depth is equal to the effective depth plus cover plus **one and one half times** the reinforcement diameter.

It remains to be checked if this slab depth satisfies the shear strength requirements given by equations (6.4) and (6.5). These equations will be critical for the interior columns supporting the central panel since these columns have the largest tributary areas. For the column common to panels A, B, D and E, say, the tributary area is calculated with these panels fully loaded while the minimum loading is applied to all other panels. The tributary lengths, L_1, L_2, L_3 and L_4 (see Fig. 6.9), can be determined accurately using the formulae in Appendix C. However, since we are only interested in the **preliminary** size of the slab, it would be excessive to use this level of accuracy to calculate the tributary lengths. For calculating the **preliminary** shear capacity of the slab using equations (6.4) and (6.5), it is reasonable to assume the following:

$L_1 = \frac{5}{8}(4) = 2.5 \, m$

$L_2 = \frac{1}{2}(4) = 2.0 \, m$

$L_3 = \frac{5}{8}(7) = 4.38 \, m$

$L_4 = \frac{1}{2}(4) = 2.0 \, m$

(the accuracy of this assumption is left to the reader to evaluate). Hence, the tributary area is:

$$(L_1 + L_2) \times (L_3 + L_4) = 4.5 \times 6.38 = 28.7 \text{ m}^2$$

Assuming a self-weight of concrete of 25 kN/m², allowing for a 75 mm floor finish of density 23 kN/m³, and assuming a weight for ceilings and services of 0.5 kN/m², gives a total permanent gravity load of:

$$\text{Perm. load} = 25 \times 0.275 + 23 \times 0.075 + 0.5$$

$$= 9.1 \text{ kN/m}^2$$

Applying the partial safety factors from Chapter 3 (1.35 for permanent and 1.5 for variable gravity load), the total maximum ultimate load per unit area on this floor is:

$$\text{ultimate load, } w = 1.35 \times 9.1 + 1.5 \times (2 + 1)$$

$$= 16.8 \text{ kN/m}^2$$

Thus:

$$\frac{1250w(\text{area supported by column})}{(\text{column perimeter} + 9h)d} = \frac{1250 \times 16.8 \times 28.7}{(4 \times 400 + 9 \times 275)232}$$

$$= 0.64 \text{ N/mm}^2$$

$$\not< 0.60 \text{ N/mm}^2$$

and

$$\frac{1250w(\text{area supported by column})}{(\text{column perimeter})d} = \frac{1250 \times 16.8 \times 28.7}{4 \times 400 \times 257}$$

$$= 1.62 \text{ N/mm}^2$$

This is less than $0.9\sqrt{f_{ck}}$ ($=4.93$) and 5 N/mm². While the first of the inequalities is not exactly satisfied, it is **almost** satisfied, and it is worth re-evaluating some of the assumptions before increasing the slab depth beyond 275 mm. The screed will be present over the top reinforcement and will provide protection against fire. Hence, the preliminary cover for fire purposes of 35 mm is excessively conservative. Adopting a cover of 20 mm has the effect of increasing the effective depth by 15 mm and reducing the calculated stress from 0.64 to 0.60 N/mm². As both inequalities are now satisfied, the flat slab depth of 275 mm is taken as the preliminary solution.

Note 1: A slab depth of 275 mm is very large and an alternative form of construction may be preferable.

Note 2: Such short spans in flat slab construction have today become unusual. Panels of up to 7 m × 8 m are not now uncommon. For such spans, preliminary design in accordance with the above criteria would give excessively deep slabs. This is due to the fact that the criteria are conservative and that punching shear can now be resisted using shear reinforcement while the criteria seem to be based on an unreinforced solution.

Proportions of ribbed slabs

To ensure adequate stiffness against bending and torsion and to allow ribbed slabs to be treated as solid slabs for the purposes of analysis, EC2 recommends that the following restrictions on size are satisfied (refer to Fig. 6.10):

1. The rib spacing must not exceed 1.5 m.
2. The rib depth, d_w, must not exceed four times the rib breadth, b_w.
3. The flange depth, d_f, must exceed the greater of one-tenth of the clear distance between ribs, b_c, and 50 mm. In the case of ribbed slabs incorporating permanent blocks (Fig. 2.19(d)), the lower limit of 50 mm may be reduced to 40 mm.

Fig. 6.10 Ribbed slab

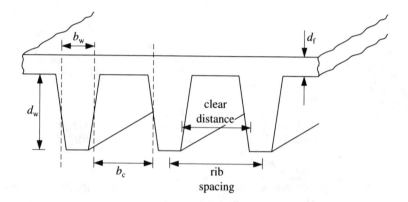

In addition to the minimum values given in Table 6.7 to satisfy fire-resistance requirements, the breadth of ribs may be governed by shear strength requirements. The method proposed in the ISE manual for the estimation of rib breadths limits the shear stress in the rib to 0.6 N/mm² for concrete with a characteristic cylinder strength of 25 N/mm² or more. The required breadth is given by:

$$b = \frac{V}{0.6d} \quad [\text{mm}] \tag{6.6}$$

where V is the maximum shear force in newtons on the rib considered as simply supported and d is the effective depth in millimetres. For characteristic cylinder strengths less than 25 N/mm², the breadth should be increased in proportion.

6.5 Reinforcement in beams and slabs

Once the section dimensions of a beam or slab are fixed and the grade of concrete has been decided upon, the steel reinforcement is the only remaining design variable. The quantity of flexural reinforcement required follows directly from the moment capacity required in the member to resist moments due to applied loads.

Preliminary analysis

At the preliminary design stage, the adequacy of the reinforcement needs only to be checked at mid-span and over the supports. Some approximate rules of thumb are given in the ISE manual for the estimation of internal moments and shear forces at these locations due to applied loads. For continuous beams and one-way spanning slabs, bending moments and shear forces may be taken from Table 6.10 if the following conditions are satisfied:

1. The variable gravity load does not exceed the permanent gravity load.
2. There are at least three spans.
3. The spans do not differ in length by more than 15 per cent of the longest span.

Table 6.10 Ultimate bending moments and shear forces for continuous beams and one-way spanning slabs where L is the span length in metres (from the ISE manual)

	Uniformly distributed load, q, acting on all spans (kN/m)	Central point load, Q, acting on all spans (kN)
Bending moment		
at support	$0.10qL^2$	$0.150QL$
at mid-span	$0.08qL^2$	$0.175QL$
Shear force	$0.65qL$	$0.65Q$

For two-way spanning slabs on linear supports, the average moment per metre width, m, can be taken as:

$$m = q\left(\frac{L_x L_y}{24}\right) \quad [\text{kNm/m}] \tag{6.7}$$

where L_x and L_y are the shorter and longer span lengths, respectively, in metres and q is the ultimate (factored) load in kilonewtons per square metre. If $L_y > 1.5L_x$, the slab should be treated as a one-way spanning slab.

For solid flat slabs, the moments per unit width in strips of slab between columns can be taken as 1.5 times those for one-way spanning slabs. For one-way spanning ribbed slabs, the bending moments at mid-span can be assessed on a width equal to the rib spacing and assuming simple supports throughout. For two-way spanning ribbed slabs where $L_y < 1.5L_x$, the average rib moment in both directions can be taken as:

$$m = q\left(\frac{L_x L_y}{24}\right)c \quad [\text{kNm/m}] \tag{6.8}$$

where c is the rib spacing. If $L_y > 1.5L_x$, the slab should be treated as one-way spanning.

Reinforcement in rectangular sections

When a rectangular section is subjected to sag moment as illustrated in Fig. 6.11(a), the top fibres are compressed and the bottom fibres are extended.

Fig. 6.11
Reinforcement
requirement in
reinforced concrete
beam: (a) beam in
bending; (b) segment of
beam in bending at ULS

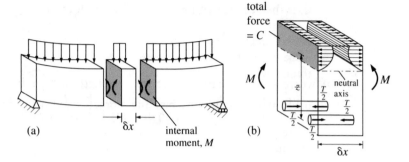

(a)   internal moment, M (b)

Concrete has good compressive strength and very low tensile strength. For design purposes, zero tensile strength can be assumed. Thus, the bending causes tension only in the reinforcing bars near the bottom of the section and compression in the concrete near the top of the section (and in the top reinforcement if any is present). For the high levels of stress that exist at ULS, the distribution of compressive stress in the concrete is of the form illustrated in Fig. 6.11(b). This parabolic distribution of stress is often approximated by an equivalent uniform (rectangular) distribution. By force equilibrium at the face illustrated in Fig. 6.11(b), we have:

compression force, C = tension force, T

Moment equilibrium at the face gives:

$$M = Tz \qquad (6.9)$$

where z is the lever arm between centres of the tension force and the compression force as shown in Fig. 6.11(b).

An estimate of the lever arm that is sufficiently accurate for detailed design is based on an assumed uniform distribution of compressive stress in the concrete (so-called equivalent rectangular stress block). Using the equivalent rectangular stress block, EC2 gives:

$$z = d(0.5 + \sqrt{0.25 - 0.88K}) \qquad (6.10)$$

(Equation (6.10) is derived later in Chapter 7.) In this equation, d is the effective depth of the section and the parameter K is given by:

$$K = \frac{M}{f_{ck}bd^2} \qquad (6.11)$$

where f_{ck} is the characteristic compressive cylinder strength of concrete at 28 days and b is the breadth of the section. Note that the units used in equations (6.10) and (6.11) must be consistent. Alternatively, for preliminary design purposes, a rough estimate of the length of the lever arm can be taken as:

$$z = 0.8d \qquad (6.12)$$

The area of tension reinforcement required to resist the internal moment,

M, is then calculated using the following formula:

$$A_s = \frac{M}{0.87f_yz} \quad (6.13)$$

where f_y is the characteristic yield strength of the reinforcement.

For relatively high moment, specifically if $K > 0.166$, compression reinforcement is required to maintain ductility in the section. For preliminary purposes, the required area of compression reinforcement, A_s', is calculated using the formula:

$$A_s' = \frac{M - 0.166f_{ck}bd^2}{0.87f_y(d - d')} \quad (6.14)$$

where d' is the effective depth from the extreme fibre in compression to the centroid of the compression steel. The required area of tension reinforcement then becomes:

$$A_s = \frac{0.24bdf_{ck}}{f_y} + A_s'$$

Reinforcement in flanged beams and ribbed slabs

The design procedure for flanged beams and ribbed slabs depends on whether the neutral axis lies within the flange or the web (Fig. 6.12). If the neutral axis lies within the flange, its location can be calculated using:

$$x = (d - z)/0.4 \quad (6.15)$$

where z is determined using equations (6.10) and (6.11) with $b = b_f$, the flange breadth. If the neutral axis does not lie within the flange, equation (6.15) is inappropriate. However, this equation can still be used to determine whether x is greater or less than the flange depth, that is $x \lessgtr h_f$.

Fig. 6.12 Alternative locations of neutral axis for a flanged section in sag: (a) within flange; (b) within web. Shaded area indicates zone of compression

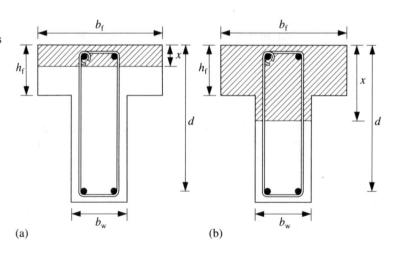

(a)   (b)

Neutral axis within flange

If the neutral axis lies within the flange for a sagging section (i.e. $x < h_f$), as shown in Fig. 6.12(a), the compression zone is rectangular and the section can be designed as a rectangular section. The shape of the tensile zone of the concrete, assumed cracked, is of no relevance in this case. It can be seen in Fig. 6.13 that the equivalent stress block specified in EC2 does not extend fully to

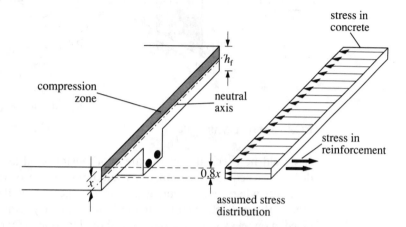

Fig. 6.13 Sagging flanged section assuming equivalent rectangular stress block

the neutral axis (the reason for this is to provide a simpler distribution in which the resultant compressive force acts at a point reasonably near to the actual resultant). It is clear from the figure that the compression zone actually remains rectangular for all $x \leq h_f/0.8$. For $x > h_f/0.8$, the assumption of rectangular section behaviour ceases to be valid since the compression zone moves into the web. The moment at which this occurs is given by:

$$M = 0.57 b_f h_f f_{ck}(d - 0.5 h_f) \tag{6.16}$$

Thus, for all sections where the applied moment, M, is less than the value obtained from equation (6.16), the area of tension steel required is derived from equations (6.10)–(6.13) using b_f in place of b. If M exceeds the value obtained from equation (6.16), the neutral axis does not lie within the web and the procedure described below is used.

Neutral axis within web

If the neutral axis lies within the web for a section in sag (Fig. 6.12(b)), the compression zone is not rectangular and the section cannot be designed using equations (6.10)–(6.13). It is possible to derive a formula similar to equation (6.13) for such a case but for preliminary design purposes this unduly complicates the procedure. It is simpler to redesign the section by increasing the web depth so that the neutral axis lies within the flange.

For flanged sections in hog (Fig. 6.14), the neutral axis generally lies in the web. Hence, the minimum tension steel requirements are found using equations (6.10)–(6.13) with $b = b_w$, the web width.

Fig. 6.14 Hogging flanged section with neutral axis in web

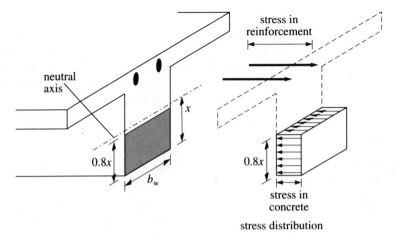

stress distribution

EC2 restrictions

It should be checked that the areas of tension steel and of compression steel do not exceed the recommended maximum value of $0.04A_c$ (except in long members where reinforcement is lapped) where A_c is the total cross-sectional area of the concrete section. In addition, secondary reinforcement must be provided in one-way spanning slabs and should be at least 20 per cent of the principal reinforcement. The spacing of principal reinforcement should not exceed the lesser of $1.5h$ and 350 mm where h is the total depth of the slab. In secondary slab reinforcement, the spacing should not exceed the lesser of $2.5h$ and 400 mm.

Example 6.5 Reinforcement in beams

Problem A four-span rectangular beam of span length 8 m is to carry a total ULS factored gravity load of 57 kN/m. A typical section of the beam is illustrated in Fig. 6.15. For this section, a preliminary depth, h, of 500 mm and a

Fig. 6.15 Rectangular section

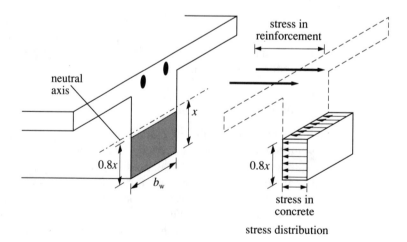

stress distribution

preliminary breadth, b, of 325 mm have been selected. Determine the quantity of longitudinal reinforcement required to resist the applied loads. For this quantity of steel, check that the beam breadth is sufficient to allow adequate bar spacing. Assume a concrete characteristic cylinder strength, f_{ck}, of 35 N/mm².

Solution From Table 6.10, the mid-span moment in the beam due to the applied load of 57 kN/m is:

$$M = 0.08qL^2 = 0.08(57)(8)^2 = 292 \text{ kNm}$$

Assuming (initially) a tension reinforcement diameter of 25 mm and, with reference to Table 6.4, the effective depth is given by:

$$d = 500 - 45 - \tfrac{1}{2}(25)$$

$$= 442.5 \text{ mm}$$

From equation (6.11), we have:

$$K = \frac{M}{f_{ck}bd^2}$$

$$= \frac{292 \times 10^6}{35 \times 325 \times 442.5^2}$$

$$= 0.131$$

As $K < 0.166$, compression reinforcement will not be required. From equation (6.10), the lever arm is:

$$z = d(0.5 + \sqrt{0.25 - 0.88K})$$

$$= 442.5(0.5 + \sqrt{0.25 - 0.88(0.131)})$$

$$= 384 \text{ mm}$$

Using high-yield reinforcement ($f_y = 460$ N/mm²), the required area of tension reinforcement is:

$$A_s = \frac{M}{0.87f_yz}$$

$$= \frac{292 \times 10^6}{0.87 \times 460 \times 384}$$

$$= 1901 \text{ mm}^2$$

One 25 mm diameter bar has a cross-sectional area of 491 mm². Thus, four 25 mm diameter bars (total area = 1963 mm²) must be provided as tension reinforcement.

The bar arrangement for this section is illustrated in Fig. 6.16. Although no compression reinforcement is required, two 10 mm diameter bars (nominal reinforcement) are provided in the top of the section to act as hangers for the links. From Table 6.6, the minimum spacing of the tension bars is 25 mm. Thus,

Fig. 6.16 Bar
arrangement for section
of Fig. 6.15

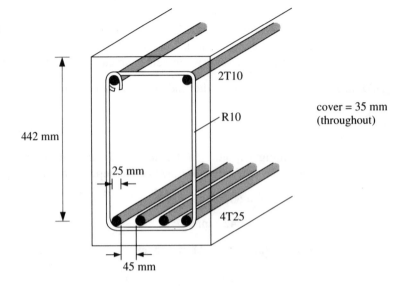

the minimum beam breadth required for this arrangement is:

$$b = 2(45) + 7(25) = 265 \, \text{mm}$$

Hence, the assumed breadth of 325 mm is satisfactory for this bar arrangement.

6.6 Preliminary sizing of columns and walls

Columns

Columns are classified in EC2 and BS8110 as being:

(a) either 'short' or 'slender'; and
(b) either 'braced' or 'unbraced'.

These distinctions are made because the strength of slender and/or unbraced columns can be substantially reduced by the horizontal deflection of the column under the applied loads. The majority of structural columns are within the short classification and slenderness effects can be ignored. For simplicity, only 'short' columns should be used at the preliminary design stage.

The criteria by which columns are classified as short or slender in EC2 are quite complex. For this reason, the BS8110 classification is more suitable for preliminary design, even if the final design is to be in accordance with EC2. BS8110 defines a short column as one for which the ratio of the effective height to the least lateral dimension does not exceed 15 for braced columns and 10 for unbraced columns. For preliminary design, the effective height can be taken as 0.85 times the clear column height for braced columns and as 1.5 times the clear height for unbraced columns.

Table 6.11 Minimum dimensions for the preliminary design of columns

Fire resistance (h)	Minimum side dimension (mm)
1	200
2	300
4	450

The minimum dimensions required for the fire resistance of rectangular columns are given in Table 6.11. A preliminary estimate of the capacity of a short column to resist axial force is given in the ISE manual as:

$$N = A_c\left(0.44f_{ck} + \frac{p}{100}(0.67f_y - 0.44f_{ck})\right) \qquad (6.17)$$

where A_c is the cross-sectional area of the column (mm^2)
f_{ck} is the characteristic cylinder strength (N/mm^2)
f_y is the characteristic strength of the reinforcement (N/mm^2)
p is the percentage of reinforcement ($100A_s/A_c$)

A satisfactory preliminary column design is one which has an ultimate load capacity, calculated using equation (6.17), greater than the total applied load, that is the sum of the ultimate loads applied at each floor level. This equation is applicable to braced frames in which axial forces, rather than moments, are the dominant criteria for column design. However, some moment occurs in all frames owing to the lack of symmetry in the arrangement of loads. To allow for this, the ultimate applied load from the floor immediately above the column being considered should be multiplied by the following factors (ISE manual):

Columns balanced in two perpendicular directions
    (Fig. 6.17(a)):   1.25
Columns balanced in only one direction
    (Fig. 6.17(b)):   1.50
Columns unbalanced in two perpendicular directions
    (Fig. 6.17(c)):   2.00

Fig. 6.17 Column types: (a) balanced in two perpendicular directions (some internal columns); (b) balanced in one direction only (external columns); (c) unbalanced in two perpendicular directions (corner columns)

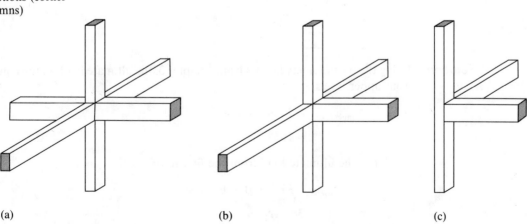

(a)   (b)   (c)

The maximum percentage of longitudinal reinforcement allowed in columns by EC2 is 8 per cent. Allowing for laps, this limits the area in regions between laps to 4 per cent. However, it is recommended to take the percentage of reinforcement, p, as 2 per cent at the preliminary stage of design. This leaves scope either to increase or decrease the amount of reinforcement in the detailed design.

Example 6.6 Preliminary sizing of columns

Problem An internal column in a multi-storey building, illustrated in Fig. 6.18, has a clear height of 5.5 m between the foundation pad and the first floor. The column carries a total characteristic permanent gravity load of 750 kN and a total characteristic variable gravity load of 450 kN from upper levels. In addition, characteristic permanent and variable gravity loads of 170 kN and 90 kN, respectively, are transferred to the column at the first floor. If the column is to have a square cross-section, select appropriate preliminary dimensions and determine the quantity of reinforcement required. Assume that the column is braced and that $f_{ck} = 35$ N/mm^2 and $f_y = 460$ N/mm^2.

Fig. 6.18 Internal ground floor column in a multi-storey structure

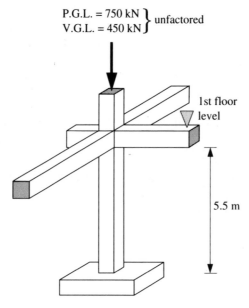

Solution Using the partial safety factors from Chapter 3, the ultimate load on the column from upper levels is:

$$N_{\text{upper levels}} = 1.35 \times 750 + 1.5 \times 450$$

$$= 1688 \text{ kN}$$

Similarly, the ultimate load from the first floor is:

$$N_{\text{first floor}} = 1.35 \times 170 + 1.5 \times 90$$

$$= 365 \text{ kN}$$

The applied loads from the first floor are factored by 1.25 (i.e. column balanced in two perpendicular directions) to compensate for bending effects. Thus, the total ultimate load on the column is:

$$N_{total} = N_{upper\ levels} + 1.25 \times N_{first\ floor}$$

$$= 1688 + 1.25 \times 365$$

$$= 2144\ kN$$

For a reinforcement percentage of 2, equation (6.17) gives:

$$N = A_c\left(0.44f_{ck} + \frac{p}{100}(0.67f_y - 0.44f_{ck})\right)$$

$$\Rightarrow \quad 2144 \times 10^3 = A_c\left(0.44(35) + \frac{2}{100}(0.67 \times 460 - 0.44 \times 35)\right)$$

$$\Rightarrow \quad A_c = \frac{2144 \times 10^3}{0.44(35) + (2/100)(0.67 \times 460 - 0.44 \times 35)}$$

$$= 100\ 866\ mm^2$$

Assuming the column is square, the side length, h, is given by:

$$h = \sqrt{100\ 866} = 318\ mm$$

This is rounded up to give a preliminary side length of 325 mm.

For the calculated depth of the column, it is necessary to confirm the initial assumption that the column is 'short'. The effective height of the column (assuming it is braced) is:

$$L_e = 0.85L = 0.85(5.5) = 4.675\ m$$

Thus:

$$\frac{L_e}{h} = \frac{4.675}{0.325} = 14.4 < 15$$

Hence, the column is indeed 'short'.

One possible solution in which 2 per cent reinforcement is provided is with the provision of eight 20 mm diameter bars as illustrated in Fig. 6.19 (total area of 2513 mm²). In addition, links are provided to give restraint against buckling of the main reinforcement.

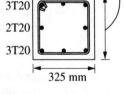

3T20

2T20

3T20

325 mm

325 mm

Fig. 6.19 Preliminary solution for column of Example 6.6

EC2 specifies that the spacing of links in columns should not exceed the minimum of:

(a) 12 times the minimum diameter of the longitudinal bars
(b) the least dimension of the column
(c) 300 mm

Walls

Walls carrying vertical loads should be designed as columns. Hence, as for columns, only non-slender walls should be used at the preliminary design stage.

Fig. 6.20 Wall with a return at the compression end

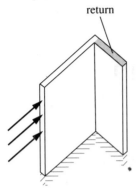

return

Shear walls should be designed as vertical cantilevers. When the dimensions are such that shear walls are not 'deep', the quantity of reinforcement required is determined in the same manner as for beams. Where a shear wall has a return at the compression end such as in Fig. 6.20, it should be designed as a flanged beam.

In general, walls should have a thickness of no less than 200 mm. This allows sufficient access for compaction hoses and, in most cases, ensures adequate strength and stiffness. EC2 specifies that the area of vertical reinforcement in walls be kept between 0.4 and 4 per cent with half the bars located on each face. Horizontal reinforcement should also be provided between the vertical reinforcement and the nearest surface. It should not be less than 50 per cent of the vertical reinforcement.

Problems

Sections 6.3 and 6.5

6.1 Suggest preliminary dimensions and maximum areas of reinforcement for a continuous rectangular beam with two 7 m spans. The total factored ULS loading (deemed to include self weight) is 30 kN/m and the fire rating is one hour.

6.2 The four-span T-beam illustrated in Fig. 6.21 is subjected to unfactored permanent and variable gravity loadings of 25 kN/m (excluding self weight) and 20 kN/m respectively. Determine preliminary values for the maximum areas of top and bottom reinforcement given a fire rating of one hour. This beam is integral with 300×300 mm columns.

Fig. 6.21 Beam of Problem 6.2

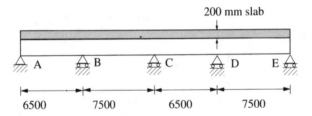

200 mm slab

| A | B | C | D | E |

| 6500 | 7500 | 6500 | 7500 |

Sections 6.4 and 6.5

6.3 For the slab of Example 6.3, make a preliminary estimate of the maximum required areas of reinforcement. The characteristic variable gravity loading of 5 kN/m² includes an allowance for non-permanent partitions.

Section 6.6

6.4 Determine preliminary dimensions and quantities of reinforcement for a reinforced concrete perimeter wall subjected to a total factored ULS load of 5000 kN uniformly distributed over its 5 m length. Of this load, 1250 kN is applied from the floor immediately above. The wall is part of a braced structure and is 4 m high.

Part III

Detailed Member Design

Introduction

Part III of this book deals with the topics traditionally covered in concrete design courses. While the 'new' topics such as qualitative design are most important, an understanding of the detailed requirements of codes of practice is also necessary.

Perhaps the most important mechanism of load transfer is flexure. This is dealt with in three chapters. Chapter 7 deals with the design of ordinary reinforced concrete to resist internal bending moment. Design for bending in prestressed concrete is dealt with in Chapter 8 including a section on partial prestressing. The internal shear force, which usually accompanies bending, is dealt with in Chapter 10 for both ordinary reinforced concrete and prestressed concrete members. Design to resist torsional moment is also treated in this chapter as there are many parallels with design for shear. As the analysis of slabs was dealt with in Part II, the design of slabs for flexure and shear is identical to that of beams as can be seen from the examples in these chapters. An exception to this rule is the design of slabs for punching shear. Punching shear in slabs is treated in a separate section in Chapter 10.

Members subject to axial force are treated in Chapter 9. This chapter includes all members which transmit load through axial force mechanisms. Thus, in addition to columns and walls, this chapter applies to tension members and to deep beams.

This part of the book is intended to complete the education of the engineer in concrete design. It also reflects the final stage in the complete design process. Thus, the designer, having gone through the stages of conceptual design and preliminary analysis and design, must now carry out a detailed analysis and, using the information in this part of the book, must complete a detailed design for all structural members.

7

Design of Reinforced Concrete Members for Bending

7.1 Introduction

Beams and slabs, by definition, transmit load by bending and must be designed to resist the resulting moment. Two limit states are generally considered. At the serviceability limit state (SLS), cracking occurs which must be limited to prevent air and moisture from reaching the reinforcement and causing corrosion. Also at SLS, deflections occur which must be kept within acceptable levels. From the design viewpoint, SLS checks are characterized by relatively low stresses. It follows that it can be reasonably assumed that the stress/strain relationship is linear, that is **stress is proportional to strain at SLS**. It is a fundamental assumption of bending theory that plane sections remain plane. Hence, the distribution of **strain through a cross-section is always linear**, as illustrated in Fig. 7.1(b). It then follows from the linear stress/strain relationship that the **stress distribution is also linear at SLS**, as illustrated in Fig. 7.1(c).

At the ultimate limit state (ULS), the safety of the beam or slab is considered. Thus, at the critical cross-sections, the capacity to resist moment is compared with the moment due to applied loads. From the design viewpoint, ULS checks are characterized by high stresses and by non-linear stress/strain relationships. The assumption that plane sections remain plane is still valid. However, **distributions of stress are generally non-linear at ULS**.

7.2 Second moments of area (linear elastic)

An internal bending moment, M, due to applied loads, generates a distribution of axial stresses within the section as illustrated in Fig. 7.1. Recall that, for **linear elastic** behaviour, the relationship between the applied moment, M, and the stress at a distance y above the neutral axis of the section, $\sigma(y)$, is given by the flexure formula:

$$\frac{M}{I} = \frac{\sigma(y)}{y} = \frac{E\varepsilon(y)}{y} \tag{7.1}$$

where $\varepsilon(y)$ is the strain at a distance y above the neutral axis and E is the

Fig. 7.1 Linear distributions of stress and strain at SLS: (a) section; (b) strain distribution; (c) stress distribution

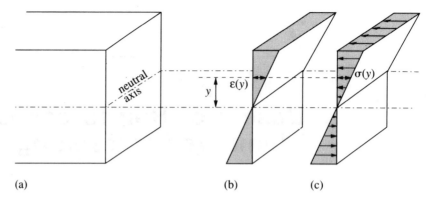

(a)   (b)   (c)

Fig. 7.2 Cross section

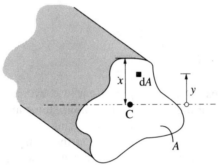

modulus of elasticity. The term I is known as the second moment of area. It is a constant which relates the applied moment at a section to the stress or strain which it causes.

For a given cross-section of area A, as illustrated in Fig. 7.2, the second moment of area about the axis containing the centroid, C, is defined by the integral:

$$I = \int_A y^2 \, dA \tag{7.2}$$

The centroid of an area is the geometric centre of the area. For a section in pure bending, the axis containing the centroid is the neutral axis of the section. **The neutral axis is defined as the axis at which strains are zero.** For the linear elastic case, the location of the neutral axis is found by taking first moments of area about any point. For this purpose, moments are usually taken about the extreme fibre in compression which, for sag moment, is the top fibre. The calculation of neutral axis location and of the second moment of area, I, is illustrated by the following examples.

Example 7.1   Rectangular section

Problem   The rectangular section of Fig. 7.3 is subjected to a sag moment of 500 kNm.

(a) Determine the neutral axis location and the second moment of area.
(b) Assuming that stress is everywhere proportional to strain, find the maximum axial tensile strain in the section given that $b = 300$ and $h = 500$.

Fig. 7.3 Section of
Example 7.1

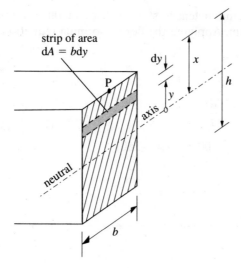

strip of area
$dA = b\,dy$

P

Solution   (a) Taking first moments of area about P in Fig. 7.3:

$$\int (\text{area} \times \text{distance from P}) = (\text{total area}) \times x$$

where x is the distance from the neutral axis to the extreme fibre in compression, which in this case is P. Hence:

$$(bh)(h/2) = (bh)x$$

$$\Rightarrow \qquad x = h/2$$

The second moment of area is calculated using eq (7.2) where y is measured from the neutral axis:

$$I = \int_{y=-h/2}^{y=h/2} y^2 \, dA$$

Referring to the strip of area, dA, in Fig. 7.3:

$$I = \int_{y=-h/2}^{y=h/2} y^2(b\,dy)$$

$$= b \int_{y=-h/2}^{y=h/2} y^2 \, dy$$

$$= b \left. \frac{y^3}{3} \right|_{-h/2}^{h/2}$$

$$= \frac{b}{3}\left[\left(\frac{h}{2}\right)^3 - \left(-\frac{h}{2}\right)^3\right]$$

$$= \frac{b}{3}\left[\frac{h^3}{8} - \left(-\frac{h^3}{8}\right)\right]$$

$$= \frac{b}{3}\frac{h^3}{4} = \frac{bh^3}{12}$$

(b) The maximum tensile stress will be at the extreme bottom fibre, that is $y = -250$ mm. Applying the flexure formula (equation (7.1)):

$$\frac{M}{I} = \frac{\sigma(y)}{y}$$

$$\Rightarrow \quad \sigma(y) = \frac{My}{I}$$

$$= \frac{(500 \times 10^6)(-250)}{(300 \times 500^3/12)}$$

$$= -40 \text{ N/mm}^2$$

Example 7.2   T-section

Problem   Find the location of the neutral axis for the T-section illustrated in Fig. 7.4. Hence calculate the second moment of area of the section about the neutral axis.

Fig. 7.4 Section of Example 7.2

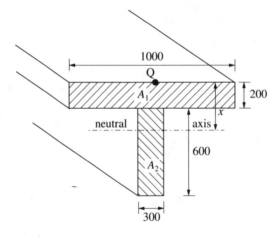

Solution   Taking moments of area about point Q in Fig. 7.4:

$$(A_1 + A_2)x = A_1(100) + A_2(600/2 + 200)$$

where $A_1 = 200 \times 1000 = 200\,000$ mm^2 and $A_2 = 600 \times 300 = 180\,000$ mm^2. Hence:

$$x = \frac{200\,000(100) + 180\,000(500)}{200\,000 + 180\,000}$$

$$= 290 \text{ mm}$$

That is, the neutral axis is 290 mm from the top fibre or 90 mm from the bottom of the flange. Hence, the second moment of area is:

$$I = \int_A y^2 \, \mathrm{d}A$$

$$= \int_{-(600-90)}^{+90} y^2(300 \, \mathrm{d}y) + \int_{90}^{290} y^2(1000 \, \mathrm{d}y)$$

$$= 300 \left. \frac{y^3}{3} \right|_{-510}^{90} + 1000 \left. \frac{y^3}{3} \right|_{90}^{290}$$

$$= 100[90^3 - (-510)^3] + 333.3[290^3 - 90^3]$$

$$= 21.225 \times 10^9 \text{ mm}^4$$

Alternatively, the second moment of area can be evaluated by summing, for all parts of the section, the second moment of area plus the product of area and the distance to the section neutral axis squared:

$$I = \frac{1000 \times 200^3}{12} + A_1(x - 100)^2 + \frac{300 \times 600^3}{12} + A_2(600/2 - 90)^2$$

$$= (667 + 7220 + 5400 + 7938) \times 10^6$$

$$= 21.225 \times 10^9 \text{ mm}^4$$

The neutral axis can be located from first principles using equilibrium of the forces parallel to the X-axis of the beam. For pure bending there are no axial forces, that is:

$$F = \int_A \sigma(y) \, \mathrm{d}A = 0 \tag{7.3}$$

Similarly equations (7.1) and (7.2) can be derived by summing moments about the neutral axis. Example 7.3 below illustrates how stresses can be calculated from first principles.

Example 7.3 **Stresses from first principles**

Problem (a) Using the equilibrium of force method (equation (7.3)), locate the neutral axis of the section of Fig. 7.5(a).

(b) Derive an expression for the stress $\sigma(y)$ at a distance y above the neutral axis.

Solution (a) As plane sections remain plane, the strain distribution for this section is as illustrated in Fig. 7.5(b). If the maximum compressive strain is ε_0 as illustrated, the strain at a distance y from the neutral axis is (by similar triangles):

$$\frac{\varepsilon(y)}{y} = \frac{\varepsilon_0}{x}$$

$$\Rightarrow \quad \varepsilon(y) = \frac{\varepsilon_0 y}{x}$$

Design of Reinforced Concrete Members for Bending

Fig. 7.5 Section of Example 7.3: (a) section; (b) strain distribution; (c) stress distribution

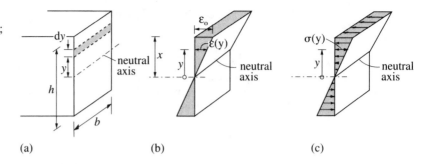

(a) (b) (c)

As stress is proportional to strain everywhere, the stress at this level is:

$$\sigma(y) = E\varepsilon(y)$$

$$= \frac{E\varepsilon_0 y}{x}$$

Thus, equation (7.3) becomes:

$$F = \int_A \sigma(y)\, dA = 0$$

$$\Rightarrow \quad \int_A \frac{E\varepsilon_0 y}{x}\,(b\, dy) = 0$$

$$\Rightarrow \quad \frac{E\varepsilon_0 b}{x} \int_{-(h-x)}^{x} y\, dy = 0$$

$$\Rightarrow \quad \frac{E\varepsilon_0 b}{x} \frac{y^2}{2}\bigg|_{-(h-x)}^{x} = 0$$

which reduces to:

$$\frac{E\varepsilon_0 b}{2x}(-h^2 + 2hx) = 0$$

As none of the constants is zero, this implies:

$$-h^2 + 2hx = 0$$

$$\Rightarrow \quad x = h/2$$

which is the centroid of the section.

(b) The sum of all moments about the neutral axis must equal the applied moment, M, that is:

$$M = \int_A \sigma(y) y\, dA$$

$$= \int_A \left(\frac{E\varepsilon_0 y}{x}\right) y\, dA$$

$$= \frac{E\varepsilon_0}{x} \int_A y^2 \, dA = E\left(\frac{\varepsilon(y)}{y}\right) \int_A y^2 \, dA$$

$$= \frac{\sigma(y)}{y} \int_A y^2 \, dA$$

Hence:

$$\sigma(y) = \frac{My}{\int_A y^2 \, dA}$$

$$= \frac{My}{I}$$

where:

$$I = \int_A y^2 \, dA$$

Uncracked sections

For concrete sections, the second moment of area considered in Examples 7.1 to 7.3 is known as the 'gross' second moment of area, I_g. In the calculation of I_g, the presence of the reinforcement is ignored and it is assumed that the entire section is available to resist bending. In the calculation of internal bending stress, however, it is sometimes useful to take account of the composite nature of reinforced concrete, that is to account for the presence of the reinforcement which is considerably stiffer than concrete. The flexure formula (equation (7.1)) is directly applicable only to homogeneous sections in bending. Therefore, in order to use it, the composite steel and concrete section is transformed into an equivalent homogeneous concrete section.

Transformed sections

Consider the two-material composite beam of Fig. 7.6(a). The two materials are bonded together so that, in bending, no slip can occur between them. The top material has an elastic modulus of E_1 and the bottom material has an elastic modulus of E_2 such that $E_1 < E_2$. If a bending moment, M, is applied to the section, the assumption that plane sections remain plane is still valid, and the strain varies linearly as illustrated in Fig. 7.6(b). The stress at any level in the top material is equal to $E_1\varepsilon$ and the stress at any level in the bottom material is given by $E_2\varepsilon$. The stress distribution corresponding to these conditions is illustrated in Fig. 7.6(c). To transform this into an equivalent homogeneous section having one elastic modulus $E = E_1$, the breadth of the bottom material is increased, as illustrated in Fig. 7.6(d). The equivalent breadth

Design of Reinforced Concrete Members for Bending

Fig. 7.6 (a) Composite section; (b) strain distribution; (c) stress distribution; (d) equivalent section with elastic modulus E_1

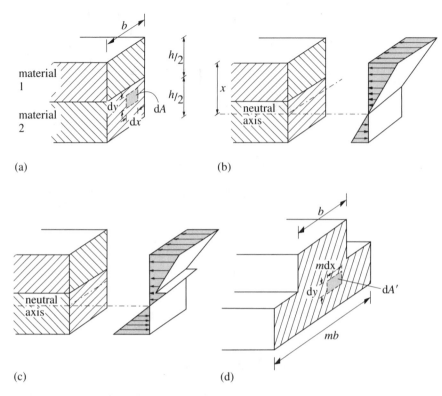

(a)

(b)

(c)

(d)

is determined by considering a force dF acting on an area $dA = dx\,dy$ of the section of Fig. 7.6(a). Thus:

$$dF = \sigma\,dA = (E_2\varepsilon)\,dx\,dy$$

The equivalent area, dA', of height dy in the section of Fig. 7.6(d) is:

$$dA' = m\,dx\,dy$$

where $m\,dx$ is the equivalent breadth for some m. In the equivalent section the force acting on this area is calculated using the equivalent area and the new modulus E_1. Hence it is:

$$dF' = (E_1\varepsilon)\,dA' = (E_1\varepsilon)m\,dx\,dy$$

Equating the forces dF and dF' gives:

$$(E_2\varepsilon)\,dx\,dy = (E_1\varepsilon)m\,dx\,dy$$

$$\Rightarrow \qquad\qquad m = E_2/E_1 \qquad\qquad\qquad\qquad (7.4)$$

Thus, for a composite section (two materials such that $E_1 \neq E_2$) of breadth b, the breadth of the section is be increased to mb over the depth of the stiffer material, where m, known as the modular ratio, is given by equation (7.4).

In the case of reinforced concrete, the equivalent transformed section is obtained by **replacing** the areas of tension and compression reinforcement, A_s and A_s' respectively, by their equivalent areas of concrete, mA_s and mA_s', where

Fig. 7.7 (a) Reinforced concrete section; (b) equivalent concrete section

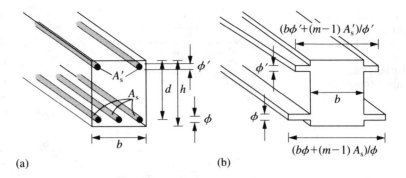

(a)   (b)

m is the ratio of the elastic modulus for steel to that for concrete:

$$m = E_s/E_c$$

Thus, for the section of Fig. 7.7(a), the equivalent section is illustrated in Fig. 7.7(b). The depth of each strip is not important as it is only the area that affects the calculations. For convenience, a depth equal to the steel diameter is assumed. The compression reinforcement displaces an area of concrete, A'_s. Thus the total equivalent area of the top strip is given by:

$$\text{(area of concrete)} + \text{(equivalent area of steel)} = (b\phi' - A'_s) + (mA'_s)$$
$$= b\phi' + (m-1)A'_s$$

where b is the breadth of the section and ϕ' is the diameter of the compression reinforcement. Similarly, the total equivalent area of the bottom strip is given by the area of the rectangular section plus $(m-1)A_s$.

Once the equivalent section is established, the calculation of I is identical to the calculation of I_g. In this case, however, I is known as the 'uncracked' second moment of area, I_u, as the possibility of the concrete cracking has not yet been considered. It is important to remember that stresses that are calculated using the transformed section refer to the less stiff material, that is concrete.

Cracked sections

In calculating I_g and I_u, the concrete is assumed to be uncracked. However, reinforced concrete sections in pure bending develop tensile cracks under very small loads. Thus, for the calculation of the stresses in a section under applied loads and for the calculation of crack widths, the second moment of area of the section must account for the cracking which has occurred. One estimate is to assume that the concrete has fully cracked; that is, that cracks have penetrated the tension region as far as the neutral axis. Thus, no tensile stress exists in the concrete in this region and the concrete in this region is effectively absent as illustrated in Fig. 7.8. There will, of course, continue to be tension in the bottom reinforcement.

As in the case of an uncracked section, we use the concept of equivalent areas of concrete when calculating the second moment of area for a cracked

Fig. 7.8 Cracked section: (a) beam in bending; (b) section X–X; (c) stress and strain distributions; (d) equivalent concrete section

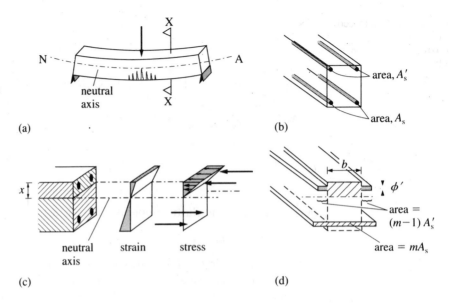

(a)

(b)

(c)

(d)

section. Thus, for the cracked section of Fig. 7.8, the equivalent section is illustrated in Fig. 7.8(d). Like the uncracked section, the total equivalent area of the top strip is given by:

$$b\phi' + (m - 1)A'_s$$

(i.e. the original area of the gross section at this level plus $(m - 1)A'_s$). In the bottom strip, however, there is no usable concrete to displace and so the total equivalent area of the strip is given by:

$$mA_s$$

Once the equivalent section is established, the calculation of I is identical in principle to the calculation of I_g or I_u. In this case, however, I is known as the 'cracked' second moment of area, I_c.

Choice of Second Moment of Area

To use I_g or I_u in the calculation of the stresses at a cracked section would give very poor results for reinforced concrete. For such calculations, only I_c gives reasonably accurate and conservative results, and it can be as little as half the value of I_g or I_u. In addition, I_c is generally used to calculate crack widths under service loads. However, for the calculation of deflection of reinforced concrete members, the use of I_c will give only approximate results as not all sections in the span will be cracked. Furthermore, parts of a beam in hog will generally have different cracked second moments of area than parts in sag. The variation in I for one span of a continuous member is illustrated in Fig. 7.9. For accurate deflection calculations, a weighted mean of I_u, I_c for the top steel and I_c for the bottom steel, known as the effective second moment of area, I_{eff}, can be used.

Fig. 7.9 Variation in I for continuous beam: (a) beam in building frame; (b) bending moment diagram; (c) deflected shape

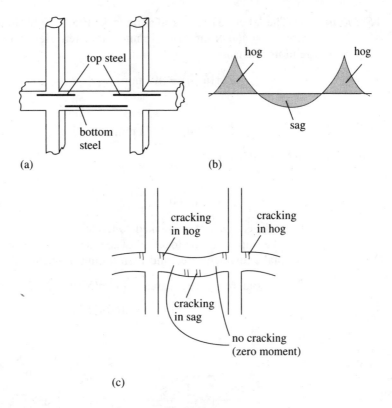

(a)

(b)

(c)

Example 7.4   Rectangular section

Problem   The rectangular beam illustrated in Fig. 7.10 is subjected to a variable uniform loading of 15 kN/m.

(a) Calculate the equivalent 'uncracked' and 'cracked' second moments of area of the section for the dimensions given. Assume $E_c = 30\,500$ N/mm² and $E_s = 200\,000$ N/mm².

(b) Determine the bending moment at the centre, B, due to the variable loading. Assuming cracked conditions and no long-term effects, calculate the stress in the concrete at the extreme top fibre at B and the stress in the tension reinforcement.

Fig. 7.10 Beam of Examples 7.4 and 7.5: (a) elevation; (b) section X–X

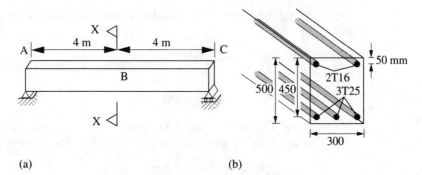

(a)

(b)

Solution (a) The label, '2T16', as illustrated in Fig. 7.10(b), indicates two high tensile bars ($f_y = 460$ N/mm^2) of 16 mm diameter. Hence the area of compression reinforcement is:

$$A'_s = 2(\pi 16^2/4) = 402 \text{ mm}^2$$

Similarly:

$$A_s = 3(\pi 25^2/4) = 1473 \text{ mm}^2$$

The modular ratio is:

$$m = \frac{E_s}{E_c} = \frac{200\,000}{30\,500} = 6.56$$

Hence the equivalent uncracked concrete section is as illustrated in Fig. 7.11(a). As in Example 7.2, the location of the neutral axis is found by first calculating the total equivalent area and then taking moments about P in Fig. 7.11(a):

$$\text{total (equivalent) area} = (300)(500) + 2235 + 8190$$
$$= 160\,425 \text{ mm}^2$$

Fig. 7.11 Equivalent concrete section of Example 7.4: (a) uncracked; (b) cracked

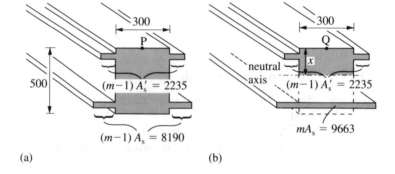

(a) (b)

Taking moments about P:

$$(160\,425)x = (300)(500)(250) + (2235)(50) + (8190)(450)$$
$$\Rightarrow \quad x = 257 \text{ mm}$$

Using the alternative method demonstrated in Example 7.2, the second moment of area is:

$$I_u = \frac{(300)(500)^3}{12} + (300)(500)(x - 250)^2 + (2235)(x - 50)^2$$
$$+ (8190)(450 - x)^2$$

Note: The second moment of area of the steel reinforcement about its own axis is small and is neglected here. Substituting for x above gives:

$$I_u = 3.533 \times 10^9 \text{ mm}^4$$

The equivalent cracked section is illustrated in Fig. 7.11(b). The equivalent total area is:

$$\text{total (equivalent) area} = (300)(x) + 2235 + 9663$$

$$= 300x + 11\,898 \text{ mm}^2$$

Taking moments about Q results in a quadratic equation in x:

$$(300x + 11\,898)x = 300(x)(x/2) + (2235)(50) + (9663)(450)$$

$$\Rightarrow \qquad 150x^2 + 11\,898x - 4\,460\,100 = 0$$

The two roots of this quadratic are $x = 137$ mm and $x = -217$ mm. Thus (discounting the negative root), the neutral axis is 137 mm below the top fibre. The cracked second moment of area, I_c, is:

$$I_c = \frac{300x^3}{12} + (300x)\left(\frac{x}{2}\right)^2 + (2235)(x - 50)^2 + (9663)(450 - x)^2$$

$$\Rightarrow \qquad I_c = 1.221 \times 10^9 \text{ mm}^4$$

(b) The moment at the centre of a simply supported beam subjected to a uniformly distributed loading, ω, is:

$$M_B = \frac{\omega L^2}{8} = \frac{15(8)^2}{8}$$

$$= 120 \text{ kNm}$$

In the long term, the creep of concrete has the effect of increasing strain and redistributing stresses from the concrete to the steel. This phenomenon is allowed for in design calculations by the use of a reduced value for the elastic modulus of concrete, E_c. As only short-term effects are being considered in this example, no such adjustments need to be made for creep. Hence the stress in the concrete at the extreme top fibre, P, is:

$$\sigma(y = x) = \frac{Mx}{I_c} = \frac{(120 \times 10^6)(137)}{(1.221 \times 10^9)}$$

$$= 13.5 \text{ N/mm}^2$$

When calculating stresses it must be remembered that the transformed section is an equivalent **concrete** section. Hence the stress in the concrete can be calculated directly but the stress in the steel cannot. At the level of the tension reinforcement

$$y = -(450 - x)$$

$$= -313 \text{ mm}$$

Hence the stress that would be in the **concrete** at this level (were it not cracked) is:

$$\sigma(y = -313) = \frac{M(-313)}{I_c} = \frac{(120 \times 10^6)(-313)}{(1.221 \times 10^9)}$$

$$= -30.8 \text{ N/mm}^2$$

The strain at this level is then $-30.8/E_c$. This strain is the same in both the steel and concrete. The steel stress is thus calculated as:

$$\sigma_s = E_s(-30.8)/E_c$$
$$= m(-30.8)$$
$$= -202 \text{ N/mm}^2$$

that is, a tensile stress of 202 N/mm².

It can be seen from Example 7.4 that the stress in reinforcement at a distance y above the neutral axis is:

$$f_s = \frac{mMy}{I_c} \qquad (7.5)$$

7.3 Elastic deflections and crack widths

Reinforced concrete members carrying lateral loads respond to these loads by bending. The moment–curvature relationship for a segment of the simply supported reinforced concrete member of Fig. 7.12(a) is illustrated in Fig. 7.12(c). It can be seen that the segment remains uncracked and has a large stiffness, EI_u, until the moment reaches the **cracking moment**, M_{cr} (point A). When this happens, the member cracks and the stiffness at the cracked section reduces to EI_c. As the load (and hence the moment) is increased further, more cracks occur and existing cracks increase in size. Eventually, the reinforcement yields at the point of maximum moment, corresponding to point C on the diagram. Above this point, the member displays large increases in deflection for small increases in moment. The service load range is between the origin and point C on the diagram and it is in this range that deflections are checked and stresses calculated.

Consider a point B within the service load range. This curvature represents the instantaneous (short-term) curvature under an applied moment, M. If the moment is sustained, however, the curvature increases with time to point D owing to the creep of the concrete. The curvature at this point is known as the long-term or sustained curvature. As deflection results from curvature, there are both instantaneous and sustained deflections which must be considered in the design of members with bending.

Limits on deflection

The deformations (or deflections) which result from bending must be limited such that they do not adversely affect the function and appearance of the member or the entire structure. To prevent the visual appearance of members from being impaired, EC2 limits their maximum deflections to span/250 under

Fig. 7.12 Moment/
curvature relationship
for beam segment:
(a) deflected shape;
(b) curvature of segment
of beam,
curvature = 1/R;
(c) moment/curvature
plot for segment of part
(b)

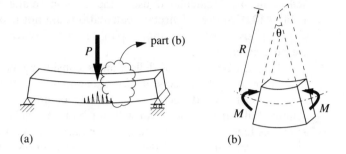

(a)

(b)

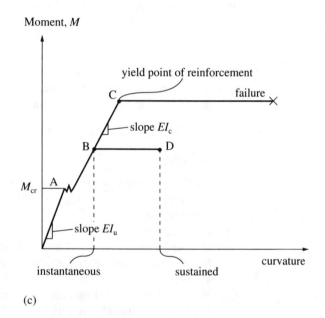

(c)

the quasi-permanent load combination of Table 3.12, that is the sustained loads. Excessive deflections may also affect the function of the structure. For instance, sagging roof slabs can result in the ponding of rainwater which can result in higher applied loads and possibly leakage.

In addition to causing detriment to the appearance and function of a member, deflections can cause damage to non-structural elements (partitions, windows, doors, etc.) which are supported by the structural members in bending. To prevent such damage, EC2 recommends an upper deflection limit of span/500 for normal circumstances.

Calculation of deflection

The deflection of reinforced concrete members can be calculated using the methods described in Chapter 4. The instantaneous deflection, δ_i, is calculated using the rare combination of applied loads given in Table 3.12. The sustained

241

deflection, δ_s, is calculated using the quasi-permanent load combination of Table 3.12. However, deflection calculations are not necessary on a 'deemed to satisfy' basis if the preliminary span/depth ratios recommended in Table 6.5 are not exceeded.

Where a calculation of deflection is considered necessary, the method of calculation must take account of a phenomenon known as 'tension stiffening' and, for sustained deflections, the effects of creep. In a beam subjected to bending, all parts of the beam will not be in a cracked state at any one time as cracks tend to occur at discrete intervals. The extra stiffness in a beam due to the uncracked portions (between cracks) is known as tension stiffening. In accounting for tension stiffening, rather than proposing a value for the effective second moment of area, I_{eff}, EC2 allows the use of the following approximate formula:

$$\delta = \zeta\delta_c + (1 - \zeta)\delta_u \tag{7.6}$$

where δ_u = deflection assuming an uncracked section (i.e. use I_u)
δ_c = deflection assuming a fully cracked section (i.e. use I_c)
ζ = distribution factor given by equation 7.7

For uncracked sections $\zeta = 0$. For the case where cracking has occurred:

$$\zeta = 1 - \beta_1\beta_2(f_{sr}/f_s)^2 \tag{7.7}$$

where β_1 = coefficient which accounts for the bond properties of the reinforcement: $\beta_1 = 1.0$ for high-bond bars (normally used); $\beta_1 = 0.5$ for plain bars
β_2 = coefficient which accounts for the duration of the loading or of repeated loading: $\beta_2 = 1.0$ for single short-term loading; $\beta_2 = 0.5$ for sustained loading or repeated loading
f_s = stress in tension steel assuming a cracked section
f_{sr} = stress in tension steel assuming a cracked section due to loading which causes initial cracking

The calculation of long-term deflection, δ_s, must take account of creep. The effect of creep is to increase the strain in the concrete with time, as illustrated in Fig. 7.13. Deflection calculations can allow for this increase in stress by

Fig. 7.13 Increase of strain with time: (a) section; (b) strain distribution; (c) strain/time relationship

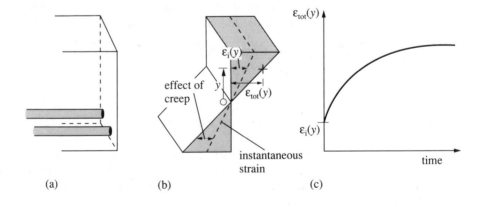

reducing the value of the elastic modulus for concrete, E_c, to an effective elastic modulus, $E_{c,eff}$. The effective modulus recommended by EC2 is given by:

$$E_{c,eff} = \frac{E_c}{1 + \varphi} \tag{7.8}$$

where φ is a creep coefficient taken from Table 7.1. It can be seen from this table that the creep coefficient is largest in young, heavily loaded concrete where the atmospheric conditions are dry. This is because creep is related to the degree of moisture present in concrete.

Table 7.1 Creep coefficient, φ, for normal weight concrete. A_c is the cross-sectional area of the concrete and u is the length of the perimeter of that area (from EC2)

Age of concrete at loading (days)	Notional size, $2A_c/u$ (mm)					
	50	150	600	50	150	600
	Dry atmospheric conditions, i.e. inside (RH = 50%)			Humid atmospheric conditions, i.e. outside (RH = 80%)		
1	5.5	4.6	3.7	3.6	3.2	2.9
7	3.9	3.1	2.6	2.6	2.3	2.0
28	3.0	2.5	2.0	1.9	1.7	1.5
90	2.4	2.0	1.6	1.5	1.4	1.2
365	1.8	1.5	1.2	1.1	1.0	1.0

Example 7.5 Deflection calculation

Problem The beam of Fig. 7.10 is subjected to a total permanent gravity loading (including self-weight) of 15 kN/m. Of the total variable gravity loading, 8 kN/m has been judged to be quasi-permanent. Find the long-term deflection at mid-span due to these loads given that the beam is located indoors and is first loaded 7 days after casting. The elastic modulus of the concrete is 30 500 N/mm^2 and the mean tensile strength is 2.6 N/mm^2. The elastic deflection at the centre of a beam of length L, due to uniform loading ω, is:

$$\delta = \frac{5}{384} \frac{\omega L^4}{EI}$$

where E is the elastic modulus and I is the second moment of area.

Solution The area of concrete in this cross-section is (approximately):

$$A_c \approx (300)(500) = 150\,000 \text{ mm}^2$$

Hence the notional size, as defined in Table 7.1, is:

$$\text{notional size} = 2\frac{A_c}{u} = \frac{2(150\,000)}{2(300 + 500)}$$

$$= 188 \text{ mm}$$

As the beam is first loaded at an age of 7 days, the value of φ is found by interpolating between the values of 2.6 and 3.1 in Table 7.1:

$$\varphi = 2.6 + \frac{(3.1 - 2.6)(600 - 188)}{(600 - 150)}$$

$$= 3.06$$

Hence the effective modulus for concrete is, from equation (7.8):

$$E_{c,\,eff} = \frac{E_c}{1 + \varphi} = \frac{30\,500}{4.06}$$

$$= 7512 \text{ N/mm}^2$$

The uncracked second moment of area is generally similar in magnitude to the gross value and it is common practice to use I_g in lieu of I_u. However, in this example, the effective concrete modulus is so small that the presence of reinforcement will have a significant stiffening effect. The equivalent transformed section for uncracked conditions is calculated using $m = E_s/E_{c,\,eff} = 26.62$ and is illustrated in Fig. 7.14(a).

Fig. 7.14 Equivalent transformed sections: (a) uncracked; (b) cracked

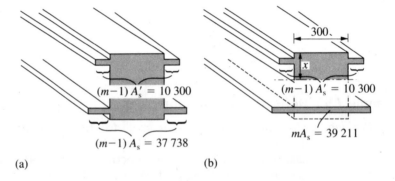

(a)

$(m-1)\,A_s' = 10\,300$

$(m-1)\,A_s = 37\,738$

(b)

300

x

$(m-1)\,A_s' = 10\,300$

$mA_s = 39\,211$

Taking moments about the top fibre gives $x = 278$ mm. Hence:

$$I_u = \frac{(300)(500)^3}{12} + (300)(500)(x - 250)^2 + (10\,300)(x - 50)^2$$

$$+ (37\,738)(450 - x)^2$$

$$= 4.894 \times 10^9 \text{ mm}^4$$

Thus the deflection, assuming an uncracked section, is:

$$\delta_u = \frac{5}{384} \frac{\omega L^4}{E_{c,\,eff} I_u} = \frac{5}{384} \left(\frac{(15 + 8)(8000)^4}{(7512)(4.894 \times 10^9)} \right)$$

$$= 33 \text{ mm}$$

The equivalent transformed section for cracked conditions is illustrated in Fig. 7.14(b). Taking moments about the top fibre gives a quadratic in x:

$$150x^2 + 49\,511x - 18\,160\,000 = 0$$

The two roots of this quadratic equation are $x = 220$ mm and $x = -550$ mm of which the only feasible root is $x = 220$ mm. Using this value for x, the cracked second moment of area is calculated as:

$$I_c = \frac{300x^3}{12} + (300x)\left(\frac{x}{2}\right)^2 + (10\,300)(x - 50)^2 + (39\,211)(450 - x)^2$$

$$I_c = 3.437 \times 10^9 \text{ mm}^4$$

Hence the deflection, assuming a cracked section, is:

$$\delta_c = \frac{5}{384}\frac{\omega L^4}{E_{c,\,eff}I_c} = \frac{5}{384}\left(\frac{(15 + 8)(8000)^4}{(7512)(3.437 \times 10^9)}\right)$$

$$= 48 \text{ mm}$$

The maximum moment in the beam is:

$$M = \frac{\omega L^2}{8} = \frac{(15 + 8)(8)^2}{8} = 184 \text{ kNm}$$

The stress in the tension steel due to this moment is given by equation (7.5):

$$f_s = mMy/I_c$$

where $y = -(450 - x) = -230$ mm. Hence:

$$f_s = \frac{(26.62)(184 \times 10^6)(-230)}{3.437 \times 10^9} = -328 \text{ N/mm}^2$$

The section first cracks when the maximum tensile stress in the concrete equals the tensile strength, that is:

$$M_{cr}y/I_u = -2.6 \text{ N/mm}^2$$

$$\Rightarrow \quad M_{cr} = \frac{(-2.6)I_u}{y} = \frac{-2.6(4.894 \times 10^9)}{-(500 - 278)} = 57 \times 10^6 \text{ Nmm}$$

The stress in the tension reinforcement due to this moment is:

$$f_{sr} = \frac{mM_{cr}y}{I_c} = \frac{(26.62)(57 \times 10^6)(-230)}{3.437 \times 10^9} = -102 \text{ N/mm}^2$$

Hence, assuming high-bond reinforcement, the distribution factor given by equation (7.7) is:

$$\zeta = 1 - \beta_1\beta_2(f_{sr}/f_s)^2$$

$$= 1 - (1)(0.5)(-102/-328)^2$$

$$= 0.952$$

The long-term deflection, from equation (7.6), is:

$$\delta = \zeta \delta_c + (1 - \zeta)\delta_u$$
$$= (0.955)(48) + (1 - 0.955)(33)$$
$$= 47 \text{ mm}$$

Limits on cracking

The formation of cracks in reinforced concrete members with bending is inevitable. This is unfortunate because flexural cracking leads to corrosion of the reinforcement, with the rate of corrosion directly related to the width of the cracks. For this reason, plus the fact that cracks are unsightly and can cause public concern, the size to which cracks develop is strictly controlled in many structures. Since the environment in the interior of buildings is usually non-severe, corrosion does not generally pose a problem and limits on crack widths will be governed by their appearance. On the other hand, for reinforced concrete structures in aggressive environments, corrosion is a problem and stringent limits are imposed on the width of cracks that are allowed to develop.

The limits recommended by EC2 take account only of durability (corrosion) requirements for different environmental conditions. The five exposure classes relating to environmental conditions defined in EC2 are given in Table 6.2. For reinforced concrete members in exposure class 1, crack widths, although relatively large, do not lead to an excessive degree of corrosion and no limit is specified. For exposure classes 2 to 4, the recommended limit on crack widths is 0.3 mm under the quasi-permanent combination of service load. No specific limit is proposed for exposure class 5.

Calculation of crack widths

Crack widths are calculated using the quasi-permanent service load combination of Table 3.12. In general, however, crack width calculations, like deflection calculations, are not necessary on a 'deemed to satisfy' basis if certain detailing rules are satisfied. Specifically, crack widths can be assumed not to exceed the limiting values if the limits on bar spacing or bar diameters, specified by EC2 and given in Table 7.2, are satisfied, and if minimum areas of reinforcement, also specified, are provided.

In specific cases where a crack width calculation is considered necessary, the following formula is specified in EC2:

$$w_k = \beta s_{rm} \varepsilon_{sm} \tag{7.9}$$

where w_k = design crack width

        s_{rm} = average final crack spacing (equation (7.11))

        ε_{sm} = mean strain in the tension steel allowing for tension stiffening and time-dependent effects (equation (7.10))

        β = coefficient relating the average crack width to the design value; $\beta = 1.7$ for sections in bending under applied loads

Table 7.2 Maximum size and spacing of high-bond bars for control of cracking (from EC2)

Steel stress* (N/mm^2)	Maximum bar spacing (mm)	Maximum bar diameter (mm)
160	300	32
200	250	25
240	200	20
280	150	16
320	100	12
360	50	10
400	–	8
450	–	6

* Steel stresses are determined using quasi-permanent loads.

The mean strain is simply the strain in the steel adjusted by the distribution factor, ζ, of equation (7.7):

$$\varepsilon_{sm} = \zeta \frac{f_s}{E_s} \tag{7.10}$$

where f_s is the stress in the tension reinforcement and E_s is the elastic modulus for steel. The average final crack spacing in millimetres is calculated using the equation:

$$s_{rm} = 50 + 0.25 k_1 k_2 \frac{\phi}{\rho_r} \tag{7.11}$$

where k_1 = coefficient which accounts for the bond properties of the reinforcement: $k_1 = 0.8$ for high-bond bars; $k_1 = 1.6$ for plain bars

k_2 = coefficient which takes account of the form of strain distribution: for bending, $k_2 = 0.5$

ϕ = bar diameter

ρ_r = effective reinforcement ratio $A_s/A_{c,eff}$, where $A_{c,eff}$ is the effective tension area of the concrete, as illustrated in Fig. 7.15

Fig. 7.15 Effective tension area of concrete: (a) beams; (b) slabs

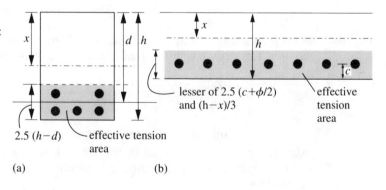

(a)   (b)

The application of these formulae is illustrated in Example 7.6. Except for term β_2 which influences the factor ζ, these equations do not appear to take account of the increase in crack size that will occur with time owing to creep. While this is not stated in EC2, it would seem prudent for sustained loading to use a reduced value for the elastic modulus of concrete such as that given in equation (7.8).

Example 7.6   Crack width calculation

Problem   Determine the crack width due to a characteristic variable gravity loading of 3 kN/m² for the strip of slab illustrated in Fig. 7.16. The elastic modulus for the concrete is 32 000 N/mm² and it has a tensile strength of 2.9 N/mm². The elastic modulus for the reinforcement is 200 000 N/mm².

Fig. 7.16 Strip of slab of Example 7.6: (a) elevation; (b) section X–X; (c) transformed section

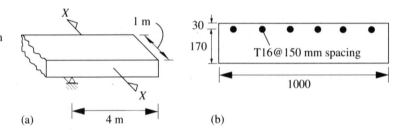

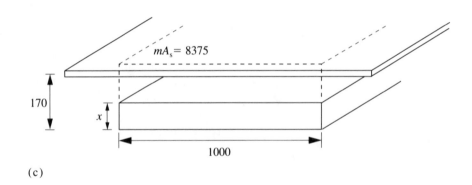

Solution   As the bars are spaced at 150 mm centres, there will, on average, be 1000/150 bars in a 1 m strip, giving the area of tension reinforcement, A_s, as:

$$A_s = \left(\frac{1000}{150}\right)\left(\frac{\pi 16^2}{4}\right) = 1340 \text{ mm}^2$$

The modular ratio, m, is 200 000/32 000 = 6.25. As the applied moment is hogging, the top of the slab is in tension as can be seen in the transformed section of Fig. 7.16(c). The neutral axis location and second moment of area can be shown to be:

$$x = 46 \text{ mm}$$

$$I_c = 161.2 \times 10^6 \text{ mm}^4$$

The maximum moment in a cantilever due to a uniform load is:

$$M = \frac{\omega L^2}{2} = \frac{(3)(4)^2}{2} = 24 \text{ kNm/m}$$

The stress in the tension reinforcement due to this moment is:

$$f_s = \frac{mM[-(170-x)]}{I_c} = \frac{(6.25)(24 \times 10^6)[-(170-46)]}{161.2 \times 10^6}$$

$$= -115 \text{ N/mm}$$

Given that the tensile strength of the concrete is 2.9 N/mm², the moment when the first cracks appear is found from:

$$-2.9 = \frac{M_{cr}[-(200-x)]}{I_u}$$

As the modular ratio is not excessively large, the neutral axis is assumed to be central (as in the gross section) and the gross second moment of area is used in lieu of the uncracked value, that is:

$$I_u \approx I_g = \frac{(1000)(200)^3}{12} = 667 \times 10^6 \text{ mm}^4$$

Note: This approximation can make a significant difference and should not be used where crack widths are close to the allowable limit.

Hence:

$$M_{cr} = \frac{-2.9(667 \times 10^6)}{-(200-100)} = 19.3 \times 10^6 \text{ Nmm}$$

and:

$$f_{sr} = \frac{mM_{cr}[-(170-x)]}{I_c} = \frac{(6.25)(19.3 \times 10^6)[-(170-46)]}{161.2 \times 10^6}$$

$$= -93 \text{ N/mm}^2$$

Thus, using equation (7.7):

$$\zeta = 1 - \beta_1\beta_2(f_{sr}/f_s)^2 = 1 - (1)(1)(-93/-115)^2$$

$$= 0.346$$

Hence using equation (7.10):

$$\varepsilon_{sm} = \frac{\zeta f_s}{E_s} = \frac{0.346 \times 115}{200\,000}$$

$$= 199 \times 10^{-6}$$

The effective tension area of the concrete $A_{c,\,eff}$ is illustrated in Fig. 7.15(b). For this example, the distance from the reinforcement to the extreme fibre in tension is $c = 30$ mm and the bar diameter is $\phi = 16$ mm. Hence the depth of the effective tension area is:

$$\text{depth} = \text{lesser of } 2.5(c + \phi/2) \text{ and } (h - x)/3$$

$$= \text{lesser of } 2.5(30 + 16/2) \text{ and } (200 - 46)/3$$

$$= 51.3 \text{ mm}$$

which gives an area of:

$$A_{c, \text{eff}} = 1000 \times 51.3 = 51\,300 \text{ mm}^2$$

The effective reinforcement ratio is thus:

$$\rho_r = 1340/51\,300$$

$$= 0.0261$$

or 2.61 per cent. Finally the average crack spacing is calculated using equation (7.11):

$$s_{\text{rm}} = 50 + 0.25k_1 k_2 \frac{\phi}{\rho_r} = 50 + 0.25(0.8)(0.5)\left(\frac{16}{0.0261}\right)$$

$$= 111 \text{ mm}$$

from which the design crack width is:

$$w_k = \beta s_{\text{rm}} \varepsilon_{\text{sm}} = 1.7(111)(199 \times 10^{-6})$$

$$= 0.038 \text{ mm}$$

7.4 Stress/strain relationships and modes of failure

The short-term stress/strain relationship for a concrete test cylinder in compression is illustrated in Fig. 7.17. The stress/strain relationship is approximately

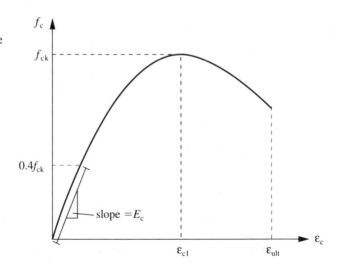

Fig. 7.17 Stress/strain relationship for concrete

linear (i.e. $f_c \approx E_c \varepsilon_c$) for low values of strain. Thus, the slope of the initial part of the diagram is reasonably constant. As the stress is increased, the stress/strain relationship becomes substantially non-linear until the point is reached where very little extra stress causes a large increase in strain. The peak in stress is known as the compressive strength. Beyond this point of maximum strength, the stress must be decreased if failure is to be prevented. The characteristic compressive strength is denoted f_{ck} (subscript k implies 'characteristic' through-out EC2) and the corresponding compressive strain is denoted ε_{c1}. Finally, at a certain ultimate strain, ε_{ult}, the concrete crushes regardless of the level of stress. Much research has been carried out to predict ε_{c1} and ε_{ult} for different grades of concrete. The design values recommended by EC2 for all concrete grades are as follows:

$$\varepsilon_{c1} = 0.002$$

$$\varepsilon_{ult} = 0.0035$$

As the stress/strain relationship for concrete is non-linear, an average slope, known as a **secant modulus**, is used in lieu of the regular concrete modulus. This is defined as the slope of the line joining the origin to the point corresponding to 40 per cent of the characteristic strength (as illustrated in Fig. 7.17). Where great accuracy is not required, EC2 allows the estimation of the secant modulus, E_c, with the formula:

$$E_c = 9500(f_{ck} + 8)^{0.33} \tag{7.12}$$

where E_c is in newtons per square millimetre and f_{ck} is the characteristic cylinder strength, also in newtons per square millimetre.

A typical stress/strain diagram for reinforcement in compression or tension is illustrated in Fig. 7.18. The relationship is reasonably linear for a considerable portion of the diagram. The elastic modulus, E_s, is equal to the slope of the linear portion of the diagram. EC2 allows a mean value of $200\,000$ N/mm^2 to be assumed for E_s. The characteristic stress, f_{yk} or f_y, at which yielding is said to occur is defined as the stress at which strain has exceeded the value predicted by the linear relationship by 0.002 (see Fig. 7.18). This nominal yield stress is

Fig. 7.18 Stress/strain relationship for reinforcement

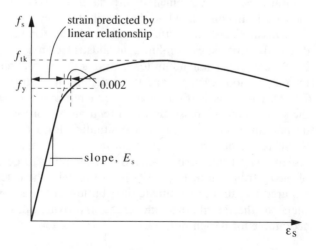

Fig. 7.19 Stress
distributions for
increasing moment:
(a) section;
(b) uncracked;
(c) cracked;
(d) non-linear;
(e) ultimate

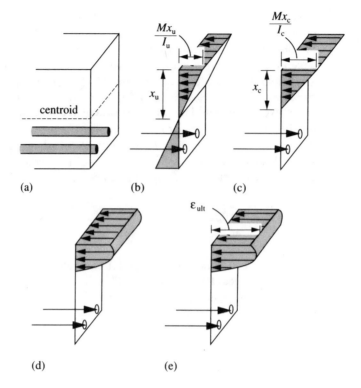

of equal magnitude in tension and compression. Beyond this point, increases in stress result in successively larger increases in strain. The ultimate tensile and compressive strengths are also equal in magnitude and are denoted by f_{tk}.

Under service loads (SLS), strains in reinforced concrete members in bending are small and both the steel and concrete stress/strain relationships remain linear. Initially, segments of a member such as illustrated in Fig. 7.12(b) are uncracked and the stress distribution is as illustrated in Fig. 7.19(b) with a maximum compressive stress of Mx_u/I_u, where x_u is the depth to the uncracked neutral axis. As the moment is increased, the section cracks and the stresses increase in magnitude while the distribution remains linear (see Fig. 7.19(c)). As loads are increased further, strains in the section become large and the stress distribution becomes non-linear as illustrated in Fig. 7.19(d). At the ultimate limit state, under maximum load, strains in the concrete at the extreme fibres in compression reach ε_{ult} and the distribution of stress is as illustrated in Fig. 7.19(e). The stress distribution in the concrete in compression has approximately the same shape as the stress/strain relationship of Fig. 7.17. However, the two do not correspond directly as the situation for a fibre of concrete in bending is different from that for concrete in a standard test cylinder. Considerable research effort has been expended in attempting to determine the exact shape of the distribution in Fig. 7.19(e). Fortunately, the calculation of the moment required to cause this ultimate distribution of stress is insensitive to the exact shape of the distribution and crude approximations of the shape are quite acceptable for design purposes.

The cracking moment, M_{cr}, was introduced in the previous section as the moment at which the concrete member first cracks. After members have cracked, the ultimate capacity to resist moment, M_{ult}, is greatly affected by the area of tension reinforcement, A_s. If this area is particularly small, the moment capacity after first cracking will be less than M_{cr} which will result in sudden failure of the member when the applied moment reaches M_{cr} (see Fig. 7.20(a)). To prevent such sudden failure, sufficient reinforcement must be provided so that:

Fig. 7.20 Moment/
curvature plots for
increasing areas of
reinforcement;
(a) $M_{ult} < M_{cr}$ (very
small A_s); (b) small A_s;
(c) under-reinforced
(typical A_s);
(d) balanced design
(large A_s);
(e) over-reinforced
(very large A_s)

$$M_{ult} > M_{cr} \qquad (7.13)$$

To calculate a conservatively high value for M_{cr}, for use in this inequality, the tensile strength with a 95 per cent probability of not being exceeded can be used:

$$f_{ct,0.95} = 0.39 f_{ck}^{2/3} \qquad (7.14)$$

The corresponding cracking moment is defined by:

$$-f_{ct,0.95} = \frac{M_{cr}[-(h-x)]}{I_u} \qquad (7.15)$$

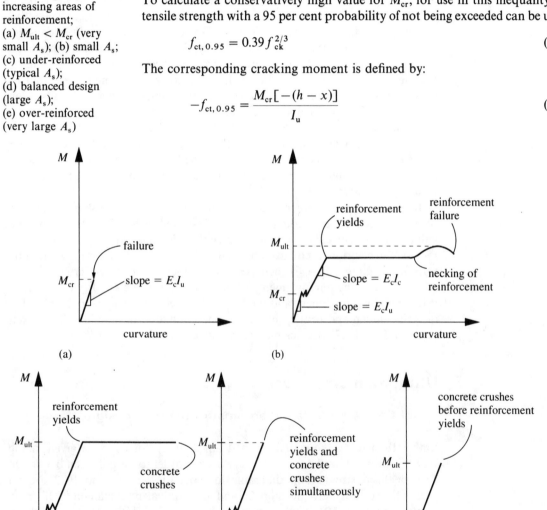

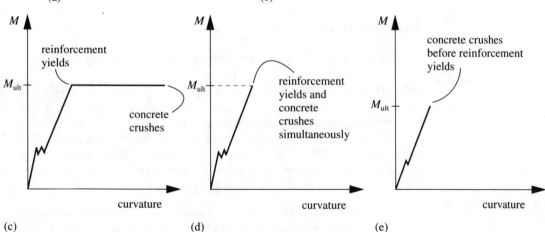

where h is the section depth and x is the distance from the neutral axis to the extreme fibre in compression. Substituting for $f_{ct,0.95}$ in equation (7.15) and rearranging gives:

$$M_{cr} = \frac{0.39 f_{ck}^{2/3} I_u}{(h - x)} \tag{7.16}$$

To ensure that M_{ult} exceeds M_{cr}, EC2 specifies a minimum value for A_s which results in an ultimate moment in excess of M_{cr}.

When the area of reinforcement, A_s, is increased so that $M_{ult} > M_{cr}$, the beam does not fail when it first cracks, but continues to deform elastically as illustrated in Fig. 7.20(b). However, as A_s is still small, the reinforcement still yields before the maximum compressive strain in the concrete reaches ε_{ult}. The plastic deformation which occurs after yielding of the reinforcement results in considerable deformation of the beam before, finally, the reinforcement fails completely. The more normal mode of failure, in practice, occurs when a larger area of tension reinforcement is present. In this case, the reinforcement yields but the maximum compressive strain in the concrete reaches ε_{ult} before it fails. Thus, failure of the beam is due to crushing of the concrete in compression, as illustrated in Fig. 7.20(c). Because the amount of reinforcement is small, the beam is termed **under-reinforced** in both these cases (Figs 7.20(b) and (c)). They are referred to as **ductile failures** because of the large plastic deformations which occur prior to collapse. Since there is ample warning of imminent failure this is highly desirable in design.

If A_s is increased further to a level where the concrete crushes at the same time as the reinforcement first yields, as illustrated in Fig. 7.20(d), the mode of failure is known as **balanced failure**. Since there is no plastic deformation of the reinforcement prior to collapse the beam acts as a brittle material giving little warning of failure. If A_s is increased still further, the concrete crushes before any yielding of the reinforcement, as illustrated in Fig. 7.20(e). In this case there is too much reinforcement and the beam is termed **over-reinforced**. Since failure is again sudden and brittle, this is an unacceptable form of failure. To prevent brittle failure, EC2 specifies maximum values for A_s.

7.5 Ultimate moment capacity

Simplified stress/strain diagrams for member design

Rather than using the actual stress/strain diagram for concrete given in Fig. 7.19(e), other idealized diagrams can be used in order to simplify member design. The idealized stress/strain diagrams for concrete specified in EC2 are the parabolic–rectangular, the bilinear and the equivalent rectangular diagrams illustrated in Fig. 7.21. The diagrams to be applied in design (shown solid) are derived from the idealized diagrams (shown dotted) by means of a reduction of all stresses in the idealized diagrams by the factor α/γ_c, where:

α = coefficient which takes account of the long-term effects on the compressive strength and of the unfavourable effects resulting from

Fig. 7.21 Simplified stress/strain diagrams for concrete (idealized shown dotted, design stress shown solid): (a) parabolic–rectangular; (b) bi-linear; (c) rectangular

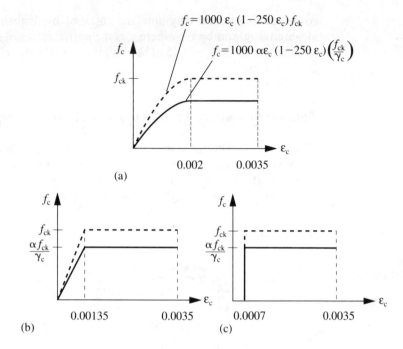

(a)

(b)   (c)

the way in which the load is applied: $\alpha = 0.85$ for the parabolic–rectangular and the bilinear diagrams; for the equivalent rectangular diagram, $\alpha = 0.85$ if the section is at least as wide at the extreme compression fibre as it is elsewhere in the compressive zone; otherwise, $\alpha = 0.8$

γ_c = partial factor of safety for concrete strength equal to 1.5

For reinforcement, the simplified stress/strain diagram adopted in EC2 is the bilinear diagram illustrated in Fig. 7.22(a) in which f_y (f_{yk} is used in EC2) is the characteristic yield strength and f_{tk} is the characteristic ultimate strength. When using this diagram, the strain in the reinforcement should be limited to 0.01. For further simplicity, the diagram can be altered so that the top branch is horizontal, as illustrated in Fig. 7.22(b). In this case, no limit on the steel strain is necessary. In both cases, the idealized diagrams are shown dotted. The

Fig. 7.22 Idealized stress/strain relationships for reinforcement (from EC2): (a) inclined top branch; (b) horizontal top branch

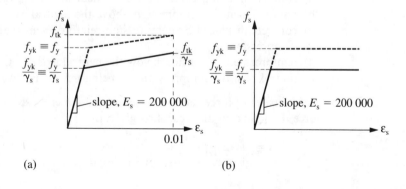

(a)   (b)

corresponding design diagrams are obtained by factoring the stresses of the idealized diagrams by $1/\gamma_s$ where γ_s is the partial factor of safety for the strength of steel. At ULS, $\gamma_s = 1.15$. The design value for the elastic modulus, E_s, is 200 000 N/mm².

General procedure for calculating ultimate moment capacity

Given the assumptions stated above, the ultimate moment capacity, M_{ult}, can be calculated for a section of specified dimensions and areas of reinforcement such as that illustrated in Fig. 7.23. In accordance with EC2, any combination of the design stress/strain diagrams for steel and concrete can be used.

Fig. 7.23 Distribution of stress and strain in typical section: (a) section; (b) linear strain distribution; (c) parabolic–rectangular stress distribution

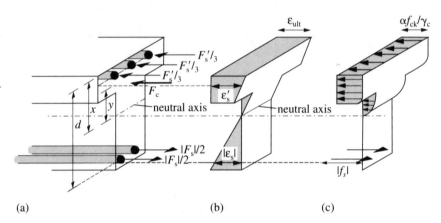

(a)   (b)   (c)

For all modes of failure (with the exception of the very under-reinforced section, a case which can be ignored), ultimate moment failure occurs when the strain in the extreme concrete fibre in compression reaches $\varepsilon_{ult} = 0.0035$, at which stage the concrete stress distribution is assumed to equal one of the distributions illustrated in Fig. 7.21 (see Fig. 7.23(c)). As in the calculation of second moments of area (Example 7.3), equilibrium of forces can then be used to determine the ultimate moment capacity, M_{ult}, of a section in bending. The different forces which act on a reinforced concrete section **in sag**, illustrated in Fig. 7.24, are as follows:

(a) concrete in compression above the neutral axis, F_c (gross section)
(b) reinforcement in compression above the neutral axis, F'_s
(c) concrete displaced by compression reinforcement that would have been in compression, F_{disp}
(d) reinforcement in tension below the neutral axis, F_s. Using the sign convention of positive compression, F_s, being tensile, is always negative.

A similar set of forces acts on a reinforced concrete section in hog. If the total force, F, acting on an area, A, is given by:

$$F = \int_A \sigma \, dA \qquad (7.17)$$

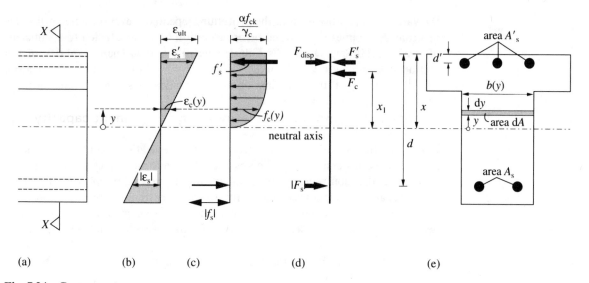

(a) (b) (c) (d) (e)

Fig. 7.24 Components of axial force in typical section: (a) elevation; (b) strain; (c) parabolic–rectangular stress distribution; (d) axial forces on section; (e) section X–X

then the compressive force in the concrete, F_c, is given by:

$$F_c = \int_0^x f_c(y)\, \mathrm{d}A = \int_0^x f_c(y) b(y)\, \mathrm{d}y \tag{7.18}$$

where $f_c(y)$ is the concrete stress at height y and $b(y)$ is the breadth of the section at that height, as illustrated in Fig. 7.24(e). By allowing later for the force that would have been in the concrete displaced by compression reinforcement, this integration can be over the gross section. The formula for $f_c(y)$ depends on which idealized stress/strain diagram for concrete is used.

The compression force in the top steel, F'_s, is equal to $A'_s f'_s$ where f'_s is the stress there. Similarly, the force in the bottom steel, F_s, is equal to the product of area and stress, $A_s f_s$ (both f_s and F_s, being tensile, will be negative). The concrete displaced by the compression reinforcement has an area A'_s and would, if present, have had a stress equal to the stress in the concrete at that level. This depends on the point at which the compression reinforcement intersects the stress distribution, but the situation illustrated in Fig. 7.24 is typical, where the stress in the displaced concrete is $\alpha f_{ck}/\gamma_c$. Hence the force is typically:

$$F_{disp} = A'_s\left(\frac{\alpha f_{ck}}{\gamma_c}\right) \tag{7.19}$$

By equilibrium, the sum of the compressive forces must equal zero, that is:

$$F_c - F_{disp} + F'_s + F_s = 0$$

$$\Rightarrow \quad \int_0^x f_c(y) b(y)\, \mathrm{d}y - F_{disp} + A'_s f'_s + A_s f_s = 0 \tag{7.20}$$

The only unknown in equation (7.20) is x. In practice, a trial and error approach is sometimes used to find the correct value. An initial value for x is assumed and the forces on the section are determined. If equation (7.20) is not satisfied,

257

the value for x is adjusted and the procedure repeated. Several refined estimates of x may sometimes be required before equation (7.20) is satisfied to a sufficient degree of accuracy. Once x is known, the ultimate moment capacity of the section is found by taking moments about any horizontal axis in the section. For instance, by taking moments about the neutral axis, the ultimate moment capacity, M_{ult}, is found to be (refer to Fig. 7.24(d)):

$$M_{ult} = F_c x_1 + F'_s(x - d') - F_{disp}(x - d') + |F_s|(d - x) \qquad (7.21)$$

Note that the tension reinforcement component is **additive** to the components due to the compressive force. This is because, while the forces act in opposite directions, the components of moment in all cases (except the displaced concrete) act anti-clockwise. The depth from the extreme fibre in compression to the tension reinforcement is known as the **effective depth** and is traditionally denoted as d (see Fig. 7.24(e)). Substituting for the forces in equation (7.21) gives:

$$M_{ult} = (x_1) \int_0^x f_c(y)b(y)\,dy + A'_s f'_s(x - d') - F_{disp}(x - d')$$
$$+ A_s|F_s|(d - x) \qquad (7.22)$$

Ultimate moment capacity of rectangular sections

For a rectangular section with specified total depth h and having constant breadth b, as illustrated in Fig. 7.25(a), equation (7.18) becomes:

$$F_c = b \int_0^x f_c(y)\,dy \qquad (7.23)$$

The term $\int_0^x f_c(y)\,dy$ is equal to the total area under the compressive stress block for the concrete. Clearly, the value of the total area is dependent on which of the distributions of Fig. 7.21 is adopted. Fortunately, while the distributions are quite different in form, the areas under each are in fact quite similar in magnitude. The simplest of these is the equivalent rectangular stress block of Fig. 7.21(c). In this case, the stress is constant for all strains in excess of 0.0007 and is zero elsewhere. When a section is on the point of collapse, the strain at the extreme fibre is ε_{ult}, as illustrated in Fig. 7.25(b). Thus, the strain increases

Fig. 7.25 Stress and strain distributions at ULS: (a) elevation and section; (b) strain; (c) equivalent rectangular stress block

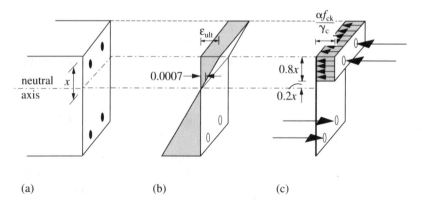

(a)   (b)   (c)

linearly from zero at the neutral axis to a maximum of ε_{ult} at the extreme fibre in compression. The strain is 0.0007 at a distance from the neutral axis of:

$$\frac{0.0007x}{\varepsilon_{ult}} = \frac{0.0007x}{0.0035} = 0.2x$$

Hence the stress distribution is as illustrated in Fig. 7.25(c). The total compressive force in the concrete is simply the product of stress and area. Taking the gross section:

$$F_c = \left(\frac{\alpha f_{ck}}{\gamma_c}\right)(0.8xb) \tag{7.24}$$

Taking the recommended values $\alpha = 0.85$ and $\gamma_c = 1.5$, equation (7.24) becomes:

$$F_c = 0.453xbf_{ck} \tag{7.25}$$

The force, F_c, acts at the centroid of the compressive stress block, that is at a distance of $0.4x$ from the top of the section.

The depth to the neutral axis, x, is calculated using equation (7.20). Substitution of equation (7.25) into equation (7.20) gives:

$$0.453xbf_{ck} - F_{disp} + A'_s f'_s + A_s f_s = 0 \tag{7.26}$$

In using equation (7.26) to calculate x, assume initially, say, that both the tension and compression reinforcements have yielded, that is $-f_s = f'_s = f_y/\gamma_s$. Once x is found, these assumptions can be checked from the strain diagram of the section. If they are incorrect, that is either or both of the reinforcements has not yielded, the procedure is repeated using the revised assumption. Having found x, the ultimate moment capacity is calculated using equation (7.22). The method is illustrated in the following examples.

Example 7.7 **Ultimate moment capacity of rectangular beam**

Problem For the beam of Fig. 7.26, calculate the value for the applied load, P, which will cause collapse (ignoring the self-weight of the beam). Use the rectangular stress distribution for concrete and the horizontal top branch in the stress distribution for the reinforcement. Assume the concrete has $f_{ck} = 35 \text{ N/mm}^2$ and the reinforcement has $f_y = 460 \text{ N/mm}^2$.

Solution The moment capacity of the beam is found from a consideration of the stress distribution in the section. Assuming that the reinforcement has yielded, the stress distribution is as illustrated in Fig. 7.27(c). Using equation (7.24), the total compressive force in the concrete is:

$$F_c = \left(\frac{\alpha f_{ck}}{\gamma_c}\right)(0.8xb) = \frac{(0.85)(35)(0.8x)(250)}{1.5}$$

$$= 3967x \text{ N}$$

Assuming that the reinforcement has yielded, the total force in it is:

$$F_s = -A_s\left(\frac{f_y}{\gamma_s}\right) = -3\left(\frac{\pi 32^2}{4}\right)\left(\frac{460}{1.15}\right)$$

$$= -965\,100 \text{ N}$$

Fig. 7.26 Beam of
Examples 7.7 and 7.10:
(a) elevation; (b) section
A–A

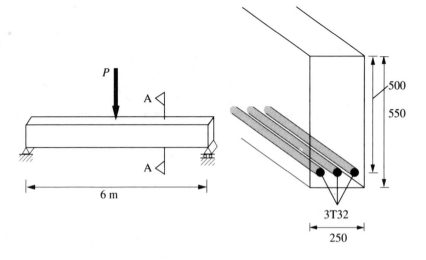

(a)   (b)

Fig. 7.27 Assumed
distributions of stress
and strain: (a) elevation
and section; (b) strain;
(c) stress

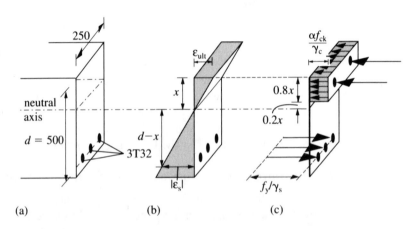

(a)   (b)   (c)

As there is no net axial force in the member, $F_c + F_s = 0$. Hence:

$$3967x - 965\,100 = 0$$

$\Rightarrow \qquad x = 243$ mm

The assumption that the reinforcement has yielded is now checked. The strain, ε_s, in the reinforcement is calculated using similar triangles (see Fig. 7.27(b)):

$$\frac{|\varepsilon_s|}{(d-x)} = \frac{\varepsilon_{ult}}{x}$$

$\Rightarrow \qquad |\varepsilon_s| = \dfrac{(500-243)(0.0035)}{243} = 0.00370$

The yield strain (from Fig. 7.22) is:

$$\frac{f_y/\gamma_s}{E_s} = \frac{460/1.15}{200\,000} = 0.002$$

Therefore the assumption that the steel has yielded is a correct one. The moment capacity is found by taking moments about the neutral axis:

$$M_{\text{ult}} = F_{\text{c}}(0.6x) + |F_{\text{s}}|(d - x)$$

but:

$$|F_{\text{s}}| = F_{\text{c}}$$

$$\Rightarrow \quad M_{\text{ult}} = |F_{\text{s}}|(0.6x + d - x)$$

$$= |F_{\text{s}}|(d - 0.4x)$$

$$= (965\,100)[500 - 0.4(243)] \text{ Nmm}$$

$$= 389 \text{ kNm}$$

The bending moment diagram for this beam is illustrated in Appendix B, No. 2. The maximum applied moment can be seen to be $Pl/4$. Thus, the beam will collapse when:

$$\frac{P_{\text{ult}}l}{4} = M_{\text{ult}}$$

$$\Rightarrow \quad P_{\text{ult}} = \frac{4M_{\text{ult}}}{l} = \frac{4(389)}{6}$$

$$= 259 \text{ kN}$$

Example 7.8   Parabolic/rectangular stress block

Problem   For the slab illustrated in Fig. 7.28 calculate the ultimate moment capacity (per unit breadth). Use the parabolic–rectangular stress block for the concrete and the horizontal top branch in the stress/strain relationship for reinforcement. Assume that $f_{\text{ck}} = 40 \text{ N/mm}^2$ and $f_{\text{y}} = 460 \text{ N/mm}^2$.

Fig. 7.28 Slab of Example 7.8: (a) elevation and section; (b) strain; (c) stress

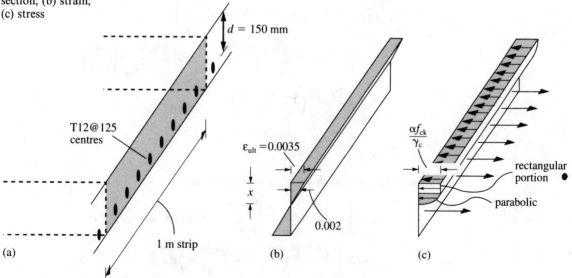

Solution The total compressive force on the concrete is:

$$F_c = F_c^1 + F_c^2$$

where F_c^1 is the compressive force in the rectangular portion of the stress distribution and F_c^2 is the compressive force in the parabolic portion (see Fig. 7.28(c)). The point on the stress distribution where the stress changes from being constant to being parabolic occurs when the strain is 0.002 (see Fig. 7.21(b)). Hence, referring to Fig. 7.28(b), the change point occurs at a distance from the neutral axis given by:

$$\text{change point} = \frac{0.002x}{0.0035} = 0.571x$$

Hence:

$$F_c^1 = \left(\frac{\alpha f_{ck}}{\gamma_c}\right)(x - 0.571x)1000$$

$$= \left(\frac{(0.85)(40)}{1.5}\right)(x - 0.571x)1000$$

$$= 9724x \text{ N}$$

The strain in the concrete at any point, $\varepsilon_c(y)$, is found from Fig. 7.28(b) (by similar triangles):

$$\frac{\varepsilon_c(y)}{y} = \frac{\varepsilon_{ult}}{x}$$

$$\Rightarrow \qquad \varepsilon_c(y) = \frac{0.0035y}{x}$$

In the parabolic region (refer to Fig. 7.21(a)):

$$f_c(y) = 1000\alpha\varepsilon_c(y)[1 - 250\varepsilon_c(y)]\left(\frac{f_{ck}}{\gamma_c}\right)$$

$$= 1000(0.85)\left(\frac{0.0035y}{x}\right)\left[1 - 250\left(\frac{0.0035y}{x}\right)\right]\left(\frac{40}{1.5}\right)$$

$$= \frac{79.33y}{x}\left(1 - \frac{0.875y}{x}\right)$$

$$= \frac{79.33y}{x} - \frac{69.41y^2}{x^2}$$

Hence the force in this region is (stress × area):

$$F_c^2 = \int_A f_c(y)\,dA = b\int_{y=0}^{y=0.571x} f_c(y)\,dy$$

$$= b\int_{y=0}^{y=0.571x}\left(\frac{79.33y}{x} - \frac{69.41y^2}{x^2}\right)dy$$

which reduces to:

$$F_c^2 = 8625x \text{ N}$$

Assuming that the reinforcement has yielded gives:

$$F_s = -A_s\left(\frac{f_y}{\gamma_s}\right) = -\left(\frac{1000}{125}\right)\left(\frac{\pi 12^2}{4}\right)\left(\frac{460}{1.15}\right) = -362\,000 \text{ N}$$

Then equilibrium of axial forces gives:

$$F_c^1 + F_c^2 + F_s = 0$$

$$\Rightarrow \qquad 9724x + 8625x - 362\,000 = 0$$

$$\Rightarrow \qquad x = 19.7 \text{ mm}$$

The corresponding steel strain is well in excess of the yield strain. To get the moment capacity, M_{ult}, moments are taken about the neutral axis. The contribution of the parabolic portion of the stress distribution to the moment capacity is found by integrating the product of force and lever arm:

$$M_c^2 = \int_A f_c(y)y \, dA$$

$$= b \int_{y=0}^{y=0.571x} f_c(y)y \, dA$$

$$= b \int_{y=0}^{y=0.571x} \left(\frac{79.33y^2}{x} - \frac{69.41y^3}{x^2}\right) dy$$

where $x = 19.7$ mm. This reduces to:

$$M_c^2 = 1.2 \text{ kNm}$$

The total moment capacity is the sum of the contributions from the parabolic and rectangular portions of the compressive block and that of the reinforcement:

$$M_{\text{ult}} = F_c^1\left(\frac{0.571x + x}{2}\right) + M_c^2 + |F_s|(d - x)$$

which reduces to 51.3 kNm for the 1 m strip.

Example 7.9 **Reinforcement not yielded**

Problem Calculate the ultimate hogging moment capacity for the section illustrated in Fig. 7.29. Use the rectangular stress distribution of Fig. 7.21(c) with $f_{ck} = 35$ N/mm². For the reinforcement, use the simplified stress/strain distribution of Fig. 7.22(b) with $f_y = 460$ N/mm².

Solution As the moment is hogging, the bottom of the section is in compression and the stress and strain distributions are as illustrated in Fig. 7.30. Provided $0.8x < 150$, this section will behave exactly as a rectangular section because the shape of the compression zone will be rectangular. Assuming therefore that $0.8x < 150$,

Fig. 7.29 Section of Example 7.9

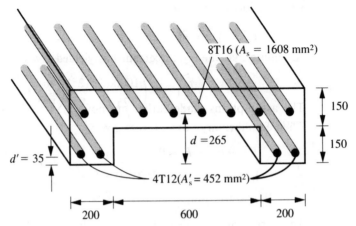

8T16 (A_s = 1608 mm²)

150

$d = 265$

150

$d' = 35$

4T12(A'_s = 452 mm²)

200   600   200

Fig. 7.30 Strain and stress distributions for section of Example 7.9: (a) elevation and section; (b) strain; (c) stress

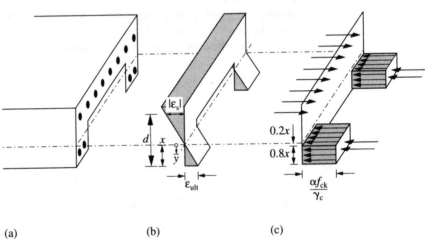

$|\varepsilon_s|$

d x

y

ε_{ult}

$0.2x$

$0.8x$

$\dfrac{\alpha f_{ck}}{\gamma_c}$

(a)   (b)   (c)

the breadth is 2 × 200 mm and the compressive force in the concrete is:

$$F_c = \left(\frac{\alpha f_{ck}}{\gamma_c}\right)(400)(0.8x) = \left(\frac{(0.85)(35)}{(1.5)}\right)(400)(0.8x) = 6347x \text{ N}$$

Assuming that the tension reinforcement has yielded implies:

$$|F_s| = A_s\left(\frac{f_y}{\gamma_s}\right) = (1608)\left(\frac{460}{1.15}\right) = 643\,200 \text{ N}$$

Assuming that the compression reinforcement has yielded implies:

$$F'_s = A'_s\left(\frac{f_y}{\gamma_s}\right) = (452)\left(\frac{460}{1.15}\right) = 180\,800 \text{ N}$$

The force that would have been present were the concrete not displaced by the compression reinforcement is given by:

$$F_{disp} = \left(\frac{\alpha f_{ck}}{\gamma_c}\right)A'_s = \left(\frac{(0.85)(35)}{1.5}\right)(452) = 8965 \text{ N}$$

By equilibrium of the forces on the section:

$$F_c - F_{disp} + F'_s = |F_s|$$

$$\Rightarrow \quad 6347x - 8965 + 180\,800 = 643\,200$$

$$\Rightarrow \quad x = 74.3 \text{ mm}$$

The assumptions made earlier are now checked. By similar triangles in Fig. 7.30(b):

$$\frac{|\varepsilon_s|}{(d-x)} = \frac{\varepsilon_{ult}}{x}$$

$$\Rightarrow \quad |\varepsilon_s| = \frac{(0.0035)[265 - 74.3)]}{74.3} = 0.0090 > 0.002$$

which implies that the tension reinforcement has yielded and the initial assumption is correct. Similarly:

$$\varepsilon'_s = \frac{(0.0035)[(74.3 - 35)]}{74.3} = 0.001\,85$$

This time, it emerges that the initial assumption that the compression reinforcement had yielded is incorrect. Thus, the value for x must be recalculated using a revised expression for F'_s:

$$\varepsilon'_s = \frac{\varepsilon_{ult}(x - 35)}{x}$$

$$\Rightarrow \quad f'_s = \frac{E_s \varepsilon_{ult}(x - 35)}{x}$$

$$\Rightarrow \quad F'_s = \frac{A'_s E_s \varepsilon_{ult}(x - 35)}{x} = \frac{(452)(200\,000)(0.0035)(x - 35)}{x}$$

$$= \frac{(316\,400)(x - 35)}{x}$$

Then equilibrium gives:

$$F_c - F_{disp} + F'_s = |F_s|$$

$$\Rightarrow \quad 6347x - 8965 + \frac{(316\,400)(x - 35)}{x} = 643\,200$$

which reduces to:

$$6.347x^2 - 336x - 11\,074 = 0$$

Solving this equation for x, the only positive root is $x = 75.9$ mm. This time, on checking, all of the assumptions are found to be correct; that is, the tension reinforcement has yielded, the compression reinforcement has not and $0.8x < 150$ mm.

The ultimate moment capacity is now calculated by taking moments about the neutral axis:

$$M_{ult} = (6347x)(0.6x) - (8965)(x - 35)$$

$$+ \frac{(316\,400)(x - 35)}{x}(x - 35) + (643\,200)(265 - x)$$

$$\Rightarrow \quad M_{ult} = 150\,\text{kNm}$$

Reinforcement required to resist moment

Up to now, all of the examples considered have involved the calculation of the moment capacity for cross-sections in which the areas of reinforcement present were known. However, the more usual problem is that the applied moment is known and the area of reinforcement required to resist it is sought, that is, we usually wish to determine the area of reinforcement necessary to provide a specified moment capacity. When there is no compression reinforcement, the procedure for calculating the required area of tension reinforcement, A_s, is as follows:

(a) Determine an expression for F_c as a function of x.
(b) Taking moments about the neutral axis gives equation (7.21) (with $F'_s = F_{disp} = 0$). Use equilibrium of axial forces (i.e. $F_c = |F_s|$) to remove F_s from this equation to give an equation relating M_{ult} to x. Solve this (by iteration if necessary) to find x.
(c) Back-substitute into equation (7.21) to find F_s and hence find A_s.

Using this procedure the formula for a rectangular section is derived below. Formulae for other sections can be derived in a similar manner.

Formula for A_s required in a rectangular section

It is assumed that the reinforcement has yielded. Steps will be taken later to ensure that the beam is under-reinforced, as illustrated in Fig. 7.20(c), and hence that this assumption is valid.

(a) If there is no compression reinforcement, then, for the rectangular section of arbitrary dimensions illustrated in Fig. 7.31:

$$F_c = \left(\frac{\alpha f_{ck}}{\gamma_c}\right)(0.8xb)$$

(b) Summing moments about the neutral axis gives the ultimate moment capacity:

$$M_{ult} = F_c(0.6x) + |F_s|(d - x)$$

But $|F_s| = F_c$ from which:

$$M_{ult} = F_c(d - 0.4x)$$

$$\Rightarrow \quad M_{ult} = F_c z \tag{7.27}$$

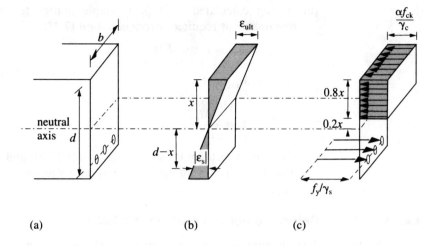

Fig. 7.31
Rectangular section
of arbitrary
dimensions:
(a) elevation and
section; (b) strain
distribution;
(c) stress
distribution

(a) (b) (c)

where z is the lever arm equal to $d - 0.4x$. Hence:

$$x = \frac{d - z}{0.4} \tag{7.28}$$

It is, in this case, more convenient to solve equation (7.27) to find z and subsequently to use equation (7.28) to find x. Thus:

$$M_{\text{ult}} = F_{\text{c}}z = \left(\frac{\alpha f_{\text{ck}}}{\gamma_{\text{c}}}\right)(0.8xb)z$$

$$= \left(\frac{\alpha f_{\text{ck}}}{\gamma_{\text{c}}}\right)(0.8)\left(\frac{d - z}{0.4}\right)bz$$

which reduces to:

$$\frac{M_{\text{ult}}}{bd^2 f_{\text{ck}}} = \frac{2\alpha}{\gamma_{\text{c}}}\left(1 - \frac{z}{d}\right)\left(\frac{z}{d}\right)$$

It is convenient to define a non-dimensional constant, K, as:

$$K = \frac{M_{\text{ult}}}{bd^2 f_{\text{ck}}} \tag{7.29}$$

Then, taking $\alpha = 0.85$ and $\gamma_{\text{c}} = 1.5$ gives:

$$\left(\frac{z}{d}\right)^2 - \left(\frac{z}{d}\right) + \frac{K}{1.133} = 0$$

$$\Rightarrow \quad \frac{z}{d} = \frac{1 \pm \sqrt{1 - 4K/1.133}}{2}$$

$$= 0.5 \pm \sqrt{0.25 - K/1.133}$$

Of the two roots, it is generally only the positive one that is valid. Hence:

$$z = d(0.5 + \sqrt{0.25 - 0.88K}) \tag{7.30}$$

267

(c) Having calculated z, it is a simple matter to calculate the area of reinforcement required. From equation (7.27):

$$M_{ult} = F_c z = |F_s| z$$

$$\Rightarrow \quad M_{ult} = A_s \left(\frac{f_y}{\gamma_s}\right) z$$

$$\Rightarrow \quad A_s = \frac{M_{ult}}{(f_y/\gamma_s) z} \tag{7.31}$$

Thus, for a rectangular section, equations (7.30) and (7.31) can be used to find the area of reinforcement required to provide a moment capacity of M_{ult}.

Example 7.10   Reinforcement in rectangular beam

Problem   The ultimate applied sag moment on the section of Fig. 7.31 is 250 kNm. If the breadth of the beam is 300 mm and the overall depth of the beam is 600 mm, use the equivalent rectangular stress block to calculate the area of tension reinforcement required to resist the applied moment. Assume a cover of 50 mm and $f_{ck} = 35$ N/mm^2.

Solution   Assuming a 10 mm diameter stirrup and a longitudinal tension reinforcement bar diameter of 25 mm, the effective depth is:

$$d = 600 - 50 - 10 - 25/2$$

$$= 527 \text{ mm}$$

From equation (7.29):

$$K = \frac{M_{ult}}{bd^2 f_{ck}} = \frac{250 \times 10^6}{(300)(527)^2(35)} = 0.0857$$

Equation (7.30) then gives:

$$z = d(0.5 + \sqrt{0.25 - 0.88K})$$

$$= (527)(0.5 + \sqrt{0.25 - 0.88(0.0857)})$$

$$= 484 \text{ mm}$$

The area of reinforcement required to resist M_{ult} is found from equation (7.31):

$$A_s = \frac{M_{ult}}{(f_y/\gamma_s) z} = \frac{250 \times 10^6}{(460/1.15)484} = 1291 \text{ mm}^2$$

3T25 gives an area of $3(\pi 25^2/4) = 1473$ mm^2, which exceeds the required area and is thus sufficient to resist the applied moment.

Flanged beams and ribbed slabs

It was briefly stated in Chapter 6 that the design of reinforced flanged beams (i.e. T-beams and L-beams) and ribbed slabs depends on the location of the

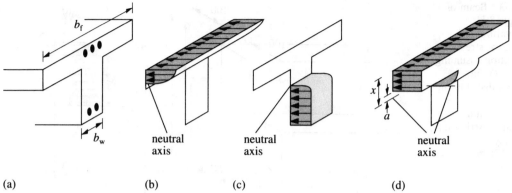

Fig. 7.32 Flanged sections: (a) elevation and section; (b) rectangular compression zone (sag); (c) rectangular compression zone (hog); (d) T-shaped compression zone (sag)

neutral axis. For light or moderately loaded beams, the neutral axis lies within the flange for a section in sag as illustrated in Fig. 7.32(b). In such circumstances, the compression zone is rectangular (assuming that the flange is rectangular) and the section can be designed as a rectangular section having a breadth equal to the effective breadth of the flange, b_f. Similarly, the neutral axis generally lies within the web for a section in hog, as illustrated in Fig. 7.32(c), and, assuming a constant web breadth, the section can be designed as a rectangular section having a breadth equal to the web breadth, b_w. (This was true of Example 7.9.) In certain cases, however, the neutral axis may not lie within the flange for sections in sag or in the web for sections in hog. In such cases, the design procedure is not as straightforward since the compression zone is no longer rectangular as illustrated in Fig. 7.32(d) and so the member cannot be treated as a simple rectangular section. From equation (7.18), the total compressive force in the concrete for the section of Fig. 7.32(d) is given by:

$$F_c = b_w \int_0^a f_c(y)\,dy + b_f \int_a^x f_c(y)\,dy \qquad (7.32)$$

Using equation (7.32) and an appropriate idealized stress block, the ultimate moment capacity for such a section can be calculated.

Example 7.11 Flanged beam

Problem Determine the ultimate moment capacity in sag of the flanged beam of Fig. 7.33 using the equivalent rectangular stress block. Take $f_{ck} = 30$ N/mm^2.

Solution Assuming the tension reinforcement to have yielded and the compressive stress block to be in the top 175 mm of the section:

$$|F_s| = \frac{A_s f_y}{\gamma_s} = \frac{(804)(460)}{1.15} = 321\,600 \text{ N}$$

$$F_c = \left(\frac{\alpha f_{ck}}{\gamma_c}\right)[(100)(0.8)x]$$

$$= \left(\frac{(0.85)(30)}{1.5}\right)[(100)(0.8)x]$$

$$= 1360x \text{ N}$$

Fig. 7.33 Beam of
Example 7.11:
(a) elevation and
section; (b) stress
distribution assuming
$0.8x < 175$ mm;
(c) stress distribution
assuming
$0.8x > 175$ mm;
(d) strain distribution
assuming
$0.8x > 175$ mm

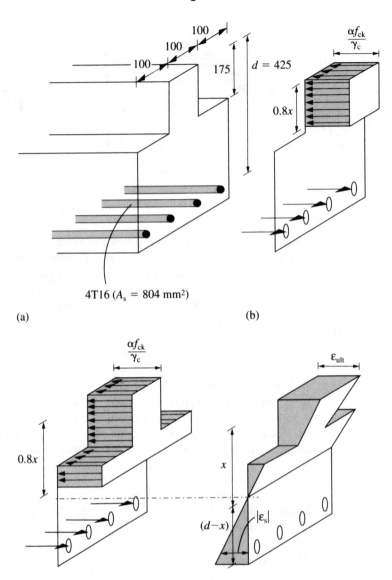

Equating F_s and F_c gives:

$$321\,600 = 1360x$$

$\Rightarrow$ $\quad x = 236$ mm

As $0.8x\,(=189$ mm$)$ exceeds 175 mm, the assumption is clearly not valid. It is next assumed that the compressive stress block extends below the top 175 mm of the section as illustrated in Fig. 7.33(c). In the definition of the coefficient, α, EC2 specified a value of 0.85 only when the section is at least as wide at the extreme compression fibre as elsewhere in the compression zone. In the distribution of Fig. 7.33(c) this is no longer the case and the lesser value of

$\alpha = 0.8$ must be adopted. Hence:

$$F_c = \left(\frac{\alpha f_{ck}}{\gamma_c}\right)[(100)(175) + (300)(0.8x - 175)]$$

$$= \left(\frac{(0.8)(30)}{1.5}\right)[(100)(175) + (300)(0.8x - 175)]$$

$$= 3840x - 560\,000$$

This time, equating F_s and F_c gives:

$$321\,600 = 3840x - 560\,000$$

$$\Rightarrow \qquad x = 230 \text{ mm}$$

which verifies the assumption regarding the extent of the compressive stress block. Now the assumption that the steel has yielded is checked. From similar triangles in Fig. 7.33(d):

$$|\varepsilon_s| = \frac{\varepsilon_{ult}(d - x)}{x} = \frac{(0.0035)(425 - 230)}{230} = 0.00296 > 0.002$$

Finally the ultimate moment capacity, M_{ult}, can be calculated by taking moments about the neutral axis:

$$M_{ult} = \left(\frac{(0.8)(30)}{1.5}\right)[(100)(175)(x - 175/2) + (300)(0.8x - 175)^2/2]$$

$$+ (321\,600)(425 - x)$$

$$\Rightarrow \qquad M_{ult} = 103 \text{ kNm}$$

Non-uniform sections

For non-rectangular beams which have a varying breadth, such as those illustrated in Fig. 7.34, the calculation of the ultimate moment capacity is somewhat more complex than for rectangular sections. The reason behind this is that, since the breadth, $b(y)$, varies with depth, the total compressive force in the concrete must be calculated directly from equation (7.18) by integration.

Fig. 7.34 Non-uniform sections

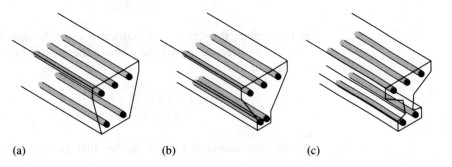

(a)   (b)   (c)

The complexity of the problem is further increased if the parabolic–rectangular stress block is used, in which the stress is not constant over the depth. The procedure is illustrated below in Example 7.12.

Example 7.12   Non-uniform section

Problem   For the section illustrated in Fig. 7.35, determine the area of tension reinforcement required to resist an applied ultimate moment of 62 kNm. Assume the bilinear stress distribution for concrete, as illustrated in Fig. 7.21(b), with $f_{ck} = 30 \text{ N/mm}^2$. Assume the simplified stress distribution for reinforcement with $f_y = 460 \text{ N/mm}^2$.

Fig. 7.35 Beam of Example 7.12: (a) elevation and section; (b) stress distribution; (c) elevation of stress distribution

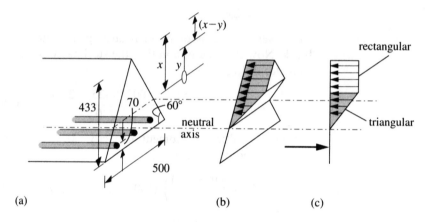

Solution   The section breadth varies from 500 mm at the base to zero at the top. Thus at a distance y above the neutral axis (i.e. $x - y$ below the top fibre), the breadth is:

$$b(y) = \frac{(x - y)\,500}{433}$$

Referring to Fig. 7.21(b), the triangular portion of the stress distribution ends at a strain of 0.001 35. This occurs at a distance y_0 above the neutral axis where:

$$y_0 = \frac{0.001\,35x}{\varepsilon_{ult}} = \frac{0.001\,35x}{0.0035} = 0.386x$$

In this region, the stress in the concrete varies linearly with strain. From Fig. 7.21(b), the stress at a distance y above the neutral axis is related to the strain at this point by:

$$f_c(y) = \left(\frac{\alpha f_{ck}}{\gamma_c}\right)\left(\frac{\varepsilon_c(y)}{0.001\,35}\right)$$

Hence the compressive force in the first portion of the stress distribution is:

$$F_{\rm c}^1 = \int_A f_{\rm c}(y)\,{\rm d}A = \int_{y=0}^{y=y_0} f_{\rm c}(y)b(y)\,{\rm d}y$$

$$= \int_{y=0}^{y=0.386x} \left(\frac{\alpha f_{\rm ck}}{\gamma_{\rm c}}\,\frac{\varepsilon_{\rm c}(y)}{0.001\,35}\right)\left(\frac{(x-y)500}{433}\right){\rm d}y$$

where $\varepsilon_{\rm c}(y)$ is given by:

$$\varepsilon_{\rm c}(y) = \frac{\varepsilon_{\rm ult}\,y}{x}$$

Hence:

$$F_{\rm c}^1 = \int_{y=0}^{y=y_0} \left(\frac{\alpha f_{\rm ck}}{\gamma_{\rm c}}\,\frac{\varepsilon_{\rm ult}\,y}{0.001\,35x}\right)\left(\frac{(x-y)500}{433}\right){\rm d}y$$

which (taking $\alpha = 0.8$) reduces to:

$$F_{\rm c}^1 = 2.65x^2$$

The force in the upper part of the section is a simple product of stress and area:

$$F_{\rm c}^2 = \left(\frac{\alpha f_{\rm ck}}{\gamma_{\rm c}}\right) \times \tfrac{1}{2}(x-y_0)\left(\frac{(x-y_0)500}{433}\right)$$

which reduces to:

$$F_{\rm c}^2 = 3.48x^2$$

The ultimate moment capacity is found by taking moments about the neutral axis:

$$M_{\rm ult} = \int_{y=0}^{y=y_0} f_{\rm c}(y)b(y)y\,{\rm d}y + \int_{y=y_0}^{y=x} f_{\rm c}(y)b(y)y\,{\rm d}y + |F_{\rm s}|(d-x)$$

$$= \int_{y=0}^{y=y_0} f_{\rm c}(y)b(y)y\,{\rm d}y + \int_{y=y_0}^{y=x} f_{\rm c}(y)b(y)y\,{\rm d}y + (F_{\rm c}^1 + F_{\rm c}^2)(d-x)$$

$$= \int_{y=0}^{y=y_0} \left(\frac{\alpha f_{\rm ck}}{\gamma_{\rm c}}\right)\left(\frac{\varepsilon_{\rm ult}\,y}{0.001\,35x}\right)\left(\frac{(x-y)500}{433}\right)y\,{\rm d}y$$

$$+ \int_{y=y_0}^{y=x} \left(\frac{\alpha f_{\rm ck}}{\gamma_{\rm c}}\right)\left(\frac{(x-y)500}{433}\right)y\,{\rm d}y$$

$$+ (2.65x^2 + 3.48x^2)(d-x)$$

This reduces to:

$$M_{\rm ult} = 2225x^2 - 3.42x^3$$

This equation may be solved iteratively to find a value of x for which $M_{\rm ult} = 62 \times 10^6$ Nmm. By trial and error it is found that $x = 201$ mm. For this value of x, the force in the reinforcement is:

$$|F_s| = F_c^1 + F_c^2$$
$$= 2.65(201)^2 + 3.48(201)^2$$
$$= 248 \text{ kN}$$

Equating this to $A_s f_y / \gamma_s$ gives:

$$A_s = \frac{248 \times 10^3}{(460/1.15)} = 620 \text{ mm}^2$$

2T20 bars give an area of 628 mm^2.

7.6 Balanced design and section ductility

Balanced design

It was stated in section 7.4 that a balanced design is one in which the concrete crushes at the same instant as the reinforcement first yields under the ultimate applied loads. In other words, the strain in the reinforcement reaches its yield value at the same time as the strain in the extreme concrete fibre in compression reaches ε_{ult}, as illustrated in Fig. 7.36. To determine accurately when this occurs, material factors of safety are omitted initially in the following calculations. Then, by similar triangles on the strain diagram:

$$\frac{\varepsilon_{ult}}{x} = \frac{f_y/E_s}{d-x} \tag{7.33}$$

Assuming $f_y = 460 \text{ N/mm}^2$ (high-yield steel), $\varepsilon_{ult} = 0.0035$ and $E_s = 200\,000 \text{ N/mm}^2$, equation (7.33) becomes:

$$\frac{0.0035}{x} = \frac{460/200\,000}{d-x}$$

$$\Rightarrow \quad \frac{0.0035}{x} = \frac{0.0023}{d-x}$$

$$\Rightarrow \quad x/d = 0.603 \tag{7.34}$$

This is true of all sections regardless of their shape. The precise amount of reinforcement which gives a balanced design (i.e. satisfies equation (7.34)) can readily be calculated by equilibrium of the axial forces acting on the section. Unlike equation (7.34), however, the **area of reinforcement** for which a design is balanced depends in the cross-section being considered. Assuming an equivalent rectangular stress block for the rectangular section of Fig. 7.37,

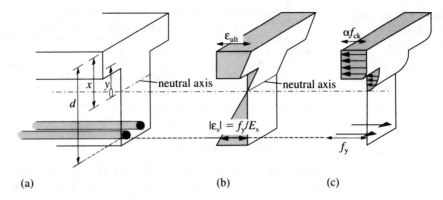

Fig. 7.36 Balanced design for typical section: (a) elevation and section; (b) strain distribution; (c) stress distribution

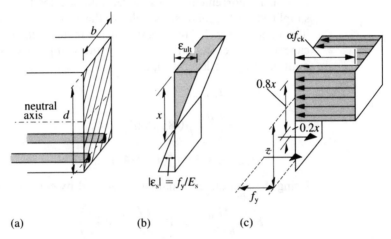

Fig. 7.37 Balanced design for singly reinforced rectangular section: (a) elevation and section; (b) strain; (c) equivalent rectangular stress block

equilibrium gives:

$$(\alpha f_{ck})(0.8xb) = A_s f_y$$

$$\Rightarrow \quad (0.85 f_{ck})[0.8(0.603d)b] = 460 A_s$$

$$\Rightarrow \quad A_s = \frac{bd f_{ck}}{1122} \ (\text{mm}^2) \tag{7.35}$$

If the area of reinforcement provided is greater than that given by equation (7.35), the beam is over-reinforced, that is the reinforcement does not yield before the concrete crushes. Similarly, if the area of reinforcement provided is less than that given by equation (7.35), the beam is under-reinforced and the reinforcement will yield before the concrete crushes. It was seen in section 7.4 that, for safety reasons, only under-reinforced beams are acceptable. To ensure that designs remain under-reinforced, EC2 recommends that the ratio, x/d, does not exceed the values given in Table 7.3. Plastic moment redistribution (refer to section 5.3) requires ductility which is affected by the degree to which the beam is under-reinforced. Even without such redistribution, beams should always be under-reinforced. However, with redistribution, beams must be well under-reinforced as can be seen from the smaller values for x/d given in the table (a small x implies large strain in the steel).

Table 7.3 Maximum values for x/d recommended in EC2

Concrete cylinder strength, f_{ck} (N/mm²)	% plastic moment redistribution						
	0	5	10	15	20	25	30
12 to 35	0.45	0.41	0.37	0.33	0.29	0.25	0.21
40 and over	0.35	0.31	0.27	0.23	0.19	0.15	0.11

Singly reinforced sections

The rectangular section illustrated in Fig. 7.37 has tension reinforcement only and is termed a singly reinforced section. For designs with $x/d < 0.603$, the tension reinforcement yields under ultimate loads. Assuming the use of the simplified rectangular stress block, the forces acting on the section are as illustrated in Fig. 7.37(c). Reintroducing material factors of safety, the moment capacity of the section is given by the compressive force (or tensile force) multiplied by the lever arm, z:

$$M_{ult} = F_c z$$

$$= \left(\frac{\alpha f_{ck}}{\gamma_c}\right)(0.8x)(b)(d - 0.4x)$$

$$\Rightarrow \quad M_{ult} = 0.453xbf_{ck}(d - 0.4x) \tag{7.36}$$

Using the non-dimensional factor defined by equation (7.29), this becomes:

$$K = \frac{M_{ult}}{bd^2 f_{ck}} = 0.453\frac{x}{d}\left(1 - 0.4\frac{x}{d}\right) \tag{7.37}$$

Recall that, for a balanced design, $x/d = 0.603$. Thus, equation (7.37) becomes:

$$K = 0.453(0.603)[1 - 0.4(0.603)]$$

$$= 0.207$$

Hence, for a balanced design, the ultimate moment capacity for the singly reinforced rectangular section is given by:

$$M_{ult} = Kf_{ck}bd^2 = 0.207f_{ck}bd^2 \tag{7.38}$$

To ensure ductile failure, the allowable values for K are calculated using the limits on x/d specified in EC2 (Table 7.3). Substituting for x/d in equation (7.37) results in the maximum values for K given in Table 7.4. Thus, the safe upper

Table 7.4 Maximum values for K in accordance with EC2

Concrete cylinder strength, f_{ck} (N/mm²)	% plastic moment redistribution						
	0	5	10	15	20	25	30
12 to 35	0.166	0.154	0.142	0.129	0.115	0.101	0.086
40 and over	0.137	0.123	0.109	0.095	0.080	0.064	0.048

limit for ultimate moment capacity, regardless of reinforcement provided, for singly reinforced rectangular sections is given by:

$$M_{ult} = K f_{ck} b d^2 \tag{7.39}$$

where the appropriate value for K is taken from Table 7.4.

Example 7.13 Minimum depth ductile section

Problem Design a rectangular singly reinforced section of breadth 350 mm and of minimum depth to resist a factored ultimate moment of 425 kNm. This moment has been reduced from an applied elastic moment of 500 kNm through plastic moment redistribution. $f_{ck} = 35$ N/mm², $f_y = 460$ N/mm².

Solution The ductility of sections increases with effective depth as a higher depth allows a lower area of reinforcement. Hence this design is governed by the requirements of ductility. A redistribution of 75 kNm, or 15 per cent, has been performed. Using Table 7.4:

$$K \leq 0.129$$

$$\Rightarrow \quad \frac{M_{ult}}{bd^2 f_{ck}} \leq 0.129$$

$$\Rightarrow \quad d \geq \sqrt{\frac{M_{ult}}{0.129 b f_{ck}}} = \sqrt{\frac{425 \times 10^6}{(0.129)(350)(35)}} = 519 \text{ mm}$$

Allowing for cover of 35 mm, a 12 mm link and a 25 mm bar diameter gives a total depth of:

$$h = 519 + 35 + 12 + 25/2 = 579 \text{ mm}$$

This is rounded up to $h = 600$ mm, giving $d = 540$ mm. Hence:

$$K = \frac{M_{ult}}{bd^2 f_{ck}} = \frac{425 \times 10^6}{(350)(540)^2(35)} = 0.119$$

The corresponding area of reinforcement is calculated using equations (7.30) and (7.31):

$$z = d(0.5 + \sqrt{0.25 - 0.88K})$$

$$= (540)(0.5 + \sqrt{0.25 - (0.88)(0.119)})$$

$$= 476 \text{ mm}$$

$$\Rightarrow \quad A_s = \frac{M_{ult}}{(f_y/\gamma_s)z} = \frac{425 \times 10^6}{(460/1.15)(476)} = 2232 \text{ mm}^2$$

5T25 gives an area of 2454 mm², which exceeds the required area and is sufficient to resist the ultimate moment.

Doubly reinforced rectangular sections

The upper limits on moment capacity presented above are only valid for singly reinforced sections. By providing compression reinforcement, greater capacities can be achieved while maintaining a ductile section. Consider first the case where the ultimate moment capacity, M_{ult}, equals the maximum value allowed for a singly reinforced section. If the area of tension reinforcement is A_{s1}, then equilibrium of axial forces dictates that:

$$F_c = \frac{A_{s1}f_y}{\gamma_s} \qquad (7.40)$$

Further, equations (7.29) to (7.31), derived using moment equilibrium, can be used to calculate A_{s1}:

$$A_{s1} = \frac{M_{ult}}{(f_y/\gamma_s)z} \qquad (7.41)$$

where:

$$z = d(0.5 + \sqrt{0.25 - 0.88K}) \qquad (7.42)$$

and:

$$K = \frac{M_{ult}}{bd^2 f_{ck}} \qquad (7.43)$$

Now, if additional tension reinforcement is provided of area A_{s2} and, at the same time, compression reinforcement of area A'_s such that:

$$A'_s f'_s = \frac{A_{s2}f_y}{\gamma_s} \qquad (7.44)$$

then, as the additional forces are self-equilibrating, equilibrium of axial forces would be satisfied without any change in x. Further, all of the tension reinforcement would have yielded as is evident from equations (7.40) and (7.44), thus maintaining the ductility of the section. The forces on this strengthened section are illustrated in Fig. 7.38(d). The new areas of reinforcement result in an increase in moment capacity of:

$$\Delta M_{ult} = A'_s f'_s(x - d') + \frac{A_{s2}f_y(d - x)}{\gamma_s} \qquad (7.45)$$

Substitution from equation (7.44) gives:

$$\Delta M_{ult} = A'_s f'_s(d - d') \qquad (7.46)$$

Hence, the area of compression reinforcement required to provide an ultimate moment capacity of $M_{ult} + \Delta M_{ult}$ is:

$$A'_s = \frac{\Delta M_{ult}}{f'_s(d - d')} \qquad (7.47)$$

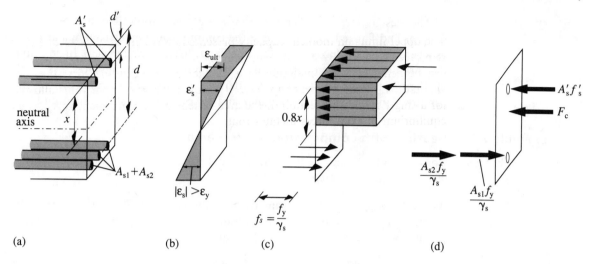

(a) (b) (c) (d)

Fig. 7.38 Ductile design for doubly reinforced rectangular section: (a) elevation and section; (b) strain distribution; (c) stress distribution; (d) axial forces

The total area of tension reinforcement required is:

$$A_s = A_{s1} + A_{s2} \tag{7.48}$$

where A_{s1} is given by equation (7.41) and, from equation (7.44):

$$A_{s2} = \frac{A'_s f'_s}{(f_y / \gamma_s)} \tag{7.49}$$

Compression reinforcement may or may not have yielded. The strain in the compression reinforcement is:

$$\varepsilon'_s = \frac{\varepsilon_{ult}(x - d')}{x} \tag{7.50}$$

Yield occurs when:

$$\varepsilon'_s = \left(\frac{f_y}{\gamma_s}\right)\frac{1}{E_s}$$

$$\Rightarrow \quad \frac{\varepsilon_{ult}(x - d')}{x} = \frac{f_y}{\gamma_s E_s}$$

Rearranging gives:

$$\frac{x}{d'} = \frac{E_s \varepsilon_{ult}}{(E_s \varepsilon_{ult} - f_y / \gamma_s)} \tag{7.51}$$

Taking $E_s = 200\,000$, $\varepsilon_{ult} = 0.0035$ and $f_y = 460$ gives:

$$\frac{x}{d'} = \frac{7}{3}$$

Therefore, if x/d' is less than 7/3, then $f'_s = E_s \varepsilon'_s$ where ε'_s is given by equation (7.50). Otherwise, $f'_s = f_y / \gamma_s$.

279

The design formulae derived above are only applicable to rectangular sections and flanged sections where the compression zone is rectangular (such as where the neutral axis lies within the flange for a section in sag). In addition, the formulae have been derived using the equivalent rectangular stress block of Fig. 7.21(c). Formulae can be derived for flanged sections where the compression zone is non-rectangular and for other stress/strain diagrams.

Example 7.14 Section with compression reinforcement

Problem For the beam of Example 7.13, determine the reinforcement requirements if the effective depth is limited to 475 mm.

Solution As for Example 7.13, the parameter K is limited to 0.129. With the reduced effective depth, the corresponding ultimate moment capacity (the maximum possible without compression reinforcement) is:

$$M_{ult} = 0.129bd^2f_{ck} = 0.129(350)(475)^2(35)$$

$$= 356 \, \text{kNm}$$

The corresponding area of tension reinforcement is found from equations (7.30) and (7.31):

$$z = d(0.5 + \sqrt{0.25 - 0.88K})$$

$$= 475(0.5 + \sqrt{0.25 - (0.88)(0.129)})$$

$$= 413 \, \text{mm}$$

$$\Rightarrow \quad A_{s1} = \frac{M_{ult}}{(f_y/\gamma_s)z} = \frac{356 \times 10^6}{(460/1.15)(413)} = 2155 \, \text{mm}^2$$

From equation (7.28), the corresponding value for x is:

$$x = \frac{d - z}{0.4} = \frac{475 - 413}{0.4} = 155 \, \text{mm}$$

Assuming a value for d' of 60 mm, the ratio x/d' then becomes 2.58. As this exceeds 7/3, compression reinforcement will yield. Hence, from equation (7.47), the required area of compression reinforcement is:

$$A'_s = \frac{\Delta M_{ult}}{f'_s(d - d')} = \frac{(425 \times 10^6 - 356 \times 10^6)}{(460/1.15)(475 - 60)} = 416 \, \text{mm}^2$$

Equation (7.49) gives the required additional area of tension reinforcement:

$$A_{s2} = \frac{A'_s f'_s}{(f_y/\gamma_s)} = A'_s = 416 \, \text{mm}^2$$

giving a total required area of tension reinforcement of $2155 + 416 = 2571 \, \text{mm}^2$.

The formula given in Chapter 6 for the preliminary design of reinforced concrete beams and slabs, namely equation (6.14), is derived from equations (7.41) and (7.47) assuming no moment redistribution has been carried out. Taking a value for K of 0.166 from Table 7.4 and substituting in equation (7.42) gives:

$$z = 0.822d$$

Then equation (7.41) gives:

$$A_{s1} = \frac{M_{ult}}{(f_y/\gamma_s)(0.822d)} = \frac{0.166bd^2 f_{ck}}{(f_y/\gamma_s)(0.822d)}$$

$$\Rightarrow \quad A_{s1} = \frac{0.232bd f_{ck}}{f_y}$$

Rounding up the constant to 0.24 and taking $A_{s2} \approx A_s'$ gives the formula for A_s presented in Chapter 6.

7.7 Anchorage length

In reinforced concrete members, the flexural strength relies on the transfer of tensile force, commonly known as **bond**, between the longitudinal reinforcement and the surrounding concrete. The quality of the bond depends on the surface pattern of the bar, the dimensions of the member and the position and inclination of the bars. Bond is achieved by the combination of adhesion and a small amount of friction between the two materials. In the case of deformed bars, a significant proportion of the bond is provided by the bearing of the concrete between the ribs. This is illustrated in Fig. 7.39 where a tensile force due to moment F_s is being transferred by the bond into the concrete. In EC2, the effect of ribbed (high-bond) bars is allowed for with a reduced bond strength. Clearly, the greater the length that the bars are embedded in the concrete, the greater the bond between the two materials.

The application of loads to a reinforced concrete member leads to bending of the member which, in turn, results in tensile forces being developed in the reinforcement. If the anchorage bond between the bars and the concrete is sufficient, the full strength of the reinforcement can be utilized. If, however, the bond is insufficient, the bar will pull out of the concrete, the tensile force will drop to zero and the member will fail. The **anchorage length**, l_b, is the length of reinforcement required to develop sufficient anchorage bond so that the full strength of the reinforcement can be used.

The design anchorage length, $l_{b,net}$, recommended for design in EC2 is given by:

$$l_{b,net} = \text{greater of} \left(\alpha l_b \frac{A_{sr}}{A_{sp}} \right) \text{and } l_{b,min} \tag{7.52}$$

where α = coefficient which accounts for shape of bars (a bar with a hooked end clearly has better anchorage): for straight bars, $\alpha = 1.0$; for curved bars where concrete cover is at least three times bar diameter, $\alpha = 0.7$

Fig. 7.39 Bond in deformed bars: (a) portion of beam (elevation and section); (b) detail around bar

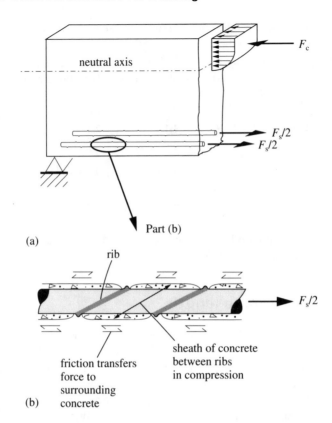

(a)

rib

(b)

friction transfers force to surrounding concrete

sheath of concrete between ribs in compression

l_b = basic anchorage length (equation (7.54))

A_{sr} = calculated area of reinforcement required for design

A_{sp} = area of reinforcement provided in design (the required area is rounded up to an integer number of bars with the result that $A_{sp} \geq A_{sr}$)

$l_{b,min}$ = minimum anchorage length: for bars in tension, $l_{b,min}$ = greater of $0.3l_b$, 10ϕ and 100 mm, where ϕ is the bar diameter; for bars in compression, $l_{b,min}$ = greater of $0.6l_b$, 10ϕ and 100 mm.

The basic anchorage length, l_b, is the straight length of bar required to anchor the force $A_s f_y$. For a bar of diameter ϕ, this force must equal the shear force developed between the bar surface and the surrounding concrete:

$$A_s f_y = (\pi \phi l_b) f_b \qquad (7.53)$$

where $(\pi \phi l_b)$ is the contact surface area and f_b is the ultimate design bond strength, that is, the maximum shear stress that can act at the interface of the two materials. This simplifies to:

$$l_b = \frac{\phi}{4} \frac{f_y}{f_b} \qquad (7.54)$$

The bond strength is dependent on the quality of the bond, with EC2 distinguishing between 'good' and 'bad' bond conditions. Good bond conditions are defined as follows:

Table 7.5 Design values of ultimate bond stress, f_b, for good bond conditions

Characteristic cylinder strength, f_{ck}	Ultimate bond stress, f_b (N/mm^2)	
	Plain bars	High-bond bars
12	0.9	1.6
16	1.0	2.0
20	1.1	2.3
25	1.2	2.7
30	1.3	3.0
35	1.4	3.4
40	1.5	3.7
45	1.6	4.0
50	1.7	4.3

1. All bars inclined at an angle of between 45° and 90° to the horizontal during casting of the member.
2. All bars inclined at an angle of between 0° and 45° to the horizontal that are either (a) placed in members whose depth does not exceed 250 mm or (b) placed in the lower half or at least 300 mm from the top of the member.

All other bond conditions are defined as poor. In conditions of good bond, the bond stress, f_b, is as given in Table 7.5. For poor bond conditions, the values in Table 7.5 should be multiplied by a factor of 0.7.

Example 7.15   Anchorage length

Problem   Determine the design anchorage length for the beam of Fig. 7.40. The beam must resist an ultimate sag moment of 190 kNm. The breadth is 300 mm, the

Fig. 7.40   Beam of Example 7.15

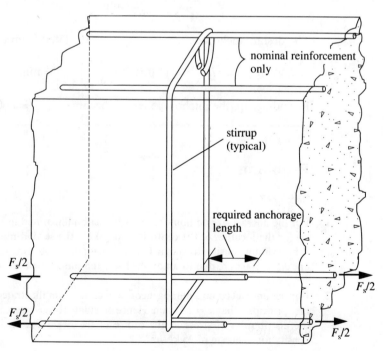

effective depth, d, is 550 mm and the total depth is 600 mm. Assume $f_{ck} = 35$ N/mm², $f_y = 460$ N/mm² and assume high-bond bars.

Solution   Given that $f_{ck} = 35$ N/mm² and high-bond bars are being used, Table 7.5 gives the ultimate bond stress, f_b, as 3.4 N/mm² (bond is good as the bars are in the lower half of the beam). The area of steel required is calculated using equations (7.29), (7.30) and (7.31). From equation (7.29):

$$K = \frac{M_{ult}}{bd^2 f_{ck}} = \frac{190 \times 10^6}{(300)(550)^2(35)} = 0.060$$

From equation (7.30):

$$z = d(0.5 + \sqrt{0.25 - 0.88K})$$
$$= (550)(0.5 + \sqrt{0.25 - 0.88(0.060)})$$
$$= 519 \text{ mm}$$

Hence using equation (7.31):

$$A_s = \frac{M_{ult}}{(f_y/\gamma_s)z} = \frac{190 \times 10^6}{(460/1.15)519} = 915 \text{ mm}^2$$

$$\Rightarrow \qquad A_{sr} = A_s = 915 \text{ mm}^2$$

2T25 provides an area of reinforcement of 982 mm².

$$\Rightarrow \qquad A_{sp} = 982 \text{ mm}^2$$

From equation (7.54), the basic anchorage length, l_b, is:

$$l_b = \frac{\phi}{4} \frac{f_y}{f_b} = \frac{25}{4} \frac{460}{3.4} = 846 \text{ mm}$$

The design anchorage length, $l_{b,net}$, is calculated from equation (7.52):

$$l_{b,net} = \alpha l_b \frac{A_{sr}}{A_{sp}} = (1.0)(846)\left(\frac{915}{982}\right) = 788 \text{ mm}$$

Rounding up, the anchorage length becomes $l_{b,net} = 800$ mm.

Problems

Section 7.2

7.1   For the uncracked and homogeneous T-section illustrated in Fig. 7.41, derive formulae, (a) for the location of the centroid and (b) for the second moment of area. Hence find the maximum tensile stress in a T-section with $b_f = 900$ mm, $b_w = 300$ mm, $h_f = 200$ mm and $h_w = 500$ mm, due to an applied sagging moment of 500 kNm.

7.2   For the uncracked and homogeneous concrete beam illustrated in Fig. 7.42, verify from first principles that $x = 183$ mm. Hence calculate the gross second moment of area. Find the distribution of stress at the centre of the beam due to its self weight. Assume a density for concrete of 25 kN/m³.

Fig. 7.41 Section of Problem 7.1

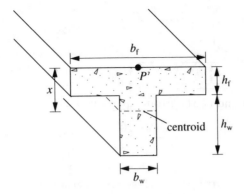

Fig. 7.42 Beam of Problem 7.2: (a) span; (b) cross-sectional dimensions

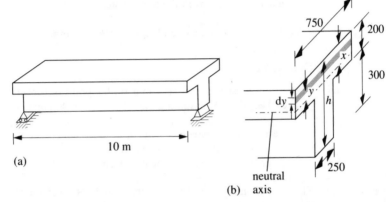

7.3 Prove that the second moment of area for an uncracked reinforced section is given by:

$$I_u = \int y^2 \, dA + (m-1)A_s(d-x)^2$$

7.4 Find, by rule, the stresses due to self weight at A for the beam illustrated in Fig. 7.43:

(a) at the top fibre in the concrete;
(b) in the reinforcement.

Assume the section to be cracked. Let $E_s = 200\,000$ N/mm^2 and $E_c = 30\,000$ N/mm^2.

Fig. 7.43 Beam of Problem 7.4: (a) elevation; (b) section X–X

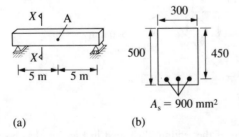

7.5 For the beam whose section is illustrated in Fig. 7.44, it can be assumed that the section is cracked and that the neutral axis is in the flange.

(a) Sketch the equivalent concrete section and calculate x.

Design of Reinforced Concrete Members for Bending

Fig. 7.44 Section of Problem 7.5

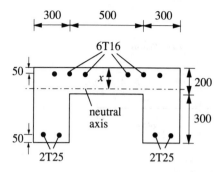

(b) Calculate the cracked second moment of area, I_c.

(c) Calculate the stress in the tension steel due to an applied moment of 150 kNm.

Take $E_s = 200\,000$ N/mm^2 and $E_c = 32\,000$ N/mm^2.

Section 7.5

7.6 The section illustrated in Fig. 7.45 is on the point of rupture (i.e. strain in concrete has just reached 0.0035). Assuming initially that $x < 400$ mm:

(a) determine if the steel has yielded;

(b) find x and verify that $x < 400$ mm;

(c) calculate the ultimate moment capacity.

The concrete has a cylinder strength, $f_{ck} = 35$ N/mm^2 and the reinforcement has a yield strength, $f_y = 460$ N/mm^2.

Fig. 7.45 Section of Problem 7.6

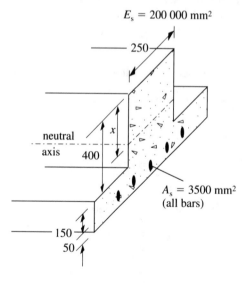

7.7 The neutral axis of the section illustrated in Fig. 7.46 has been found to be located at $x = 280$ mm. Check this result and calculate the ultimate moment capacity of the section.

7.8 Repeat 7.7 using the simplified stress–strain relationships of Figs 7.21(c) and 7.22(b).

Fig. 7.46 Problem
7.7: (a) section;
(b) stress/strain
diagram for steel;
(c) stress/strain
diagram for
concrete

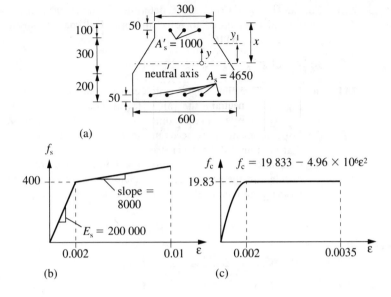

(a)

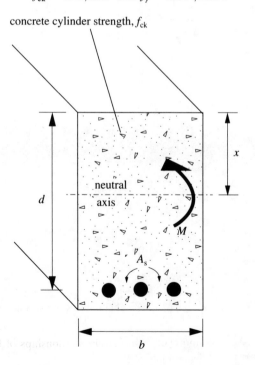

(b)　　　　　　　　(c)

7.9　(a) For the section illustrated in Fig. 7.47 derive, from first principles, a formula giving x/d in terms of $M/bd^2 f_{ck}$. Use the parabolic rectangular stress block for the concrete and the horizontal top branch in the stress/strain relationship for the reinforcement. Hence derive a formula for the area of reinforcement required to resist the applied moment.

　　(b) Determine the area of reinforcement required to resist a moment of 300 kNm in a rectangular section of breadth, 300 mm, effective depth to reinforcement, 500 mm, $f_{ck} = 40$ N/mm^2 and $f_y = 460$ N/mm^2.

Fig. 7.47 Section
of Problem 7.9

concrete cylinder strength, f_{ck}

287

7.10   For the section of Problem 7.6, determine from first principles the area of reinforcement that would be required to resist an applied ultimate moment of 300 kNm given that $f_{ck} = 35$ N/mm^2 and $f_y = 460$ N/mm^2.

Section 7.6

7.11   (a) For the T-section singly reinforced with mild steel ($f_y = 250$ N/mm^2) illustrated in Fig. 7.48, determine the value of x at which the design is balanced. Calculate the corresponding ultimate moment capacity of the section. Assume $f_{ck} = 40$ N/mm^2.
   (b) For a moment 20 per cent in excess of the balanced value, calculate x and show that the reinforcement has not yielded.
   (c) Calculate the area of compression reinforcement required to render the section ductile and the area of tension reinforcement required to resist the higher applied moment, and verify that the section is now ductile.

Fig. 7.48
Section of
Problem 7.11

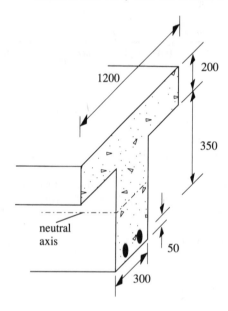

8

Design of Prestressed Concrete Members for Bending

8.1 Introduction

Loads which are carried in a member by bending are effectively transferred through it as compressive and tensile forces, as illustrated in Fig. 8.1. Owing to the low tensile strength of concrete, steel reinforcement must be provided in all structural concrete members subject to such bending forces to control tensile cracking and, ultimately, to prevent failure.

In ordinary reinforced concrete, steel bars are placed within the tension zones of concrete members to carry the internal tensile forces across flexural cracks to the supports. In such a member, the applied moment is resisted by compression of the uncracked portion of the concrete section and by tension in the reinforcing bars. It should be remembered that this form of reinforcement does not **prevent** the development of tensile cracks in members. Thus, it is only by limiting the magnitude of the strain in the bars that the cracks are prevented from becoming excessively large.

One problem with such ordinary reinforced concrete is that the presence of cracks can lead to the corrosion of the reinforcement due to its exposure to water and to chemical contaminants. Corrosion is generally only a problem for structures in aggressive exterior environments (bridges, marine structures, etc.) and is not critical in the majority of buildings. A further effect of cracking of ordinary reinforced members is the substantial loss in stiffness which occurs after cracking. The second moment of area of the cracked section, I_c, is far less

Fig. 8.1 Stresses due to bending in a beam

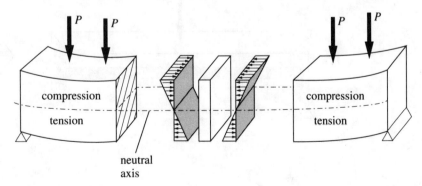

neutral axis

than the second moment of area before cracking, I_u (see section 7.2). Thus, allowing cracks to develop can cause a large increase in the deformation of the member.

Prestressed concrete is an alternative form of reinforced concrete. In prestressed concrete, compressive stresses are introduced into a member to reduce or nullify the tensile stresses which result from bending due to the applied loads. The compressive stresses are generated in a member by tensioned steel anchored at the ends of the members and/or bonded to the concrete.

Consider the simply supported member of Fig. 8.2(a), of cross-sectional area, A. The linear elastic bending stress distribution due to the applied load at a given section of the member is illustrated in the figure. When the maximum tensile stress, $\sigma_{app,\,t}$, exceeds the tensile strength of the concrete, a crack forms in the section. If there is no reinforcement, the crack will propagate until it causes failure. The introduction of a steel tendon, running along the centroid of the member, tensioned to a force of P and anchored at its ends (Fig. 8.2(b)),

Fig. 8.2 Stress distribution in prestressed concrete beam: (a) stress due to applied load; (b) stress due to prestress; (c) stress due to applied load and prestress

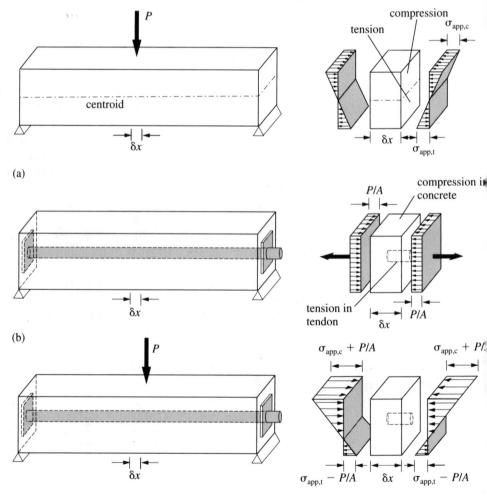

creates the stress distribution illustrated in that figure which, when combined with the stress due to applied loading (Fig. 8.2(a)), gives the distribution illustrated in Fig. 8.2(c). The maximum tensile stress in the concrete is reduced by an amount P/A and the maximum compressive stress is increased by the same amount. By applying a sufficiently large prestress force, P, the tensile stresses can be reduced below the tensile strength of the concrete.

In ways similar to this, the development of cracks in all prestressed concrete members can be controlled or prevented completely. Therefore, prestressed members are stiffer and more durable (i.e. less susceptible to corrosion) than the equivalent reinforced members. However, the magnitude of P is restricted by the need to keep the maximum compressive stress less than the compressive strength of the concrete. For the member of Fig. 8.2, the maximum compressive stress, $\sigma_{app,c} + P/A$, must remain less than the maximum allowable compressive stress of the concrete at all sections.

8.2 Prestressing methods and equipment

Prestress is normally applied to members by steel strands tensioned using hydraulic jacks at one or both ends of the member. The tensioning operation can be performed either: (a) before the concrete is cast, in which case the member is classed as **pre-tensioned**; or (b) after the concrete is cast, in which case the member is classed as **post-tensioned**.

Pre-tensioning

The pre-tensioning process involves three basic stages, each of which is illustrated in Fig. 8.3. In the first stage, the steel strands are placed in a casting bed, stressed to the required level and anchored between two supports (Fig. 8.3(a)). The concrete is then cast around the strands and allowed to set (Fig. 8.3(b)). During this curing stage, the strands bond to the surrounding

Fig. 8.3 Pre-tensioning: (a) stage 1, steel strands are tensioned; (b) stage 2, concrete is cast; (c) stage 3, strands are cut

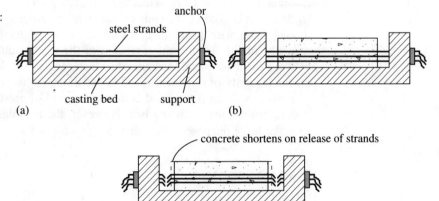

291

concrete. When the concrete has developed sufficient compressive strength, the strands are released from the supports (Fig. 8.3(c)). Immediately after the release, the strands attempt to contract. Owing to their bond with the concrete, this prestress contraction force is transferred to the concrete, thus forcing the concrete into compression.

Pre-tensioning is most commonly employed where many similar precast members are required. It is generally only carried out off site at precasting factories which have permanent casting beds. Thus, the size and weight of pre-tensioned members are limited by the transportation requirements.

Post-tensioning

The post-tensioning process also involves three fundamental stages, which are illustrated in Fig. 8.4 for a simple beam. In the first stage of the process, the concrete is cast around a hollow duct (Fig. 8.4(a)). After the concrete has set,

Fig. 8.4 Post-tensioning: (a) beam is cast; (b) beam is prestressed

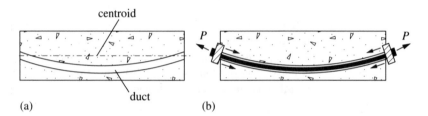

a tendon, consisting of a number of strands, is pushed through the duct (alternatively, the tendon can be placed in the duct before casting). Thus, unlike in pre-tensioned members, the tendon in post-tensioned members can be fixed in any desired linear or curved profile. By varying the eccentricity of the tendon from the centroid, the maximum effectiveness of a constant prestressing force can be utilized by applying the prestress only where it is required. Once the concrete has achieved sufficient strength in compression, the tendon is jacked from one or both ends using hydraulic jacks, thus putting the concrete into compression (Fig. 8.4(b)). When the required level of prestress is achieved, the tendon is anchored at the ends of the member. After anchorage, the ducts are usually filled with grout (a fine cement paste) under pressure. The grout is provided mainly to prevent corrosion of the tendon but it also forms a bond between the tendon and the concrete which reduces the dependence of the beam on the integrity of the anchor and hence improves its robustness.

Post-tensioning is the most common method of prestressing *in situ* because it does not require a casting bed. However, the technique is also used off site to make large purpose-built individual precast units.

Prestressing steel

The steel used for prestressed concrete comes in the form of either cold-drawn high-strength **wire** or high-strength alloy steel **bars**. The use of solid high-yield

Fig. 8.5 Stress/strain relationship for prestressing wire

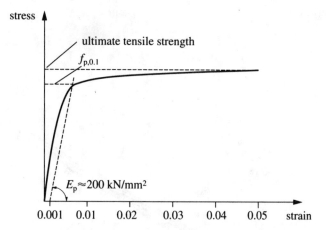

bars, however, is generally limited as they do not have the flexibility to be profiled along the length of the member. High tensile steel wire is by far the more widely used material for both pre-tensioning and post-tensioning.

The short-term stress–strain relationship for a typical wire specimen is illustrated in Fig. 8.5. Failure generally occurs at a strain somewhere between 0.04 and 0.06. The **ultimate characteristic tensile strength**, f_{pk}, of most manu-factured wires is approximately 1800 N/mm² (compared with 460 N/mm² for ordinary high-yield reinforcement) and the elastic modulus may usually be taken to be about 190 000 N/mm². It can be seen from the diagram that there is no definite elastic yield point. For this reason, the concept of **proof stress** is used as an equivalent yield stress. The most frequently used proof stress is the 0.1 per cent proof stress, $f_{p,\,0.1}$, which is the point on the stress–strain curve which intersects with a straight line drawn at an angle equal to the elastic modulus starting from 0.1 per cent strain (see Fig. 8.5).

High-strength steel wire, which comes in a range of diameters from 3 mm to 7 mm, does not generally have sufficient strength to be used singly for prestressing purposes. Thus, for most prestressing applications, several wires are twisted together to form a **strand**. The wires in a strand are spun in a helical form around a central straight wire, as illustrated in Fig. 8.6. Most manu-facturers supply strands made up of seven spun wires, as illustrated in the figure, although five-wire and nineteen-wire strands are also supplied by some manufacturers. The performance characteristics of strands differ slightly to that of the wire from which they are made up owing to the straightening of the spun wires when in tension. The strength properties of typical strands are listed in Table 8.1. In current practise, only the two largest strand sizes listed in the table are commonly used.

In post-tensioned concrete, it is common to group many strands together to form a **cable** or **tendon**, as illustrated in Fig. 8.7. A complete prestressing tendon can be made up of as many strands as are needed to carry the required tension, with all the strands enclosed in a single duct. In addition, large structures may have many individual tendons running parallel to each other along the length of the member.

Most codes of practice place restrictions on the maximum stress which can be applied to strands during the prestressing process. In particular, EC2

Fig. 8.6 Prestressing
strand: (a) elevation;
(b) section through
seven-wire strand
(anchor in background).
Photographs by
B. Dempsey

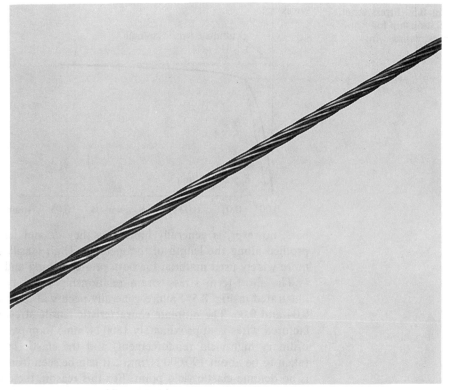

(a)

(b)

Table 8.1 Properties of
a typical seven-wire
strand (adapted from
Lin and Burns (1982))

Nominal diameter (mm)	Nominal area of strand (mm^2)	Breaking strength (kN)	0.1% proof load (kN)
6.35	23.22	40.0	34.0
7.94	37.42	64.5	54.7
9.53	51.61	89.0	75.6
11.11	69.68	120.1	102.3
12.50	93.00	164.0	139.0
15.70	150.00	265.0	225.0

Fig. 8.7 Prestressing
tendon. Photograph
courtesy of CCL Systems

recommends that the stress in the strands during jacking should not exceed the lesser of $0.8f_{pk}$ and $0.9f_{pk, 0.1}$, where f_{pk} is the **characteristic value** of the ultimate strength and $f_{pk, 0.1}$ is the **characteristic value** of the 0.1 per cent proof stress. In addition, EC2 restricts the stress in the tendons immediately after they have been anchored in the concrete (not all the stress applied during jacking remains after anchorage owing to losses – see section 8.6). At this stage, the stress in the tendons should not exceed the lesser of $0.75f_{pk}$ and $0.85f_{pk, 0.1}$.

Prestressing equipment

For both pre-tensioning and post-tensioning of concrete members, specialist equipment is required for stressing the steel and/or anchoring the stressed steel to the concrete. A wide variety of systems has been developed for these purposes, many of which are patented by their manufactuters. A detailed knowledge of each system is not generally necessary for design purposes since all systems achieve the same end result. Having said that, however, a general knowledge is needed so that sufficient space is allowed for anchoring and jacking equipment. For this reason, some of the different systems are presented below.

The tensioning of the steel is usually achieved by mechanical jacking using hydraulic jacks. In pre-tensioning, the jacks pull the steel against the supports of the casting beds. The strands in pre-tensioned members are often stressed individually using small jacks, such as that illustrated in Fig. 8.8(a). In post-tensioning, the jacks pull the steel against the hardening concrete member itself. As the strands are usually grouped in tendons, large multi-strand jacks, such as the one illustrated in Fig. 8.8(b), are often used to tension all the strands in the tendon simultaneously.

In pre-tensioning, anchorage of the strand is provided by the bond between the strand and the hardened concrete. However, before the prestress is transferred to the concrete, temporary anchors are required to hold the ends of the strands while they are being tensioned. One of the most popular methods of anchoring the ends of the strands in the casting bed is the wedge grip (Fig. 8.9). Wedge grips are also used to grip each strand during post-tensioning of a tendon and to hold the strands permanently in the tendon anchor afterwards. Figure 8.10 illustrates one such anchor. The bearing plate on this anchor transmits the force in the strands to the main body of the assembly which in turn transmits the force to the surrounding concrete.

8.3 Basis of design

The design of prestressed members is generally governed by limits on tensile and compressive stress in service rather than by their strength at the ultimate limit state. Normal practice, therefore, is first to produce an initial design (i.e. to choose an appropriate prestress force and tendon location) which satisfies the limitations on service stresses. This design is then checked at the ultimate limit state to ensure that it satisfies strength requirements. If it is found that

Fig. 8.8 Hydraulic
jacks: (a) single-strand
jack; (b) multi-strand
jack. Photographs
courtesy of CCL Systems

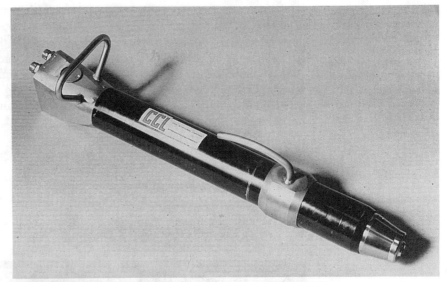

(a)

(b)

Fig. 8.9 Wedge grip anchorage: (a) wedge grip assembly (photograph by B. Dempsey); (b) wedge grips as end anchors in pre-tensioning facility (photograph courtesy of CCL Systems)

(a)

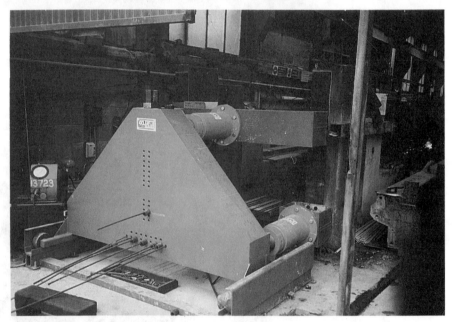

(b)

the member has insufficient strength against bending, ordinary high-yield reinforcing bars can be provided to increase its moment capacity.

Owing to the techniques employed in the construction of prestressed concrete members, two critical loading conditions arise where the stresses in the concrete must be checked against specified permissible values. The first condition, known as the **transfer condition**, occurs immediately on transfer of the prestress force to the concrete. At this stage, the concrete is still relatively young and its compressive strength has not reached its full design value. The stresses which are acting on a member during the transfer condition are prestress and stress due to the moment, M_0, induced by the applied loads which are present at transfer. Often, the only load which is present at transfer is the

Fig. 8.10 Anchorage assembly after casting. Photograph courtesy of CCL Systems

self-weight of the member. Thus, the induced transfer moment, M_0, is often equal to the moment due to member self-weight. The second condition which must be checked is the **service (SLS) condition**. This condition is reached when the concrete has matured to its full strength and the full service loads are being applied to the member. At this stage, the applied prestress force has been reduced from its initial magnitude, P, due to losses which have taken place in the concrete and the steel (prestress losses are discussed in detail in section 8.6). The total loss of prestress force between transfer and service is generally in the region of 10–20 per cent. The stresses which are acting on the member during the service condition are prestress and stress due to the moment, M_s, induced by all applied permanent and variable loads (including self-weight).

Concrete stress limits

In the design of prestressed members, limits are generally placed on the tensile and compressive stresses at transfer and service to ensure that the member remains serviceable. Tensile stress limits may be imposed to ensure that cracking does not occur in the member while compressive stress limits may be imposed to control microcracking and to prevent excessive loss of prestress due to creep. Unlike BS8110, EC2 does not lay down compulsory permissible stresses and the choice of concrete stress limits is left to the discretion of the designer. The advantage of this approach is that it gives the designer freedom to select concrete stress limits, if any, which he/she feels are most appropriate for a particular structure.

To prevent the occurrence of large deformations and substantial losses in prestress due to creep of the concrete, EC2 does, however, recommend that compressive stresses should not be allowed to exceed $0.45f_{ck,0}$ at transfer and $0.45f_{ck}$ at service, where $f_{ck,0}$ and f_{ck} are the characteristic compressive cylinder strengths of the concrete at transfer and service, respectively.

For the design of **fully prestressed members**, in which no cracking of the member is permitted, the tensile stress may be limited to the tensile strength of the concrete. Some designers will further restrict the design by not allowing any tensile stresses whatsoever to develop in the member when designing for the service condition and allow only very small tensile stresses to develop when designing for the transfer condition.

In the case of **partially prestressed members,** in which cracks are allowed to form at service, stress limits are not usually specified directly for the service condition. Instead, limits are placed on the width to which cracks are allowed to open. For this reason it has been found that it is often easier initially to design partially prestressed members for strength at the ultimate limit state and then to check that the transfer and service conditions are satisfied.

The design of partially prestressed members is discussed in more detail in section 8.9. The remainder of this section and sections 8.4 and 8.5 deal exclusively with fully prestressed members where stress limits are specified.

Example 8.1   Simple beam with tendon at centroid

Problem   The simply supported beam of Fig. 8.11 is post-tensioned by applying a prestress force of 1900 kN at the centre of the section. The permissible stresses are (compression positive):

Compression: $0.45f_{ck,0}$ at transfer
$0.45f_{ck}$ at service

Tension: $-1\,\text{N/mm}^2$ at transfer
$0\,\text{N/mm}^2$ at service

The concrete is designed to achieve a compressive cylinder strength of $35\,\text{N/mm}^2$ at transfer and $50\,\text{N/mm}^2$ at service. The permanent gravity load (excluding self-weight) is 8 kN/m of which 4 kN/m is present at transfer. In addition, there is a variable gravity load of 15 kN/m. Assuming a 20 per cent

Fig. 8.11 Beam of Example 8.1

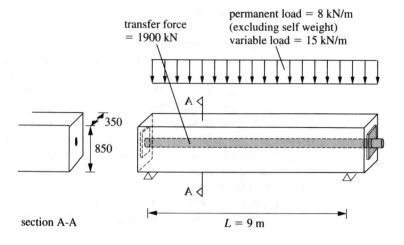

section A-A

$L = 9\,\text{m}$

loss in prestress between transfer and service, check if the level of prestress applied satisfies the given stress limits at mid-span.

Solution Transfer condition check

At transfer, self-weight and 4 kN/m of permanent gravity load are present. The cross-sectional area of the beam is given by:

$$A = 350 \times 850 = 297\,500 \text{ mm}^2$$

Hence, assuming concrete density of 25 kN/m^3 the applied load at transfer is:

$$q_0 = 25(0.298) + 4 = 11.44 \text{ kN/m}$$

from which the transfer moment is:

$$M_0 = q_0 L^2/8 = 11.44(9)^2/8 = 116 \text{ kNm}$$

The elastic bending stress, σ, caused by an applied moment, M, is given by:

$$\sigma(y) = My/\text{I} \tag{8.1}$$

where y is the distance from the centroid to the point where the stress is required and I is the second moment of area of the section. The stresses at the extreme top fibre ($y = y_t$) and the extreme bottom fibre ($y = y_b$) are given by:

$$\sigma(\text{top}) = \frac{M_0 y_t}{I} = \frac{M_0}{Z_t} \tag{8.2}$$

and:

$$\sigma(\text{bottom}) = \frac{M_0 y_b}{I} = \frac{M_0}{Z_t} \tag{8.3}$$

where Z_t ($=I/y_t$) is the section modulus for the top of the section and Z_b ($=I/y_b$) is the section modulus for the bottom of the section. Note that Z_b is negative since y is negative below the centroid. Hence, the stress at the extreme bottom fibre is negative indicating that it is tensile as expected. For this example, the magnitudes of the stresses in the extreme fibres due to M_0 are equal since $y_t = -y_b$.

For uncracked prestressed concrete members, it has been found that sufficient accuracy in calculating bending stresses is achieved using the gross second moment of area, I_g, for I in equation (8.1). The gross second moment of area of the member of Fig. 8.11 is:

$$I_g = \frac{bh^3}{12} = \frac{350 \times (850)^3}{12}$$
$$= 17.91 \times 10^9 \text{ mm}^4$$

and the section moduli are:

$$Z_t = -Z_b = \frac{17.91 \times 10^9}{425} = 42.15 \times 10^6 \text{ mm}^3$$

The stresses in the concrete at transfer are due to the prestress force plus the transfer moment, M_0. In this example, the prestress force at transfer exerts a uniform axial stress of P/A. Hence, at the extreme top and bottom fibres, the stress due to prestress is:

$$\sigma_t = \sigma_b = \frac{P}{A} = \frac{1900 \times 10^3}{297\,500}$$

$$= 6.39 \text{ N/mm}^2$$

At the extreme top fibre, there is a compressive stress due to M_0 of:

$$\frac{M_0}{Z_t} = \frac{116 \times 10^6}{42.15 \times 10^6} = 2.75 \text{ N/mm}^2$$

Similarly, the stress at the extreme bottom fibre due to M_0 is -2.75 N/mm^2.

The stress distributions due to prestress and applied transfer moment are summarized in Fig. 8.12. It can be seen from the figure that the total stress is compressive throughout the section with a maximum value of 9.14 N/mm^2 at the top fibre. The maximum permissible compressive stress at transfer is $0.45 f_{ck,\,0} = 0.45(35) = 15.75$ N/mm^2. Hence, the amount of prestress applied satisfies the stress requirements at transfer.

Fig. 8.12 Distributions of stress at transfer: (a) prestress; (b) applied loads; (c) total

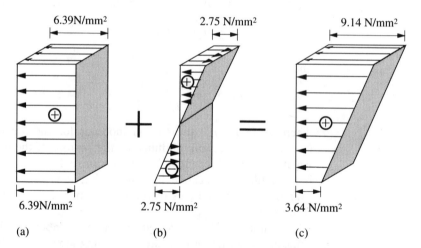

6.39N/mm²	2.75 N/mm²	9.14 N/mm²
6.39N/mm²	2.75 N/mm²	3.64 N/mm²
(a)	(b)	(c)

Service condition check

The stresses in the concrete at service are due to the prestress force (after all losses have occurred) plus the service moment. At service, the full permanent and variable gravity loads apply (including load present at transfer). Hence, the applied service loading is:

$$q_s = 25(0.298) + 8 + 15 = 30.44 \text{ kN/m}$$

from which the service moment at mid-span is:

$$M_s = q_s L^2/8 = 30.44(9)^2/8 = 308 \text{ kNm}$$

The extreme fibre stresses due to prestress are reduced from their magnitudes at transfer by 20 per cent and are given by:

$$\sigma_t = \sigma_b = 0.8 \frac{P}{A} = 0.8 \frac{1900 \times 10^3}{297\,500}$$

$$= 5.11 \text{ N/mm}^2$$

Hence, the total stress at the extreme top fibre is:

$$\sigma_t + \frac{M_s}{Z_t} = 5.11 + \frac{308 \times 10^6}{42.15 \times 10^6} = 12.42 \text{ N/mm}^2$$

and the total stress at the extreme bottom fibre is:

$$\sigma_b + \frac{M_s}{Z_b} = 5.11 + \frac{308 \times 10^6}{-42.15 \times 10^6} = -2.20 \text{ N/mm}^2$$

The total stress distribution due to the service loads and prestress is summarized in Fig. 8.13. The maximum permissible compressive stress at service is $0.45 f_{ck} = 0.45(50) = 22.5 \text{ N/mm}^2$ which is satisfactory. However, there is a

Fig. 8.13 Distributions of stress at service: (a) prestress; (b) applied loads; (c) total

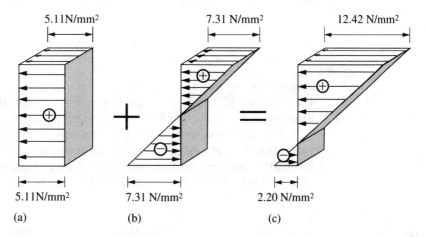

5.11N/mm²   7.31 N/mm²   12.42 N/mm²

5.11N/mm²   7.31 N/mm²   2.20 N/mm²

(a)   (b)   (c)

tensile stress of 2.20 N/mm². Hence, the amount of prestress supplied fails to satisfy the stress requirements at service and the design must be revised.

Effect of tendon location

It was seen in Example 8.1 that a tendon placed at the centroid of a member exerting a force, P, on the concrete creates a uniform distribution of stress, equal to P/A, across the section. If, instead, a straight tendon is located at an eccentricity e above the centroid of the member, as illustrated in Fig. 8.14, an eccentric force is applied to the concrete. The application of this eccentric force, P, is equivalent to applying a concentric axial force, P, and a sag bending

Fig. 8.14 Eccentric straight tendon

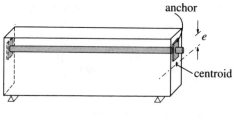

Fig. 8.15 Eccentric prestress: (a) actual prestress force; (b) equivalent force and moment at centroid

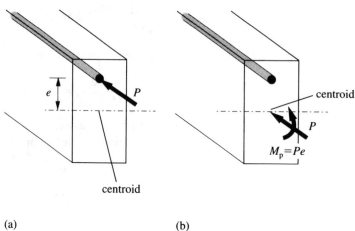

(a)   (b)

moment, M_p, at any given section as illustrated in Fig. 8.15. For a determinate beam such as this, the moment due to prestress is simply the product of prestress force and eccentricity, that is $M_p = Pe$ (for indeterminate beams, refer to section 8.7). The axial force component creates a uniform axial stress distribution, of magnitude P/A, at any given section, as illustrated in Fig. 8.16(a). The bending component, however, creates a triangular stress distribution at any given section (Fig. 8.16(b)). The stress at the extreme fibres can be determined using equations

Fig. 8.16 Stress distributions due to eccentric prestress: (a) axial component; (b) bending component; (c) total

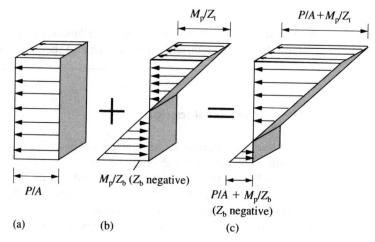

(a)   (b)   (c)

(8.2) and (8.3). Hence, the total stress due to prestress at the top fibre is:

$$\sigma_t = \frac{P}{A} + \frac{M_p}{Z_t} \qquad (8.4)$$

Similarly, at the bottom fibre:

$$\sigma_b = \frac{P}{A} + \frac{M_p}{Z_b} \qquad (8.5)$$

where Z_b is negative.

The overall stress distribution due to the axial component and the bending component of the prestress is illustrated in Fig. 8.16(c).

Example 8.2   **Simple beam with eccentric tendon**

Problem   The simply supported member of Fig. 8.11 is now post-tensioned by applying an eccentric prestress force of 1500 kN at 150 mm below the centroid of the member (i.e. $e = -150$). The prestress loss, the design concrete strengths and the applied loading are the same as for Example 8.1. Check if the stresses at mid-span are within the limits given in Example 8.1.

Solution   As before, the adequacy of the member must be checked at both the transfer condition and at the service condition.

Transfer condition check

As the beam is determinate, $M_p = Pe$, and the stress at the extreme top fibre due to prestress is:

$$\sigma_t = \frac{P}{A} + \frac{Pe}{Z_t}$$

$$= \frac{1500 \times 10^3}{297\,500} + \frac{(1500 \times 10^3)(-150)}{42.15 \times 10^6}$$

$$= 5.04 - 5.34$$

$$= -0.30 \text{ N/mm}^2$$

At the bottom fibre, the stress due to prestress is:

$$\sigma_b = \frac{P}{A} + \frac{Pe}{Z_b}$$

$$= \frac{1500 \times 10^3}{297\,500} + \frac{(1500 \times 10^3)(-150)}{-42.15 \times 10^6}$$

$$= 5.04 + 5.34$$

$$= 10.38 \text{ N/mm}^2$$

From Example 8.1, the stress at the extreme top and bottom fibres due to the applied loads at transfer is ± 2.75 N/mm². The total stress distribution at

Fig. 8.17 Transfer stress distributions at mid-span: (a) axial component of prestress; (b) bending component of prestress; (c) applied loads; (d) total

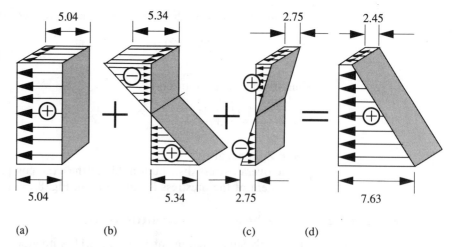

(a)   (b)   (c)   (d)

mid-span is illustrated in Fig. 8.17. It can be seen that this distribution complies with the specified stress limits.

Service condition check

As before, the stress at the extreme fibres due to the applied service loads is $\pm 7.31 \, \text{N/mm}^2$. Owing to losses, the stress due to the axial and bending components of the prestress is reduced to 80 per cent of its value at transfer. The total stress distribution at mid-span is illustrated in Fig. 8.18. It can be seen that, unlike the member of Example 8.1, the service stress limits are satisfied for this example, even with a reduction in the prestress force.

Fig. 8.18 Service stress distributions at mid-span: (a) axial component of prestress; (b) bending component of prestress; (c) applied loads; (d) total

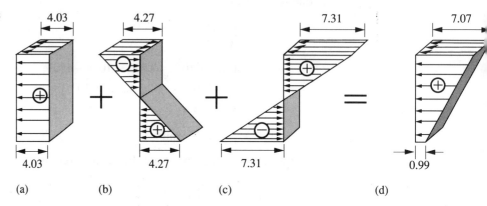

(a)   (b)   (c)   (d)

Thus, the advantage of prestressing using eccentric tendons is that not only does it reduce the tensile stresses due to the applied loads but it also prevents the compressive stresses from becoming excessively large.

Example 8.3   Simple beam with eccentric tendon

Problem   For the beam of Example 8.2, check if the stresses at the sections over the supports are within the limits specified in Example 8.1.

Fig. 8.19 Stress distribution at supports: (a) transfer stresses; (b) service stresses

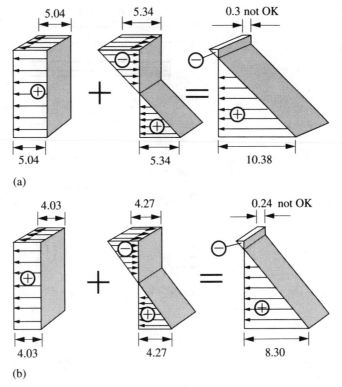

(a)

(b)

Solution   Consider the stresses at a section over one of the supports. There is no stress at this section due to the applied loads at either transfer or service, that is the bending moments M_0 and M_s are zero. However, the eccentric prestress force does cause bending at this section. The total stress distributions at transfer and service are illustrated in Fig. 8.19. It can be seen from the figure that a tensile stress is developed in the extreme top fibre at service. Therefore, this member fails to comply with the specified stress limits over the support.

Profiled tendons and debonding

In order to take advantage of the added effectiveness of eccentric prestress forces in post-tensioned members, while at the same time avoiding problems similar to that of Example 8.3 above, tendons are generally **draped** or **profiled**. For example, an appropriate tendon profile for the member of Fig. 8.11 is illustrated in Fig. 8.20. The maximum eccentricity of the tendon occurs at mid-span, where the tensile stress due to the applied loads is greatest. Over the supports the tendon is located at the centroid and so the prestress force is purely concentric

Fig. 8.20 Draped tendon

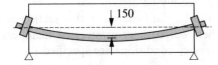

Fig. 8.21 Debonded strands: (a) elevation; (b) section A–A, stress distribution when one group of strands is bonded; (c) section B–B, stress distribution when both groups of strands are bonded

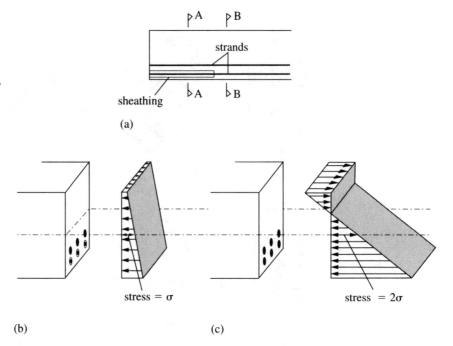

(a)

(b) stress = σ

(c) stress = 2σ

and does not generate tensile stresses over the support. For pre-tensioned members, a technique known as **debonding** is often used in practice. In debonding, a plastic duct or tape is placed around some strands near the supports, as illustrated in Fig. 8.21, to prevent them from bonding with the concrete. Thus, these strands are ineffective over the length of the duct – no prestress is applied by them. This reduction in the prestress force can prevent problems of excessive prestress, as illustrated in Example 8.3.

8.4 Minimum section moduli

In the design of prestressed members, it is often necessary to choose an initial trial section in order to estimate the load due to self-weight. The more accurate this initial section size is, the fewer iterations of the design process will be required later. In such cases, a reliable preliminary section can often be determined by a consideration of the stress requirements at transfer and service. In this section, expressions are derived in terms of the minimum values for the section moduli, Z_t and Z_b, which are required to satisfy the stress limits at critical sections. From these minimum section moduli, the minimum dimensions of the member can readily be determined.

As stated previously, the material in this section is only applicable to fully prestressed members or partially prestressed members where stress limits for the transfer and service conditions are specified.

Sign conventions

The sign conventions used in this and subsequent sections are as follows:

(a) Applied sag moments are considered positive and applied hog moments are negative.
(b) The eccentricity, e, of prestress tendons above the centroid of the member is positive and below the centroid of the member is negative.
(c) Compressive stresses are considered positive and tensile stresses are considered negative.
(d) The distance from the centroid is positive upwards. Hence, the numerical value for Z_b $(= I/y_b)$ is always negative.

These sign conventions must be observed whenever using the expressions derived below.

Section modulus requirements

Let p_{0min} and p_{0max} be the minimum and maximum permissible stresses at transfer, respectively. Similarly, let p_{Smin} and p_{Smax} denote the minimum and maximum permissible stresses at service, respectively. Hence, for the transfer condition:

$$p_{0min} \leq \sigma_t + \frac{M_0}{Z_t} \leq p_{0max} \tag{8.6}$$

and:

$$p_{0min} \leq \sigma_b + \frac{M_0}{Z_b} \leq p_{0max} \tag{8.7}$$

where σ_t and σ_b are the stresses at the extreme top and bottom fibres due to the prestress at transfer and M_0 is the transfer moment.

At service, the transfer prestress, σ_t, drops to $\rho\sigma_t$ owing to the losses which occur between transfer and service. The factor ρ is equal to the ratio of the prestress force at service to the prestress force at transfer (typically $\rho = 0.75$ to 0.90). Thus, for the service condition:

$$p_{Smin} \leq \rho\sigma_t + \frac{M_s}{Z_t} \leq p_{Smax} \tag{8.8}$$

and:

$$p_{Smin} \leq \rho\sigma_b + \frac{M_s}{Z_b} \leq p_{Smax} \tag{8.9}$$

where M_s is the total applied moment at service.
By rearranging inequalities (8.6) to (8.9), the limits on the stresses due to

prestress in the extreme top and bottom fibres are:

$$
\left(
\begin{array}{c}
P_{0min} - \dfrac{M_0}{Z_t} \\[2ex]
\dfrac{1}{\rho}\left(P_{Smin} - \dfrac{M_s}{Z_t}\right)
\end{array}
\right)
\leq \sigma_t \leq
\left(
\begin{array}{c}
P_{0max} - \dfrac{M_0}{Z_t} \\[2ex]
\dfrac{1}{\rho}\left(P_{Smax} - \dfrac{M_s}{Z_t}\right)
\end{array}
\right)
\tag{8.10}
$$

$$
\left(
\begin{array}{c}
P_{0min} - \dfrac{M_0}{Z_b} \\[2ex]
\dfrac{1}{\rho}\left(P_{Smin} - \dfrac{M_s}{Z_b}\right)
\end{array}
\right)
\leq \sigma_b \leq
\left(
\begin{array}{c}
P_{0max} - \dfrac{M_0}{Z_b} \\[2ex]
\dfrac{1}{\rho}\left(P_{Smax} - \dfrac{M_s}{Z_b}\right)
\end{array}
\right)
\tag{8.11}
$$

Sometimes a valid solution does not exist for the specified stress limits; that is, in some cases the lower limit on σ_t or σ_b may exceed the upper limit.

Top fibre

In the case of the stress due to prestress at the extreme top fibre (inequality (8.10)), the two lower limits must be less than each of the upper limits. Clearly, $P_{0min} < P_{0max}$ and $P_{Smin} < P_{Smax}$. However, in order for a valid solution to exist, the following two inequalities must also be satisfied:

$$
P_{0min} - \frac{M_0}{Z_t} \leq \frac{1}{\rho}\left(P_{Smax} - \frac{M_s}{Z_t}\right)
$$

and:

$$
\frac{1}{\rho}\left(P_{Smin} - \frac{M_s}{Z_t}\right) \leq P_{0max} - \frac{M_0}{Z_t}
$$

From these inequalities we get, respectively:

$$
\frac{1}{Z_t}\left(\frac{M_s}{\rho} - M_0\right) \leq \frac{P_{Smax}}{\rho} - P_{0min}
$$

and:

$$
\frac{1}{Z_t}\left(M_0 - \frac{M_s}{\rho}\right) \leq P_{0max} - \frac{P_{Smin}}{\rho}
$$

Hence:

$$
Z_t \geq \frac{M_s - \rho M_0}{P_{Smax} - \rho P_{0min}}
\tag{8.12}
$$

and:

$$
Z_t \geq \frac{\rho M_0 - M_s}{\rho P_{0max} - P_{Smin}}
\tag{8.13}
$$

Inequalities (8.12) and (8.13) represent two lower limits on the elastic section modulus. If these are not satisfied, then no level of prestress can result in a satisfactory design at the top fibre. Therefore, the greater of the two values for Z_t from the two expressions should be used to determine an initial section size. It follows from the sign convention that, in general, inequality (8.12) is more stringent for sections in sag (i.e. M_0 and M_s are positive) and inequality (8.13) is more stringent for sections in hog.

Bottom fibre

In the case of the stress due to prestress at the extreme bottom fibre (inequality (8.11)), a valid solution exists only if the following two inequalities are satisfied:

$$p_{0min} - \frac{M_0}{Z_b} \leq \frac{1}{\rho}\left(p_{Smax} - \frac{M_s}{Z_b}\right)$$

and:

$$\frac{1}{\rho}\left(p_{Smin} - \frac{M_s}{Z_b}\right) \leq p_{0max} - \frac{M_0}{Z_b}$$

Rearranging these inequalities we get:

$$\frac{1}{Z_b}\left(\frac{M_s}{\rho} - M_0\right) \leq \frac{p_{Smax}}{\rho} - p_{0min}$$

and:

$$\frac{1}{Z_b}\left(M_0 - \frac{M_s}{\rho}\right) \leq p_{0max} - \frac{p_{Smin}}{\rho}$$

As Z_b is negative and since multiplying both sides of an inequality by a negative number changes its sign, these inequalities become:

$$Z_b \leq \frac{M_s - \rho M_0}{p_{Smax} - \rho p_{0min}} \tag{8.14}$$

and:

$$Z_b \leq \frac{\rho M_0 - M_s}{\rho p_{0max} - p_{Smin}} \tag{8.15}$$

Thus, Z_b must be more negative than the right-hand sides of inequalities (8.14) and (8.15) or:

$$|Z_b| \geq \frac{\rho M_0 - M_s}{p_{Smax} - \rho p_{0min}} \tag{8.16}$$

and:

$$|Z_b| \geq \frac{M_s - \rho M_0}{\rho p_{0max} - p_{Smin}} \tag{8.17}$$

Inequalities (8.16) and (8.17) represent the minimum required absolute value for the elastic section modulus for the extreme bottom fibre. As for Z_t, the greater value from the two expressions should be used to determine an initial section size. It has been found that, in general, inequality (8.17) is more stringent for sections in sag (i.e. M_0 and M_s are positive) and inequality (8.16) is more stringent for sections in hog. When using inequalities (8.12), (8.13), (8.16) and (8.17) for the preliminary design of prestressed concrete flexural members, it should be remembered that satisfaction of the inequalities only guarantees that there exists a level of prestress which would result in all stress limits being satisfied. There are sometimes situations in practice where it is not possible to find a practical arrangement of strands that achieves the required level of prestress. Nevertheless, these inequalities are, for most design situations, an extremely useful aid in the preliminary sizing of prestressed concrete members.

Example 8.4   Preliminary sizing of post-tensioned beam

Problem   The single-span, simply supported post-tensioned beam of Fig. 8.22 carries a characteristic permanent gravity load of 5 kN/m (not including self-weight) and a characteristic variable gravity load of 20 kN/m, as illustrated in the figure. Given that $\rho = 0.75$, $p_{0min} = -2.5 \text{ N/mm}^2$, $p_{Smin} = -2.0 \text{ N/mm}^2$ and $p_{0max} = p_{Smax} = 20 \text{ N/mm}^2$, determine an appropriate rectangular section for the member (self-weight of concrete = 25 kN/m^3).

Fig. 8.22   Post-tensioned beam of Example 8.4

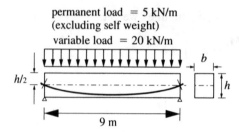

permanent load = 5 kN/m
(excluding self weight)
variable load = 20 kN/m

$h/2$

b

h

9 m

Solution   The minimum section requirements for this beam are calculated for the critical section at the mid-point of the member (i.e. where the bending moment is largest). In order to determine the applied bending moments, a first estimate of the self-weight must be made. With reference to Table 6.5, a span/depth ratio of 1:15 is adopted implying a depth of 9000/15 = 600 mm. The minimum sensible section breadth is 250 mm. Therefore, assume a trial section having breadth 250 mm and depth 600 mm. For this section:

$$\text{self-weight, } q_{sw} = 25 \times (0.25 \times 0.60)$$
$$= 3.75 \text{ kN/m}$$

It is assumed that only the self-weight portion of the permanent gravity load will be present at transfer. Thus, M_0, at mid-span, is:

$$M_0 = q_{sw}L^2/8 = 3.75(9)^2/8 = 38 \text{ kNm}$$

The total bending moment, M_s, at mid-span due to the combined action of all service loads is:

$$M_s = q_{tot}L^2/8 = [(3.75 + 5 + 20)(9)^2]/8$$
$$= 291 \text{ kNm}$$

From inequality (8.12):

$$Z_t \geq \frac{M_s - \rho M_0}{p_{Smax} - \rho p_{0min}} \geq \frac{291 - 0.75(38)}{20 - 0.75(-2.5)} \times 10^6$$
$$\Rightarrow \qquad Z_t \geq 12.0 \times 10^6 \text{ mm}^3$$

From equation (8.17):

$$|Z_b| \geq \frac{M_s - \rho M_0}{\rho p_{0max} - p_{Smin}} \geq \frac{291 - 0.75(38)}{0.75(20) + 2.0} \times 10^6$$

$$\Rightarrow \quad |Z_b| \geq 15.44 \times 10^6 \text{ mm}^3$$

Since the section is to be rectangular, the magnitude of Z_t and Z_b will be equal. Therefore, we use the larger of the limits, namely that for Z_b, to calculate the section dimensions:

$$|Z_b| = |Z_t| = \frac{I_g}{y} = \frac{bh^3/12}{h/2} = \frac{bh^2}{6}$$

Hence:

$$bh^2/6 \geq 15.44 \times 10^6 \text{ mm}^3$$

Keeping with the initial assumption, let $b = 250$ mm. Hence:

$$250h^2/6 \geq 15.44 \times 10^6 \text{ mm}^3$$

$$\Rightarrow \quad h \geq 609 \text{ mm}$$

Therefore, adopt a rectangular section, 250 mm × 650 mm. This is greater than the original section estimate and hence actual self-weight is greater than calculated. It is worth noting that self-weight does not significantly affect the required section modulus (as higher self-weight allows a higher level of prestress to be applied without violation of the stress limits at transfer). Thus, a recalculation with the revised self-weight loading will give an almost identical result.

8.5 Prestressing force and eccentricity

Once a cross-section has been chosen for a prestressed member which satisfies inequalities (8.12), (8.13), (8.16) and (8.17), the next step is to determine how much prestress force to apply, and where to locate it, at each section along the length of the member. It was shown in section 8.3 that prestress can be applied most efficiently by placing the tendons at an eccentricity and, in the case of post-tensioned members, by varying the eccentricity of the tendons along the length of the member. For a statically determinate member with an eccentric prestress force, the stresses due to prestress in the extreme fibres are given by:

$$\sigma_t = \frac{P}{A} + \frac{Pe}{Z_t} \tag{8.18}$$

$$\sigma_b = \frac{P}{A} + \frac{Pe}{Z_b} \tag{8.19}$$

where P is the prestress force at transfer. It can be seen from these expressions that the stress levels depend on the magnitudes of both P and e, that is a large stress in the extreme top fibre can be achieved using a large prestressing force and small eccentricity or vice versa. Thus, the process of choosing an appropriate prestress force, P, and tendon location (i.e. eccentricity, e) can prove to be quite difficult.

In this section a method is presented whereby the most appropriate value of P and e can be chosen for each critical section with relative ease. However, the method uses expressions based on inequalities (8.10) and (8.11) and therefore can only be applied to fully prestressed members or partially prestressed members where stress limits are specified in place of allowable crack widths.

The stress due to prestress (given by equations (8.18) and (8.19)) is limited by inequalities (8.10) and (8.11), derived in section 8.4. Thus, for the extreme top fibre:

$$\sigma_{tmin} \leq \sigma_t \leq \sigma_{tmax} \tag{8.20}$$

where:

$$\sigma_{tmin} = \text{greater of } \left(p_{0min} - \frac{M_0}{Z_t} \right) \text{ and } \frac{1}{\rho}\left(p_{Smin} - \frac{M_s}{Z_t} \right)$$

and:

$$\sigma_{tmax} = \text{lesser of } \left(p_{0max} - \frac{M_0}{Z_t} \right) \text{ and } \frac{1}{\rho}\left(p_{Smax} - \frac{M_s}{Z_t} \right)$$

Similarly, for the extreme bottom fibre:

$$\sigma_{bmin} \leq \sigma_b \leq \sigma_{bmax} \tag{8.21}$$

where:

$$\sigma_{bmin} = \text{greater of } \left(p_{0min} - \frac{M_0}{Z_b} \right) \text{ and } \frac{1}{\rho}\left(p_{Smin} - \frac{M_s}{Z_b} \right)$$

and:

$$\sigma_{bmax} = \text{lesser of } \left(p_{0max} - \frac{M_0}{Z_b} \right) \text{ and } \frac{1}{\rho}\left(p_{Smax} - \frac{M_s}{Z_b} \right)$$

Substitution from equations (8.18) and (8.19) for σ_t and σ_b in the above inequalities yields:

$$\sigma_{tmin} \leq \frac{P}{A} + \frac{Pe}{Z_t} \leq \sigma_{tmax} \tag{8.22}$$

and:

$$\sigma_{bmin} \leq \frac{P}{A} + \frac{Pe}{Z_b} \leq \sigma_{bmax} \tag{8.23}$$

Dividing inequalities (8.22) and (8.23) by P then gives:

$$\frac{1}{P}\sigma_{tmin} \leq \frac{1}{A} + \frac{e}{Z_t} \leq \frac{1}{P}\sigma_{tmax} \tag{8.24}$$

and:

$$\frac{1}{P}\sigma_{bmin} \leq \frac{1}{A} + \frac{e}{Z_b} \leq \frac{1}{P}\sigma_{bmax} \tag{8.25}$$

Inequalities (8.24) and (8.25) can be subdivided into the following four expressions:

$$\frac{1}{P}\sigma_{tmin} \leq \frac{1}{A} + \frac{e}{Z_t} \tag{8.26}$$

$$\frac{1}{A} + \frac{e}{Z_t} \leq \frac{1}{P}\sigma_{tmax} \tag{8.27}$$

$$\frac{1}{P}\sigma_{bmin} \leq \frac{1}{A} + \frac{e}{Z_b} \tag{8.28}$$

$$\frac{1}{A} + \frac{e}{Z_b} \leq \frac{1}{P}\sigma_{bmax} \tag{8.29}$$

Each of the inequalities (8.26) to (8.29) represents a linear relationship between e and $1/P$ which can be illustrated on a plot of e versus $1/P$. This form of graphical representation of the limits on the prestress force and eccentricity is attributed to Magnel and hence it is commonly known as a **Magnel diagram**. The point where the lines cross the e-axis can be found by setting $1/P$ equal to zero in each inequality. Similarly, the point where each line crosses the $1/P$-axis is found by setting e equal to zero in each inequality. Knowing these two intersection points for each line, the lines can readily be drawn on the plot.

The plot of Fig. 8.23(a) illustrates the line represented by inequality (8.26). In fact, the inequality represents a half-plane bounded by the line on which the stress limits are just satisfied, as illustrated in Fig. 8.23(b). To determine which half-plane represents the inequality, the origin is substituted into the inequality. In this case, substitution of $1/P = 0$ and $e = 0$ into inequality (8.26) gives:

$$0 \leq 1/A + 0$$

As this is true (A is positive), the correct half-plane is the one containing the origin. Figures 8.23(c)–(e) illustrate the effect of introducing the half-planes representing each of the three remaining inequalities. The shaded area in the resulting plot, Fig. 8.23(e), is the zone in which all the stress limits are satisfied and is called the **feasible zone**. Therefore, any combination of $1/P$ and e that falls within the feasible zone constitutes a valid solution. Usually, to save on prestressing steel, a solution is chosen which corresponds to a low value of P but sufficiently away from the edge of the feasible zone to allow some latitude in tendon eccentricity (see Fig. 8.23(e)).

It is often helpful to draw in two further boundary lines on the Magnel diagram which represent the maximum eccentricities which are physically possible. These lines, which may be more stringent than the inequalities of the feasible zone, ensure that any chosen value of e is physically within the section at all points and does not violate the requirements for cover. The lines, known as **physical limits**, are horizontal on a plot of e versus $1/P$, as illustrated in Fig. 8.24. To ensure adequate corrosion resistance EC2 recommends that, for pre-tensioned members, the cover should not be less than twice the diameter of the strand, while for post-tensioned members the cover should not be less than the diameter of the duct. In addition, the cover should not be less than the values specified in Table 8.2.

Fig. 8.23 Construction of Magnel diagram

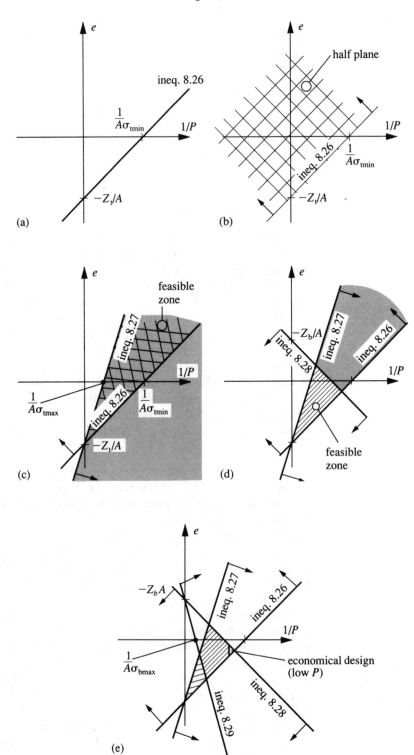

Fig. 8.24 Magnel diagram with physical limits: (a) cross-section; (b) Magnel diagram

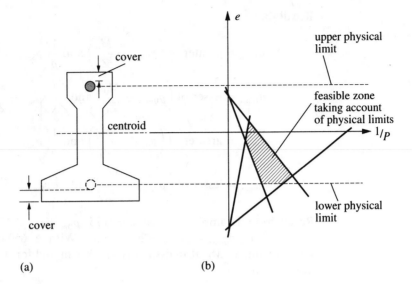

(a)   (b)

Table 8.2 Minimum cover requirements for prestressed concrete. Exposure classes correspond to those of Table 6.2 (from EC2)

Exposure class	Minimum cover (mm)
1	25
2a	30
2b	35
3	50
4a	50
4b	50
5a	35
5b	40
5c	50

Example 8.5   Prestress force and eccentricity

Problem   Using the stress limits given in Example 8.4, plot the Magnel diagram for the critical section at mid-span for the member of Fig. 8.22 and choose an appropriate prestress force at transfer (use the section size calculated in Example 8.4). Provide for a minimum of 100 mm from the centre of the tendon to the surface of the beam.

Solution   From Example 8.4, a rectangular section, 250 mm wide by 650 mm deep, is adequate for the applied loads. For this section:

$$A = 250 \times 650 = 162\,500 \text{ mm}^2$$

and:

$$Z = Z_t = -Z_b = bh^2/6$$
$$= 250 \times (650)^2/6 = 17.60 \times 10^6 \text{ mm}^3$$

317

Recall that:

$$\sigma_{tmin} = \text{greater of}\left(p_{0min} - \frac{M_0}{Z_t}\right) \text{ and } \frac{1}{\rho}\left(p_{Smin} - \frac{M_s}{Z_t}\right)$$

$$\sigma_{tmax} = \text{lesser of}\left(p_{0max} - \frac{M_0}{Z_t}\right) \text{ and } \frac{1}{\rho}\left(p_{Smax} - \frac{M_s}{Z_t}\right)$$

$$\sigma_{bmin} = \text{greater of}\left(p_{0min} - \frac{M_0}{Z_b}\right) \text{ and } \frac{1}{\rho}\left(p_{Smin} - \frac{M_s}{Z_b}\right)$$

$$\sigma_{bmax} = \text{lesser of}\left(p_{0max} - \frac{M_0}{Z_b}\right) \text{ and } \frac{1}{\rho}\left(p_{Smax} - \frac{M_s}{Z_b}\right)$$

Recall from Example 8.4 that $\rho = 0.75$, $p_{0min} = -2.5 \text{ N/mm}^2$, $p_{Smin} = -2.0$ N/mm^2 and $p_{0max} = p_{Smax} = 20 \text{ N/mm}^2$. With a 650 mm deep section the revised value of M_0 at mid-span is 41.1 kN m and for M_s is 294 kN m. Hence, at mid-span:

$$\left(p_{0min} - \frac{M_0}{Z_t}\right) = \left(-2.5 - \frac{41.1}{17.60}\right) = -4.84$$

$$\frac{1}{\rho}\left(p_{Smin} - \frac{M_s}{Z_t}\right) = \frac{1}{0.75}\left(-2.0 - \frac{294}{17.60}\right) = -24.95$$

Therefore:

$$\sigma_{tmin} = \text{greater}(-4.84, -24.95) = -4.84 \text{ N/mm}^2$$

Similarly, σ_{tmax}, σ_{bmin} and σ_{bmax} are calculated to be:

$$\sigma_{tmax} = 4.39 \text{ N/mm}^2$$

$$\sigma_{bmin} = 19.61 \text{ N/mm}^2$$

$$\sigma_{bmax} = 22.33 \text{ N/mm}^2$$

Substitution for Z and these stress limits into the general inequalities (8.26) to (8.29) yields the Magnel diagram for the critical section at mid-span illustrated in Fig. 8.25. On this plot, the physical limits are drawn in at a distance (375–100) mm on each side of the origin (which is at the centre of the section). From the plot, it can be seen that the maximum value for $1/P$ in the feasible zone, $(1/P)_{max}$, is approximately equal to $83 \times 10^{-8} \text{ N}^{-1}$ (i.e. $P_{min} = 1200$ kN) and this corresponds to an eccentricity of approximately -180 mm (which is within the physical limits of the member). To allow for some latitude in eccentricity, a higher value for P is chosen, say $P = 1300$ kN. These combinations of P and e can be substituted into equations (8.18) and (8.19) to determine the stresses due to prestress. These can in turn be substituted into inequalities (8.10) and (8.11) to check that the stresses are indeed within the specified limits.

A Magnel diagram can be plotted as e versus P rather than e versus $1/P$. The e–P plot for this example is illustrated in Fig. 8.26. Notice that, unlike the

Fig. 8.25 Magnel
diagram for Example 8.5

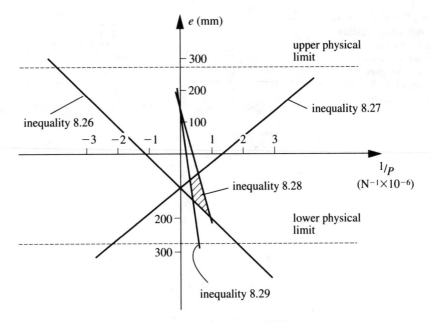

Fig. 8.26 e versus P
version of Magnel
diagram

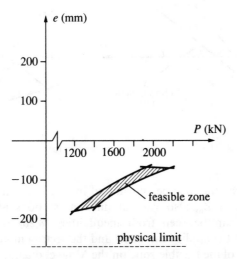

e versus $1/P$ plot of Fig. 8.25, this Magnel diagram is made up of curves rather
than lines which represent the four inequalities (8.26) to (8.29).

General procedure for post-tensioned members

Up to this point we have concentrated on the prestress force and eccentricity
requirements of a simply supported beam at the single critical section where
the moments are maximum. However, in more typical post-tensioned beams,
the critical bending moments will vary from one section to the next within each
span. In addition, many post-tensioned beams have a variable depth over their

Design of Prestressed Concrete Members for Bending

Fig. 8.27 Construction of longitudinal feasible zone: (a) elevation and longitudinal feasible zone; (b) Magnel diagram at A–A; (c) Magnel diagram at B–B

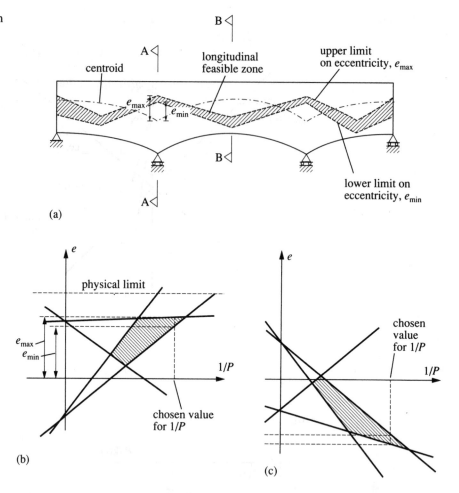

(a)

(b)

(c)

length, such as the member illustrated in Fig. 8.27. In such cases, the section moduli, Z_t and Z_b, also vary at different sections along the length of the member. It can be seen from inequalities (8.26) to (8.29) that both the magnitude of the applied moments and the section moduli determine the shape and location of the feasible zone on the Magnel diagram. Thus, for the member of Fig. 8.27(a) there is a different Magnel diagram for each section considered along its length, as illustrated in Figs 8.27(b) and (c). However, when one tendon is to be used throughout the length of the beam, the force, P, will be the same for each section except for small differences due to losses. Thus, it will often be the case that these sections will have conflicting requirements for P and e. For such members, an appropriate prestress force and tendon profile which satisfies the prestress requirements at all sections can be sought by observing the following three-step process:

1. Choose a prestress force, P, which is within the feasible zone of each Magnel diagram. If this is not possible, the section size of the member must be revised so that it is possible to find an appropriate value for P.

2. For the chosen prestress force, calculate the two limits e_{min} and e_{max} on the tendon eccentricity for each Magnel diagram, as illustrated in Fig. 8.27(b). Plot the locations of these eccentricity limits on a longitudinal section through the beam (Fig. 8.27(a)). Lines are then drawn to join the limits for adjacent sections as illustrated in the figure. These tendon limits constitute a **longitudinal feasible zone** within which the tendon must lie in order to satisfy the stress limits at each section. (Identifying discrete sections and joining them with lines is of course an approximation.)

3. A tendon profile is chosen such that its centroid lies within the longitudinal feasible zone.

The total number of sections along the length of a member which are considered is a matter for engineering judgement. In general, only a few sections (at points of maximum and minimum bending moment and one or two points in between) need to be considered in order to find an acceptable tendon profile. The following example illustrates each step of the basic post-tensioned design process.

Example 8.6   Tendon profile design

Problem   For the post-tensioned beam of Fig. 8.28, the prestress requirements are to be checked at the four numbered sections. Using the Magnel diagrams given

Fig. 8.28 Beam of Example 8.6

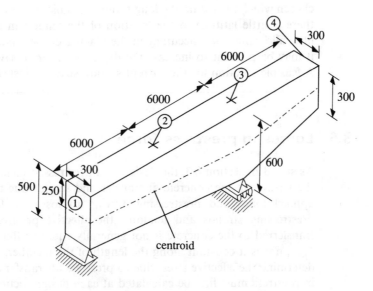

in Fig. 8.29, determine an appropriate prestress force and tendon profile. Differences in prestress force between sections generally occur in post-tensioned beams owing to friction losses. However, in short beams such as this one, such differences tend to be small and, for the purposes of this example, can be ignored. Provide for a minimum distance of 80 mm from the centroid of the tendon to the surface of the beam.

Solution   Step 1–choose value for *P*

By inspection of the four Magnel diagrams of Fig. 8.29, the most suitable value for *P* (i.e. the minimum which is within the feasible zone of each diagram) is found to be approximately 1650 kN, governed by section no. 3. It is interesting to note from Fig. 8.29(d) that this force is very close to the **maximum** possible at section no. 4.

Step 2–determine longitudinal feasible zone

For a prestress force of 1650 kN, the eccentricity limits at each section are, from Fig. 8.29:

$$\text{section no. 1:} \quad -68 \text{ mm} \le e \le 68 \text{ mm}$$
$$\text{section no. 2:} \quad -128 \text{ mm} \le e \le -45 \text{ mm}$$
$$\text{section no. 3:} \quad 172 \text{ mm} \le e \le 184 \text{ mm}$$
$$\text{section no. 4:} \quad -5 \text{ mm} \le e \le 5 \text{ mm}$$

The corresponding longitudinal feasible zone is illustrated in Fig. 8.30.

Step 3–choose suitable tendon profile

A tendon profile, made up of a continuous series of lines and parabolas, is chosen which fits within the longitudinal feasible zone of Fig. 8.30(a). Note that there is little latitude in the location of the tendon in this design and hence little allowance for inaccuracy in the placing of the tendon duct on site. Thus, it may be prudent to increase the depth (by 50 mm, say) at sections 3 and 4. A feasible solution for the current section sizes is illustrated in Fig. 8.30(b).

8.6   Losses in prestress force

As stated in section 8.3, the design of a prestressed member involves checking the stresses in the concrete at transfer and service due to the combination of applied loads and prestressing. Owing to losses of force which occur in prestressing strands and tendons, the **effective** prestress force, *P*, which is transferred to the concrete is not generally equal to the applied jacking force, P_{jack}, nor is it constant along the length of the member. Therefore, in order to determine the effective stress due to prestress at transfer and service, the losses in prestress must first be calculated at each design section.

The losses which occur in prestressed members can be divided into two groups in accordance with the time when they occur. Losses which occur prior to the point in time when stress is first felt by the concrete are collectively known as **pre-transfer** losses. Thus, the prestress force at section *i* at transfer, P_i, is the jacking force, P_{jack}, minus the pre-transfer losses for that section, that is:

$$\text{pre-transfer loss of force at section } i = P_{\text{jack}} - P_i \tag{8.30}$$

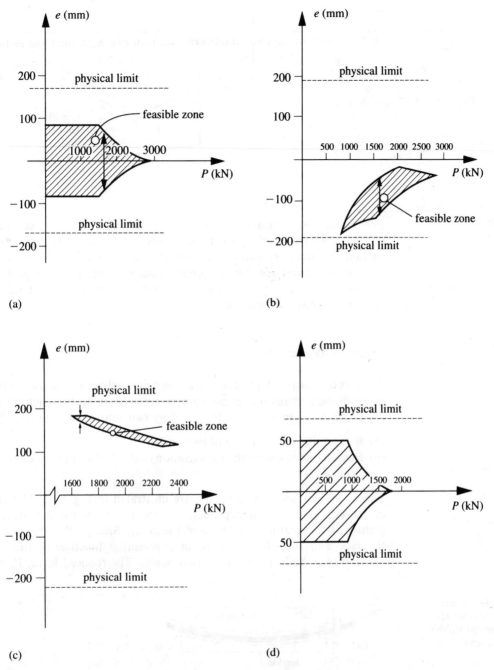

Fig. 8.29 Magnel diagrams for sections of Example 8.6: (a) section no. 1; (b) section no. 2; (c) section no. 3; (d) section no. 4

Losses which occur after the prestress is transferred to the concrete are collectively known as **post-transfer** losses. If the prestress force at section i is reduced to a final value of $\rho_i P_i$, after all losses have occurred, then the total post-transfer loss is equal to:

$$\text{post-transfer loss of force at section } i = P_i - \rho_i P_i \qquad (8.31)$$

323

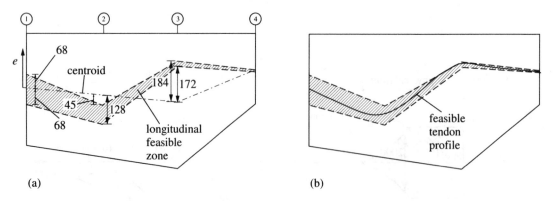

(a)   (b)

Fig. 8.30 Longitudinal feasible zone for Example 8.6: (a) longitudinal section showing feasible zone; (b) feasible zone with tendon profile

Pre-transfer losses result from elastic shortening of the concrete and, in the case of post-tensioned members, from friction between the tendons and the surrounding ducts. Post-transfer losses are caused by relaxation of the steel and by creep and shrinkage of the concrete. In addition, in the case of post-tensioned members, there is a post-transfer loss due to slippage of the tendons at the anchorage known as draw-in loss.

Friction losses

The loss of prestress force due to friction arises **only** in post-tensioned members in which the prestressing tendons are surrounded by ducts. When such tendons are profiled, the loss of force is caused by two sources of friction:

(a) friction due to curvature of the tendon
(b) friction due to unintentional variation of the duct from its prescribed profile or 'wobble'.

These two sources of friction loss are illustrated in Fig. 8.31. The magnitude of the curvature loss is dependent on the extent of the curvature (i.e. the greater the curvature, the greater the loss). Specifically, the loss between the jack and the ith section is an exponential function of the change in angle, θ (Fig. 8.32), between the two points. The reduced force, P_i, at section

Fig. 8.31 Friction losses: (a) curvature friction; (b) wobble friction (exaggerated)

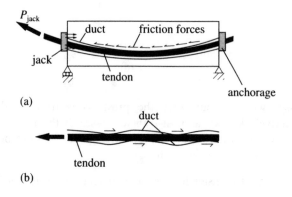

(a)

(b)

Fig. 8.32 Friction
curvature loss

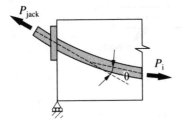

i relative to the jacking force, P_{jack}, is given by:

$$P_i = P_{\text{jack}}\, e^{-\mu\theta} \qquad (8.32)$$

where μ is a friction coefficient which depends on the surface roughness of
the duct and θ is the aggregate change in slope in radians between the jack
and section i. Thus, if the tendon slopes first upwards and then downwards,
the corresponding changes in angle are additive. From equation (8.32), the
curvature loss between the two sections is:

$$\text{curvature loss} = P_{\text{jack}} - P_i$$
$$= P_{\text{jack}}(1 - e^{-\mu\theta}) \qquad (8.33)$$

EC2 recommends, in the absence of more exact data from manufacturers,
that μ for strands be taken as 0.19.

The magnitude of the wobble loss at a section depends not on the
curvature of the tendon but on the distance of the section from the jacking end.
The reduced force, P_i, at section i relative to the jacking force, is given by:

$$P_i = P_{\text{jack}}\, e^{-\mu k x} \qquad (8.34)$$

where k is a wobble coefficient which depends on the quality of workmanship, the distance between tendon supports, the degree of vibration used in
placing the concrete and the type of duct. The term x is the distance (in
metres) from the jack to section i. From equation (8.34), the wobble loss is:

$$\text{wobble loss} = P_{\text{jack}} - P_i$$
$$= P_{\text{jack}}(1 - e^{-\mu k x}) \qquad (8.35)$$

The value for k is generally in the range 0.005 to 0.01. EC2 suggests that
design values for k be taken from technical approval documents.

From equations (8.33) and (8.35), the total loss in prestress force at
section i due to friction is equal to:

$$\text{total friction loss} = P_{\text{jack}}(1 - e^{-\mu(\theta + k x)}) \qquad (8.36)$$

Example 8.7 Friction losses

Problem For the post-tensioned bridge deck of Fig. 8.33, determine the total friction
loss at sections 2 and 3. Assume that the deck is jacked from both ends
to a tension of 40 000 kN. Take $\mu = 0.20$ and $k = 0.01$. All tendons have
the same profile for which details are given in Fig. 8.34 for one-half of the deck.

325

Design of Prestressed Concrete Members for Bending

Fig. 8.33 Post-tensioned bridge deck: (a) typical cross-section; (b) elevation showing profile of tendons

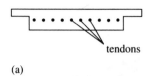

tendons

(a)

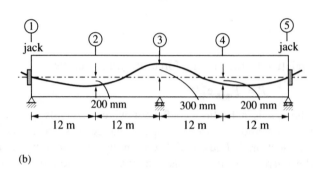

(b)

Fig. 8.34 Profile details (x and y in mm)

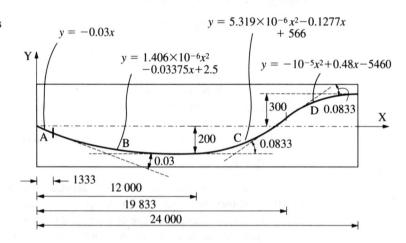

Solution   The slope at section no. 1 is found by differentiation of the equation for line A:

$$y = -0.03x$$

$$\Rightarrow \quad \frac{dy}{dx} = -0.03$$

The slope in radians is $\tan^{-1}(-0.03) = -0.03$ (the angle is very small). The slope at section no. 2 is found from the equation for parabola B:

$$y = 1.406 \times 10^{-6}x^2 - 0.033\,75x + 2.5$$

$$\Rightarrow \quad \frac{dy}{dx} = 2.812 \times 10^{-6}x - 0.033\,75$$

$$\Rightarrow \quad \frac{dy}{dx}(x = 12\,000) = (2.812 \times 10^{-6})(12\,000) - 0.033\,75$$

$$= 0$$

Hence the change in slope between sections 1 and 2 is 0.03. The total friction loss at section 2 is thus:

$$\text{total friction loss} = P_{\text{jack}}(1 - e^{-\mu(\theta + kx)})$$
$$= 40\,000(1 - e^{-0.2(0.03 + 0.01(12))})$$
$$= 40\,000(1 - 0.970)$$
$$= 1182 \text{ kN}$$

Where parabolas C and D meet, the slope is found from the equation for parabola C:

$$y = 5.319 \times 10^{-6}x^2 - 0.1277x + 566$$
$$\Rightarrow \quad \frac{dy}{dx} = 10.638 \times 10^{-6}x - 0.1277$$
$$\Rightarrow \quad \frac{dy}{dx}(x = 19\,833) = (10.638 \times 10^{-6})(19\,833) - 0.1277$$
$$= 0.0833$$

By differentiation of the equation for parabola D, the slope at section no. 3 can be shown to be zero. Thus the aggregate change in angle between sections 1 and 3 is:

$$\text{agg. change in angle} = 0.03 + 0.0833 + 0.0833$$
$$= 0.1966$$

Hence, the total friction loss over the central support is:

$$\text{total friction loss} = 40\,000(1 - e^{-0.2(0.1966 + 0.01(24))})$$
$$= 40\,000(1 - 0.916)$$
$$= 3345 \text{ kN}$$

Elastic shortening losses

As the prestress is transferred to the concrete, the concrete undergoes elastic shortening which reduces the length of the member. This can cause a slackening of the strand which results in a loss of prestress force.

Pre-tensioned members

In pre-tensioned members all the strands are stressed prior to the casting of the concrete and are released after the concrete has set. Upon release of the strands, a force P_{jack} is applied to the concrete. For a statically determinate member with eccentric strands, the stress distribution in the concrete due to prestress is illustrated in Fig. 8.35. At the level of the strands $y = e$, and the

Fig. 8.35 Elastic
shortening loss. Note:
all the stress
distributions are shown
positive. In practice e
and M_0 would generally
be of opposite sign

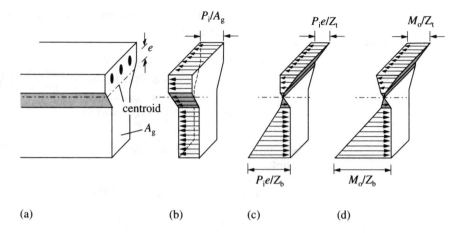

(a) (b) (c) (d)

stress in the concrete adjacent to the strands is:

$$f_c = \frac{P_{jack}}{A_g} + \frac{(P_{jack}e)e}{I_g} + \frac{M_0e}{I_g}$$

$$= P_{jack}\left(\frac{1}{A_g} + \frac{e^2}{I_g}\right) + \frac{M_0e}{I_g}$$

This causes a strain in the concrete of:

$$\varepsilon_c = \frac{P_{jack}}{E_c}\left(\frac{1}{A_g} + \frac{e^2}{I_g}\right) + \frac{M_0e}{E_cI_g}$$

This reduces the strain in the strands by the same amount, resulting in a loss
of stress in the strands of:

loss of stress $= E_p\varepsilon_c$

$$= P_{jack}\frac{E_p}{E_c}\left(\frac{1}{A_g} + \frac{e^2}{I_g}\right) + E_p\left(\frac{M_0e}{E_cI_g}\right)$$

where E_p is the modulus of elasticity of the strands. From the above expression,
the loss of prestress force due to elastic shortening in pre-tensioned members is:

$$\text{elastic shortening loss of force} = P_{jack}A_p\frac{E_p}{E_c}\left(\frac{1}{A_g} + \frac{e^2}{I_g}\right) + A_pE_p\left(\frac{M_0e}{E_cI_g}\right)$$

(8.37)

where A_p is the total cross-sectional area of the strands.

Elastic shortening is **pre-transfer**. This is because the loss occurs before the
concrete feels the full stress. The development of stress with time in the split
second after a pre-tensioned strand is cut is illustrated in Fig. 8.36(a) for one
strand and in Fig. 8.36(b) for a four-strand system. It can be seen that the full
jacking stress (before loss) is never actually applied to the concrete.

Fig. 8.36 Development of elastic shortening loss with time: (a) one strand; (b) four strands

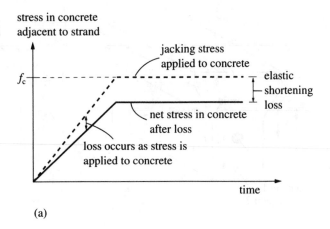

(a)

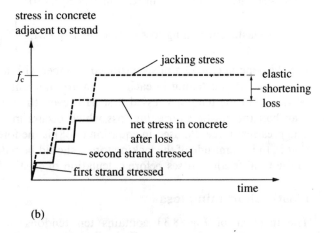

(b)

Post-tensioned members

For post-tensioned members with a single tendon in which all the strands of the tendon are stressed simultaneously (using a multistrand jack), elastic shortening of the concrete occurs during the jacking process. However, since the deformations occur during the stressing process, additional force can be applied at the jack to compensate for loss. Thus, on the anchoring of the tendon in such cases, there is no elastic shortening loss of prestress.

For post-tensioned members with more than one tendon which are stressed sequentially, there is a loss in prestress due to elastic shortening. Consider the member of Fig. 8.37 which has two tendons. Assume tendon 1 is stressed first to a force P_1 and that the transfer moment, M_0, is also applied at this time. Additional force is applied to tendon 1 during stressing to compensate for the elastic shortening loss due to this prestress and to M_0. However, when tendon 2 is then stressed to a force P_2, there is a loss of force in tendon 1 because tendon 2 causes the concrete adjacent to tendon 1 to be shortened. The stress in the concrete adjacent to tendon 1 due to tendon 2 is:

$$f_{c1} = \frac{P_2}{A_g} + \frac{(P_2 e_2)e_1}{I_g}$$

Fig. 8.37 Elastic
shortening of
post-tensioned beam

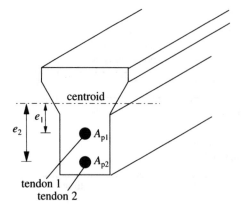

Therefore, the loss of force in tendon 1 is equal to:

$$\text{elastic shortening loss in tendon 1} = P_2 A_{p1} \frac{E_p}{E_c}\left(\frac{1}{A_g} + \frac{e_1 e_2}{I_g}\right) \qquad (8.38)$$

For the general case of a post-tensioned member with many tendons, the losses due to elastic shortening in each tendon are different with the maximum loss occurring in the tendon stressed first. However, the total loss is normally less than half the elastic shortening loss which occurs in pre-tensioned members. The prestress force, P_2, used in equation (8.38) is the force net of friction losses; that is, the magnitude of the jacking forces should be reduced by amounts equal to the total friction losses before calculation of elastic shortening losses.

Example 8.8   Elastic shortening loss

Problem   The member of Fig. 8.33 contains ten tendons of identical profile each comprising 20 strands of area 150 mm². Given that multi-strand jacks are used to stress the tendons in sequence and that the total jacking force is 40 000 kN, determine the total loss due to elastic shortening at the central support and at mid-span. Take $E_c = 35\,000$ N/mm², $E_p = 190\,000$ N/mm² and ignore secondary effects (section 8.7). For the whole bridge, the geometric properties are $I_g = 0.6 \times 10^{12}$ mm⁴ and $A_g = 5.2 \times 10^6$ mm².

Solution   From Fig. 8.33, the eccentricity of the tendons over the central support is 300 mm. From Example 8.7, the total friction loss at this section is equal to 3345 kN. Thus, the net total force for all tendons at the central support of the beam is $40\,000 - 3345 = 36\,655$ kN. When the first tendon is jacked, no elastic shortening loss occurs. However, when the second tendon is jacked, the force after friction losses is $(36\,655/10)$ kN and there is a loss in the first tendon (from equation (8.38)) of:

$$\text{elastic shortening loss in tendon 1} = \frac{36\,655 \times 10^3}{10} A_{p1} \frac{E_p}{E_c}\left(\frac{1}{A_g} + \frac{e^2}{I_g}\right)$$

$$(8.39)$$

where e is the eccentricity of all of the tendons at the central support and A_{p1}

is the cross-sectional area of tendon 1. When the third tendon is jacked, it causes a loss of equal magnitude in tendon 2 and doubles the loss in tendon 1. This process continues up to the stressing of the tenth tendon which causes losses in all the other nine tendons. Thus, the total elastic shortening loss in tendon 1 is nine times that given by equation (8.39), the total loss in tendon 2 is eight times that value, and so on. The total elastic shortening loss for all of the tendons is:

$$\frac{36\,655 \times 10^3}{10}\, A_{p1}\, \frac{E_p}{E_c}\left(\frac{1}{A_g} + \frac{e^2}{I_g}\right)(9 + 8 + 7 + 6 + 5 + 4 + 3 + 2 + 1)$$

$$= \left(\frac{36\,655 \times 10^3}{10}\right)(20 \times 150)\,\frac{190\,000}{35\,000}$$

$$\times \left(\frac{1}{5.2 \times 10^6} + \frac{(300)^2}{0.6 \times 10^{12}}\right)(45)$$

$$= 919\,537\ \text{N}$$

$$= 920\ \text{kN}$$

Similarly, the elastic shortening loss at the centre of the first span is found to be 737 kN.

From the results of Examples 8.7 and 8.8, the applied prestress at transfer can be calculated at both mid-span and central support. By definition, this is the force net of pre-transfer losses. At mid-span, the friction loss is 1182 kN and the elastic shortening loss is 737 kN. Thus:

$$P_2 = 40\,000 - (1182 + 737) = 38\,081\ \text{kN}$$

Over the central support, the friction loss is 3345 kN and the elastic shortening loss is 920 kN. Thus:

$$P_3 = 40\,000 - (3345 + 920) = 35\,735\ \text{kN}$$

Draw-in losses

After jacking the post-tensioned members to the required force, P_{jack}, the tendons are released from the jacks and the force is applied directly through the anchorages to the concrete. When wedge anchors are used, this process results in a loss of prestress due to slippage or 'draw-in' of the strands at the anchors, as illustrated in Fig. 8.38. In some instances, the movement at the anchors, Δ_s, can be up to 10 mm. Fortunately, the friction between the tendons and the ducts ensures that this loss of prestress does not extend very far from the region of the anchor in most cases, particularly in longer members.

Consider the post-tensioned member of Fig. 8.39(a). If there were no friction between the tendon and the duct (i.e. if the duct were very slippery), the loss of prestress force due to draw-in of the wedges would be constant over the entire

Fig. 8.38 Draw-in of wedges at anchors: (a) before anchoring; (b) after anchoring

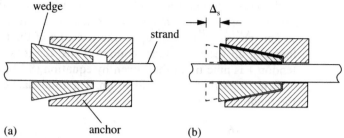

(a) anchor (b)

Fig. 8.39 Draw-in loss: (a) beam elevation; (b) variation in applied prestress force – no friction; (c) variation in applied prestress force – with friction

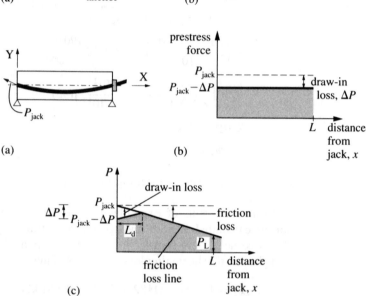

length of the member, as illustrated in Fig. 8.39(b), and would be given by:

$$\Delta P = \frac{\Delta_s}{L} E_p A_p \qquad (8.40)$$

where L is the length of the member. Consider now the real case of a similar member with friction. The variation in loss due to friction can be estimated as a linear function such as, for example, that illustrated in Fig. 8.39(c). The draw-in of the tendon results in a loss of strain of ε_{jack} at the anchorage. However, because of friction, this loss decreases along the length of the member to zero at a distance L_d from the jack. Thus, the total shortening of the tendon is:

$$\Delta_s = \tfrac{1}{2}\varepsilon_{jack}L_d$$

and the loss of force at the anchor is:

$$\Delta P = \varepsilon_{jack}E_p A_p$$

$$\Rightarrow \qquad \Delta P = \left(\frac{2\Delta_s}{L_d}\right)E_p A_p$$

As illustrated in Fig. 8.39(c), the slope of the draw-in line is equal to that of the friction loss line, but of opposite sign, that is:

$$\frac{\Delta P/2}{L_d} = \frac{P_{jack} - P_L}{L}$$

$$\Rightarrow \quad \Delta P = 2\frac{(P_{jack} - P_L)}{L}L_d \tag{8.41}$$

Equating these two expressions for ΔP gives the extent of draw-in losses:

$$L_d = \sqrt{\frac{\Delta_s E_p A_p}{(P_{jack} - P_L)/L}} \tag{8.42}$$

The magnitude of the draw-in loss at the anchorage, ΔP, is then given by equation (8.41).

Example 8.9   Draw-in loss

Problem   Upon anchoring, the strands of the post-tensioned member of Fig. 8.33 undergo a draw-in of 6 mm. Determine the extent of the draw-in loss and its magnitude at mid-span.

Solution   As this beam is jacked from both ends, the variation in prestress force after friction losses but prior to draw-in is assumed to be bilinear, as illustrated in Fig. 8.40. From Example 8.7, the total friction loss at section no. 3 is 3345 kN.

Fig. 8.40 Distribution of prestress force in beam of Example 8.9

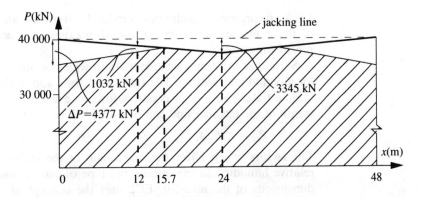

The friction loss will be assumed to vary linearly from the jack to this point, that is, the slope of the distribution of prestress force is taken as:

$$slope = \left(\frac{P_{jack} - P_L}{L}\right) = \left(\frac{3\,345\,000}{24\,000}\right)$$

$$= 139.4 \text{ N/mm}$$

Note: This is a rather crude assumption, and can sometimes result in significant inaccuracy.

From equation (8.42), the extent of the draw-in loss is:

$$L_d = \sqrt{\frac{\Delta_s E_p A_p}{(P_{jack} - P_L)/L}}$$

$$= \sqrt{\frac{6 \times 190\,000 \times (200 \times 150)}{139.4}}$$

$$= 15\,663 \text{ mm}$$

$$= 15.7 \text{ m}$$

From equation (8.41), the draw-in loss at the anchors is:

$$\Delta P = 2L_d \left(\frac{P_{jack} - P_L}{L} \right)$$

$$= 2(15.7)(139.4)$$

$$= 4377 \text{ kN}$$

From interpolation in Fig. 8.40, the draw-in loss at section no. 2 ($L = 12$ m) is given by:

$$\text{draw-in loss} = 4377(15.7 - 12)/15.7 = 1032 \text{ kN}$$

and the net force after draw-in is $40\,000 - 3345/2 - 1032 = 37\,295$ kN.

Time-dependent losses

Losses in prestress which occur gradually over time are caused by shortening of the concrete due to creep and shrinkage and due to relaxation of the prestressing steel.

Shrinkage is a time-dependent strain which occurs as the concrete sets and for a period after setting. The shrinkage strain approaches a final value, ε_{sh}, at infinite time. The loss of prestress force is generally constant over the entire length of the member and can be estimated as:

$$P_{sh} = \varepsilon_{sh} E_p A_p \tag{8.43}$$

The factors which affect the magnitude of the shrinkage strain include the relative humidity during curing, the type of cement used and the shape and dimensions of the member. EC2 uses the concept of notional size which is defined as twice the ratio of the cross-sectional area, A_g, to the perimeter length of the section, u. Values of final shrinkage strains, suggested by EC2 for the design of prestressed members, are given in Table 8.3.

When maintained at a constant tensile strain, steel gradually loses its stress with time due to relaxation. In stringed musical instruments, for example, relaxation could contribute to strings going out of tune. This phenomenon is caused by the realignment of the steel fibres under stress. The extent of the loss of stress in prestressing strands due to relaxation is determined by the stress to which the steel is tensioned, the ambient temperature and the class of steel. The loss of prestress which is caused by steel relaxation can range from

Table 8.3 Final shrinkage strains for prestressed concrete. Linear interpolation between the values given is permitted (from EC2)

Location of member	Relative humidity (%)	Notional size, $2A_g/u$ (mm)	
		≤ 150	600
Inside	50	0.60×10^{-3}	0.50×10^{-3}
Outside	80	0.33×10^{-3}	0.28×10^{-3}

3 per cent to about 12 per cent of the initial force. Precise values for relaxation losses are usually specified by the manufacturers of prestressing steel. In the absence of such information, EC2 suggests that loss due to relaxation be based on the 1000 hour values given in Fig. 8.41. The values should be trebled to determine the final relaxation losses.

Fig. 8.41 Relaxation of prestressing steel (from EC2)

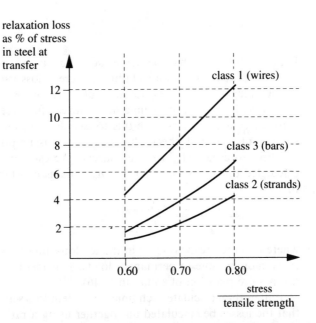

The phenomenon of creep is essentially the same as that of relaxation. The distinction is that relaxation refers to the loss of stress under constant strain while creep is the increase of strain which occurs at constant stress. For the purposes of prestressed concrete, relaxation occurs in the steel while creep occurs in the concrete. Creep of concrete is unpredictable and can be quite substantial in prestressed members where the stress is kept constant for the design life of the structure. As for shrinkage, creep in the concrete allows the prestressing strands to slacken which results in the loss of force.

Like shrinkage strain, creep strain increases with time and approaches a final value ε_∞, at infinite time. Creep strain in concrete is usually measured in terms of the creep coefficient, φ, which is the ratio of creep strain to linear elastic strain, that is:

$$\varphi = \frac{\varepsilon_\infty}{\sigma_c/E_c} \tag{8.44}$$

where ε_∞ is the final creep strain and σ_c is the elastic stress in the concrete. The magnitude of the creep coefficient depends on the size of the member, the relative humidity and on the age (or strength) of the concrete when it is first loaded. Values of the final creep coefficient, recommended by EC2, for the calculation of the total creep strain in prestressed members (and non-prestressed members) are given in Table 7.1. The final creep strain, ε_∞, at the level of the tendons is then given by:

$$\varepsilon_\infty = \frac{\varphi\sigma_c}{E_c} \tag{8.45}$$

where σ_c is the stress in the concrete at the level of the tendons due to prestress and all other **permanent** loads. From equation (8.45), the loss in prestress force due to creep is:

$$\text{creep loss} = A_p E_p \varepsilon_\infty = A_p \frac{E_p}{E_c} \varphi\sigma_c \tag{8.46}$$

It should be noted that the magnitude of σ_c in equation (8.46) is a function of the time-dependent losses including the creep loss itself and so the calculation of the creep loss involves iteration of equation (8.46). For such an iterative technique, it is usual to assume first a force after creep loss of about $0.9P_i$ (i.e. a 10 per cent creep loss), use this to calculate the creep loss, add in the other sources of loss and check that the initial assumption was (approximately) satisfied. Alternatively, a more conservative estimate of the creep loss can be calculated on the assumption that σ_c remains constant at its maximum value, that is:

$$\sigma_c = \frac{P}{A_g} + \frac{(Pe + M_{\text{perm}})e}{I_g} \tag{8.47}$$

where P is the prestress force in the tendons after pre-transfer losses and M_{perm} is the moment due to permanent loads. This method is conservative and does not require iteration of equation (8.46).

Rather than calculate each time-dependent loss separately, EC2 recommends that the losses be calculated all together using a rather expansive formula:

$$\Delta\sigma_{p,\text{tot}} = \frac{\varepsilon_{sh}E_p + \Delta\sigma_{p,r} + m\varphi\sigma_c}{1 + \dfrac{mA_p}{A_g}\left(1 + \dfrac{A_g e^2}{I_g}\right)(1 + 0.8\varphi)}$$

where $\Delta\sigma_{p,\text{tot}}$ = loss of stress in steel due to creep, shrinkage and relaxation
ε_{sh} = final shrinkage strain taken from Table 8.3
E_p = modulus of elasticity of the prestressing steel
$\Delta\sigma_{p,r}$ = loss of stress in the tendons at the design section due to relaxation (see below)
m = modular ratio, E_p/E_c
φ = final creep coefficient taken from Table 7.1
σ_c = stress in the concrete at the level of the tendons due to permanent loads plus prestress (EC2 specifies the use of equation (8.47) for the calculation of σ_c in this formula)

A_p = area of prestressing steel

A_g, I_g = gross area and second moment of area of section

e = eccentricity of tendons from centroid of section

The loss of stress in the tendons due to relaxation of the tendons, $\Delta\sigma_{p, r}$, is derived from Fig. 8.41, with the ratio of steel stress/characteristic tensile strength (σ_p/f_{pk}) calculated using:

$$\sigma_p = \sigma_{pi} - 0.3\Delta\sigma_{p, tot}$$

where σ_{pi} is the initial stress in the tendons due to the prestress force, P, and permanent loads. It can be seen that the above expression for σ_p is a function of $\Delta\sigma_{p, tot}$ which is, in turn, a function of σ_p. However, for simplicity, the term $0.3\Delta\sigma_{p, tot}$ in the above expression may be ignored and σ_p becomes:

$$\sigma_p = \sigma_{pi} = \frac{P}{A_p} + \frac{E_p}{E_c} \frac{M_{perm}e}{I_g}$$

The following example illustrates the application of the EC2 formula.

Example 8.10 **Time-dependent losses**

Problem For the member of Fig. 8.33, determine the total loss of prestress force at section no. 2 due to all effects. In addition to the data given in Examples 8.7–8.9, the following additional values may be assumed: final creep coefficient $\varphi = 2.0$, final shrinkage strain $\varepsilon_{sh} = 0.28 \times 10^{-3}$, moment at section no. 2 due to permanent loads $M_{perm} = 5744$ kNm, characteristic tensile strength of tendons $f_{pk} = 1770$ N/mm².

Solution The stress in the tendons, σ_p, is approximated by:

$$\sigma_p = \frac{P_2}{A_p} + \frac{E_p}{E_c} \frac{M_{perm}e}{I_g}$$

$$= \frac{37\,295 \times 10^3}{(10 \times 20 \times 150)} + \frac{190\,000}{35\,000} \frac{(5744 \times 10^6)(-200)}{0.6 \times 10^{12}}$$

$$= 1233 \text{ N/mm}^2$$

Hence the ratio of stress to characteristic strength is:

$$\sigma_p/f_{pk} = 1233/1770 = 0.697$$

From Fig. 8.41, the loss of stress due to relaxation is 2.5 per cent of the initial stress in the tendons due to prestress. Trebling this gives:

$$\Delta\sigma_{p, r} = 3 \times 0.025 \frac{P_2}{A_p} = 0.075 \frac{37\,295 \times 10^3}{(10 \times 20 \times 150)}$$

$$= 93.2 \text{ N/mm}^2$$

The stress, σ_c, in the concrete at the level of the tendons due to permanent loads plus prestress is calculated using equation (8.47).

$$\sigma_c = \frac{P_2}{A_g} + \frac{(P_2 e + M_{perm})e}{I_g}$$

$$= \frac{37\,295 \times 10^3}{5.2 \times 10^6} + \frac{(37\,295 \times 10^3(-200) + 5744 \times 10^6)(-200)}{0.6 \times 10^{12}}$$

$$= 7.75\ \text{N/mm}^2$$

The total time-dependent loss of stress in the tendons is:

$$\Delta\sigma_{p,\,tot} = \frac{\varepsilon_{sh}E_p + \Delta\sigma_{p,\,r} + m\varphi\sigma_c}{1 + \dfrac{mA_p}{A_g}\left(1 + \dfrac{A_g e^2}{I_g}\right)(1 + 0.8\varphi)}$$

$$= \frac{0.28 \times 10^{-3}(190\,000) + 93.2 + (190\,000/35\,000)(2.0)(7.75)}{1 + \dfrac{19\,000}{35\,000} \times \dfrac{(200 \times 150)}{5.2 \times 10^6}\left(1 + \dfrac{5.2 \times 10^6(-200)^2}{0.6 \times 10^{12}}\right)(1 + 0.8 \times 2)}$$

$$= 207.8\ \text{N/mm}^2$$

Thus, the total loss of force at section no. 2 due to time-dependent effects is:

$$\text{total time loss} = 207.8 \times (10 \times 20 \times 150)$$
$$= 6234\ \text{kN}$$

From the results of Examples 8.9 and 8.10, the applied prestress at service, $\rho_2 P_2$, can now be calculated at that section. The draw-in loss is 1032 kN and the total time-dependent loss is 6234 kN. Thus:

$$\rho_2 P_2 = P_2 - 1032 - 6234$$
$$= (40\,000 - 3345/2) - 1032 - 6234$$
$$= 31\,061\ \text{kN}$$

Hence, the ratio of the prestress force at service to the prestress force at transfer, ρ_2, is:

$$\rho_2 = 31\,061/(40\,000 - 3345/2) = 0.81$$

8.7 Secondary effects of prestress

Up to now, the moment due to prestress, M_p, has been assumed to equal the product of the prestress force and the eccentricity, Pe. However, this assumption is only true for statically determinate prestressed structures. For statically indeterminate structures, such as that of Fig. 8.33 (upon which the last four examples have been based), M_p consists of two components, **primary moments** (Pe) and **secondary moments**. Secondary moments arise from reactions developed at supports during prestressing. For example, in the beam of Fig. 8.42(a), stressing the tendon causes the beam to sag and results in a downward deflection at B. It also generates the distribution of moment, known as primary moment, illustrated in Fig. 8.42(b). If this tendency to deform is prevented, such as by

Fig. 8.42 Effect of indeterminacy on prestress: (a) deformed shape of simply supported beam; (b) distribution of moment for (a), and distribution of primary moment for (c); (c) two-span beam (indeterminate); (d) deformed shape of two-span beam; (e) distribution of secondary moment; (f) total distribution of prestress moment, M_p, equal to sum of (b) and (e)

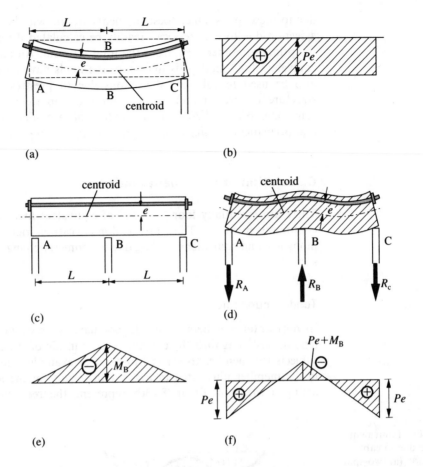

the central support in the two-span beam of Figs 8.42(c) and (d), a reaction is caused at B which generates the distribution of bending moment illustrated in Fig. 8.42(e). The moments in this distribution are known as secondary or parasitic moments. The total distribution of moment for this beam is the sum of the primary and secondary distributions and is illustrated in Fig. 8.42(f). Since statically determinate members do not provide restraint to imposed deformations, secondary moments only apply to indeterminate structures. Although these moments are termed 'secondary', they cannot be ignored as they can often have a substantial effect on the distribution of stress in a continuous member.

The magnitude of secondary moments in a given member depends on a number of factors, including tendon curvature and geometry changes in non-prismatic members. For a qualitative understanding of secondary effects, it is useful to consider the concept of concordant tendons as described, for example, by Abeles and Bardhan-Roy (1981). Here, we consider one practical method for the determination of the prestress moment, M_p, in an indeterminate member known as the **equivalent load method**. In this method, the forces applied to the concrete by the prestress are represented as externally applied loads. The structure is then analysed to obtain the total (primary plus secondary) moments

339

due to these loads. Once these moments are known, the secondary moment at a section due to prestress can readily be calculated if required by subtracting the primary moment, Pe, from the total moment. An added advantage of the equivalent load method is that, once the loads have been evaluated, they can also be used to calculate deflections due to prestress (whether or not the structure is determinate). This is important in slender members which are sometimes pre-cambered in anticipation of large deflections due to prestress and permanent gravity loads.

Calculation of equivalent loads

There are essentially four 'sources' of equivalent loading due to prestress in continuous members. These are: tendon curvature, friction loss, end forces and moments, and equivalent loading due to geometry changes in beams of variable section.

Tendon curvature

In horizontal members which do not have a linear profile, the prestressing tendon pushes against the concrete on the inside of the curve and, as a result, subjects the member to vertical forces. For example, tensioning of the tendon in the member illustrated in Fig. 8.43(a) results in the forces ω_1, ω_2 and ω_3 acting on the beam. Figure 8.43(b) represents the free-body diagram for a small

Fig. 8.43 Equivalent loading due to cable curvature: (a) two-span beam; (b) segment of beam

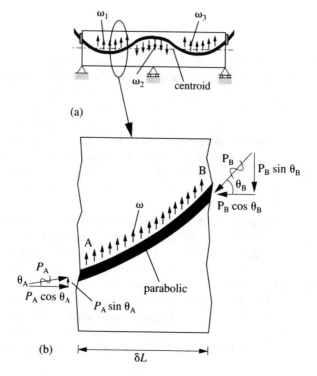

portion of the member of Fig. 8.43(a) over which the inclination of the tendon changes from θ_A to θ_B. Referrring to Fig. 8.43(b) and assuming that $P_A = P_B$, vertical equilibrium is satisfied with a total equivalent force, F, equal to:

$$F = P_A(\sin \theta_B - \sin \theta_A) \tag{8.48}$$

However, P_A and P_B are not equal as a result of friction and so the mean value of P_A and P_B is used in equation (8.48) to yield:

$$F = \frac{P_A + P_B}{2}(\sin \theta_B - \sin \theta_A) \tag{8.49}$$

It should be noted that the expression for F, given by equation (8.49), does not exactly satisfy equilibrium of vertical forces as:

$$P_B \sin \theta_B - P_A \sin \theta_A \neq \frac{P_A + P_B}{2}(\sin \theta_B - \sin \theta_A)$$

(refer to Fig. 8.43(b)). However, F is as given by equation (8.49) in order to satisfy equilibrium of moments as will be seen from Example 8.11. If the tendon profile takes the shape of a parabolic curve, the equivalent loading will be uniformly distributed over the length of the parabola. Thus, the equivalent uniformly distributed loading, ω, over the length δL is given by:

$$\omega = \frac{P_A + P_B}{2\delta L}(\sin \theta_B - \sin \theta_A) \tag{8.50}$$

If a tendon profile changes shape sharply, such as in the member of Fig. 8.44, the equivalent loading can be represented by a concentrated load rather than by a uniformly distributed load, as illustrated in that figure. The magnitude of the force, F, at the point where the slope changes is given by:

$$F = (P_A + P_B) \sin \theta \tag{8.51}$$

Fig. 8.44 Sudden change in tendon shape

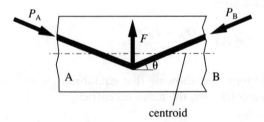

centroid

Friction losses

When the prestressing forces in adjacent design sections are not equal, there is a consequent change in the moment due to prestress, M_p. For example, in the segment of beam illustrated in Fig. 8.45(a), the moment changes from $P_A e$ at A to $P_B e$ at B. This change in prestress can be modelled using equivalent, equal

Fig. 8.45 Equivalent loading due to friction losses: (a) segment with constant eccentricity; (b) free body diagram (constant eccentricity); (c) segment with variable eccentricity

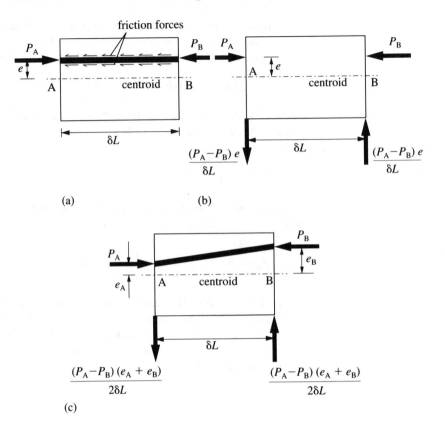

and opposite vertical forces, at A and B, of magnitude:

$$\frac{(P_A - P_B)e}{\delta L} \tag{8.52}$$

where P_A, P_B and δL are illustrated in Fig. 8.45(a). The free-body diagram, which includes these forces, is illustrated in Fig. 8.45(b). Taking moments about A in this figure gives:

$$P_A e = P_B e + \frac{(P_A - P_B)e}{\delta L} \delta L$$

which is indeed true, showing that equilibrium is satisfied. When forces are at different eccentricities, the mean eccentricity is used and the magnitude of the forces becomes:

$$\frac{(P_A - P_B)(e_A + e_B)}{2\delta L} \tag{8.53}$$

as illustrated in Fig. 8.45(c). This is an approximation and, for accuracy in the results, the length δL between points where the force is calculated should be reasonably small (less than a quarter of the span, say).

End forces and moments

In members with profiled tendons, the external force applied to the concrete by the prestress at the anchors is often inclined at an angle and/or located away from the centroid of the member. Consider the member illustrated in Fig. 8.46(a). The force, P, at the anchors can be resolved into three components: a horizontal force ($P \cos \theta$), a vertical force ($P \sin \theta$) and a moment ($P \cos \theta$)e. As the angles involved are generally small, the horizontal force can be approximated by P, and the moment by Pe, as illustrated in Fig. 8.46(b). As for equivalent loads due to tendon curvature, an average value for prestress force should be used when it is varying due to friction.

Fig. 8.46 End forces and moments: (a) beam elevation; (b) segment of beam at anchor

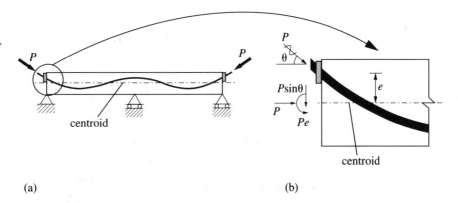

(a)   (b)

Example 8.11   Equivalent loads due to prestress

Problem   For the beam illustrated in Fig. 8.47, calculate the equivalent loading due to prestress. Verify that moment equilibrium is (approximately) satisfied at points B and C. The equation for the parabola (in metres) is:

$$y = -0.8\left[\frac{x}{L} - \left(\frac{x}{L}\right)^2\right]$$

Fig. 8.47 Prestressed beam of Example 8.11: (a) elevation; (b) equivalent loading due to tendon curvature (kN/m); (c) equivalent loading due to friction losses (kN); (d) equivalent loading due to end forces (kN)

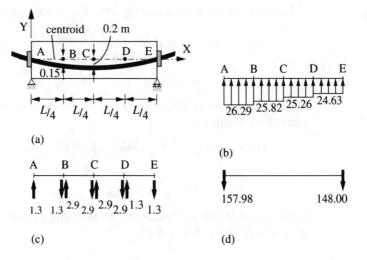

343

where the length, L, is 12 m. The prestress forces at transfer are 2400, 2350, 2300, 2250 and 2200 kN at points A, B, C, D and E respectively.

Solution The slope of the tendon is found by differentiation of the equation for the parabola:

$$\frac{dy}{dx} = -0.8\left(\frac{1}{L} - \frac{2x}{L^2}\right)$$

Hence the slopes at A, B and C are:

$$\text{at A:} \quad \frac{dy}{dx}(x = 0) = -0.0667$$

$$\text{at B:} \quad \frac{dy}{dx}\left(x = \frac{L}{4}\right) = -0.0333$$

$$\text{at C:} \quad \frac{dy}{dx}\left(x = \frac{L}{2}\right) = 0$$

At A:

$$\sin \theta_A = \sin(\tan^{-1}(-0.0667))$$
$$= -0.0665$$

while, at B, the corresponding value is:

$$\sin \theta_B = -0.0333$$

Hence, the equivalent loads due to curvature between A and B are of magnitude:

$$\left(\frac{P_A + P_B}{2}\right)\frac{(\sin \theta_B - \sin \theta_A)}{\delta L} = \left(\frac{2400 + 2350}{2}\right)\frac{(0.0665 - 0.0333)}{3}$$

$$= 26.29 \text{ kN/m}$$

Similarly, the intensity between B and C, C and D and between D and E have been calculated and are as illustrated in Fig. 8.47(b).

The eccentricity at A is clearly zero. The eccentricity at B is:

$$y(x = L/4) = -0.8\left[\frac{1}{4} - \left(\frac{1}{4}\right)^2\right]$$

$$= -0.15 \text{ m}$$

Hence the equivalent forces at A and B due to friction losses between these points have magnitude:

$$\frac{(P_A - P_B)(e_A + e_B)}{2\delta L} = \frac{(2400 - 2350)(0 - 0.150)}{2 \times 3}$$

$$= -1.3 \text{ kN}$$

Similarly the forces are calculated at each end of each segment of beam and are as illustrated in Fig. 8.47(c).

Finally the equivalent loading due to end forces is calculated. In this case, there are end forces at A and E. Taking the average prestress force for the end segments, we get a force at A of:

$$\frac{(P_A + P_B)}{2} \sin \theta_A = \frac{(2400 + 2350)}{2} (-0.0665)$$

$$= 157.98 \text{ kN}$$

This and the force at E, found similarly, are illustrated in Fig. 8.47(d). All equivalent loading due to prestress on the beam is included in the free-body diagrams of Fig. 8.48. Taking moments about B in the diagram of Fig. 8.48(a) gives:

$$M_B = -(157.98 - 1.3) \times 3 + (26.29 \times 3) \times 1.5$$

$$= -352 \text{ kNm}$$

Fig. 8.48 Free body diagrams showing total equivalent loading due to prestress

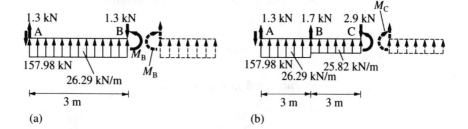

(a)   (b)

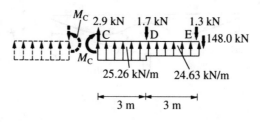

(c)

As the beam is determinate, this equals the product of prestress force and eccentricity at B:

$$M_B = 2350 \times (-0.150)$$

$$= -353 \text{ kNm}$$

Similarly, taking moments about C in the free-body diagram of Fig. 8.48(b) gives:

$$M_C = -(157.98 - 1.3) \times 6 + 1.7 \times 3$$

$$+ (26.29 \times 3) \times 4.5 + (25.82 \times 3) \times 1.5$$

$$= -464 \text{ kNm}$$

Alternatively, taking moments about C in the free-body diagram of Fig. 8.48(c) gives $M_C = -455$ kNm. The product of prestress force and eccentricity at C is $(2300 \times -0.200) = -460$ kNm. Thus, the moment, as calculated using the equivalent loading, approximately equals the actual moment due to prestress, M_p.

Geometry change

The depth of prestressed members is frequently increased over internal supports in order to increase the second moment of area at points of peak moments. Figure 8.49(a) illustrates a typical member with variable depth. In such non-prismatic members, the depth of the centroid below a horizontal datum line, D, varies along its length. This deviation has the effect of generating secondary bending moments in the member. The loading equivalent to a change in the depth, δD, of the centroid between two design sections is a pair of equal and opposite vertical forces of magnitude:

$$\frac{(P_A + P_B)\delta D}{2\delta L} \tag{8.54}$$

as illustrated in Fig. 8.49(b).

Fig. 8.49 Equivalent loading due to geometry change: (a) beam of varying cross-section; (b) equivalent forces due to geometry change

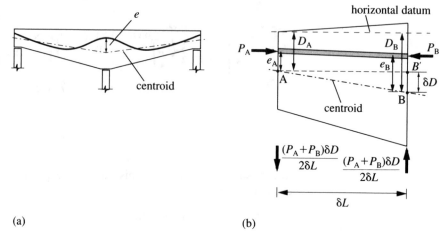

Analysis of the member under these loads, using a hand method such as moment distribution or a computer-based method, gives the total moments due to prestress at transfer. The moments due to prestress at service are then derived by reducing the transfer moments by an amount equal to the post-transfer losses. Simple factoring in this way is less accurate than calculating the service moments due to prestress using the equivalent loads. However, it is generally sufficiently accurate for design purposes.

8.8 Ultimate moment capacity of prestressed concrete

Once a prestressed member has been designed to satisfy the transfer and service stress limits, it must then be checked for other serviceability limit states (i.e. deflection and, in the case of partially prestressed members, crack widths) and the ultimate limit states. The ultimate limit-state checks consist of ensuring that the member has adequate strength in bending, shear and torsion under ultimate loads. This section deals only with the design of prestressed members for bending since the behaviour and design of prestressed members in shear and torsion is treated in Chapter 10.

As for ordinary reinforced concrete members, the ultimate moment capacity of a prestressed section is most readily calculated by equilibrium of forces acting at the section. Consider the section of Fig. 8.50(a), which has a single tendon, bonded to the concrete at an eccentricity e from the centroid of the section. Failure at the section under ultimate loads occurs when the strain in the extreme fibre in compression reaches its ultimate value, ε_{ult}. In the absence of applied loads, the strain in the concrete is that caused by the prestress force (after all losses), ρP. The distribution of strain due to prestress for the section of Fig. 8.50(a) is illustrated in Fig. 8.50(b). The compressive strain in the concrete due to prestress, ε_{ce}, at the level of the prestressing tendon is given by:

$$\varepsilon_{ce} = \frac{1}{E_c}\left(\frac{\rho P}{A_g} + \frac{\rho M_p e}{I_g}\right) \tag{8.55}$$

where M_p is the moment due to prestress. In a determinate structure, $M_p = Pe$. In an indeterminate structure, secondary effects will, in general, change M_p from Pe. However, plastic hinges may have formed in which case the structure could once again be treated as determinate. In the absence of a more detailed study, a conservative section design can be found by considering both possible values for M_p.

As the tendon in Fig. 8.50 is at a positive eccentricity, the moment due to prestress is positive (i.e. sagging) and the corresponding applied moment would normally be negative (i.e. hogging) as prestress is generally placed so as to oppose applied moment. However, the less usual situation of positive eccentricity

Fig. 8.50 Strain and stress distributions for positive applied moment and eccentricity: (a) section; (b) strain due to prestress; (c) total strain (prestress + applied); (d) total stress

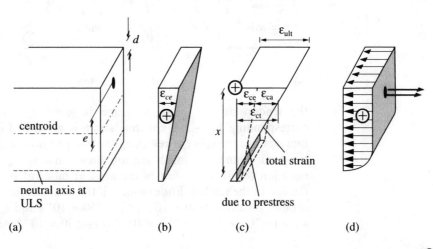

and positive moment is first considered in Fig. 8.50(c). The corresponding ultimate stress distribution is illustrated in Fig. 8.50(d). The strain in the **tendon** due to prestress alone at ULS is (contraction positive):

$$\text{tendon strain due to prestress} = -\frac{\rho P}{A_p E_p} \qquad (8.56)$$

where A_p is the area of the prestressing tendon and E_p is its modulus of elasticity. When only prestress is present, the strain in the concrete adjacent to the tendon is ε_{ce}. When the ultimate moment is applied, the distribution of strain changes from this base level to that illustrated in Fig. 8.50(c). Thus, the compression in the concrete adjacent to the tendon increases by the strain due to applied moment, ε_{ca}, and the tendon slackens accordingly to an ultimate strain of:

$$\varepsilon_{pu} = -\frac{\rho P}{A_p E_p} + \varepsilon_{ca} \qquad (8.57)$$

As the total strain in the concrete, ε_{ct}, can be determined by similar triangles and the strain due to prestress is known, it is convenient to express the tendon strain ε_{pu} as:

$$\varepsilon_{pu} = -\frac{\rho P}{A_p E_p} + \varepsilon_{ct} - \varepsilon_{ce} \qquad (8.58)$$

The more usual case, where moments due to prestress and applied loads are of opposite sign, is illustrated in Fig. 8.51. As before, the strain in

Fig. 8.51 Strain and stress distributions for negative applied moment and positive eccentricity: (a) section; (b) strain due to prestress; (c) total strain (prestress + applied); (d) total stress

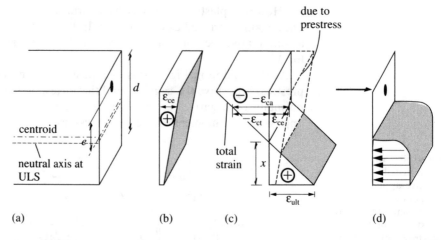

the tendon due to prestress alone is given by equation (8.56) and the corresponding base-level concrete strain is given in Fig. 8.51(b). The applied moment has the effect of reversing the compression in the concrete adjacent to the tendon and causing a tension there, that is ε_{ca} is negative (tensile) in equation (8.57). The effect of the applied moment is, in fact, to **increase** the tension in the tendon. For example, if the strain in the tendon due to prestress is (after losses) -4850×10^{-6}, $\varepsilon_{ce} = 350 \times 10^{-6}$ and $\varepsilon_{ct} = -200 \times 10^{-6}$, then $\varepsilon_{ca} = (-200 \times 10^{-6} - 350 \times 10^{-6} =) -550 \times 10^{-6}$ and the ultimate strain will be:

$$\varepsilon_{pu} = -4850 \times 10^{-6} + (-550 \times 10^{-6})$$

$$= -5400 \times 10^{-6}$$

It is sometimes more convenient to express equation (8.58) in terms of absolute values of strains. Thus, when prestress and applied moment are of opposite sign, equation (8.58) can be written as:

$$\varepsilon_{pu} = -\frac{\rho P}{A_p E_p} - |\varepsilon_{ct}| - |\varepsilon_{ce}| \qquad (8.59)$$

By considering similar triangles in Fig. 8.51(c), the total ultimate strain, ε_{ct}, in the concrete at the level of the tendon is found to be (contraction positive):

$$\varepsilon_{ct} = -\frac{\varepsilon_{ult}(d - x)}{x} \qquad (8.60)$$

EC2 recommends that ε_{ult} be taken as 0.0035. Hence, equation (8.60) becomes:

$$\varepsilon_{ct} = -\frac{0.0035(d - x)}{x} \qquad (8.61)$$

Once the magnitude of the strain in the steel at ULS has been established using equation (8.59), the ultimate moment capacity of a section, M_{ult}, is determined by taking moments about, say, the neutral axis. (Note that the neutral axis at ULS is not located at the centroid as it is at SLS.) EC2 allows any of the stress–strain relationships for concrete of Fig. 7.21 to be used for the calculation of the compressive force at the section. The following example illustrates the steps involved in the calculation of M_{ult}.

Example 8.12   Ultimate moment capacity

Problem   For the section illustrated in Fig. 8.52, determine the ultimate capacity to resist sag moment. Assume the EC2 equivalent rectangular stress block for concrete. The moduli of elasticity are $E_p = 190\,000$ N/mm^2 and $E_c = 35\,000$ N/mm^2, and the characteristic strengths are $f_{pk} = 1770$ N/mm^2 and $f_{ck} = 40$ N/mm^2. The

Fig. 8.52 Section of Example 8.12

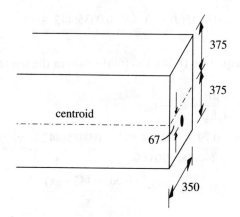

centroid

375

375

67

350

area of tendon is $A_p = 2400 \text{ mm}^2$ and it is stressed to a prestress force of 3023 kN at transfer. The post-transfer losses are 21 per cent implying a loss ratio, ρ, of 0.79.

Solution   The stress and strain distributions are illustrated in Fig. 8.53. The gross second moment of area is given by:

$$I_g = \frac{350 \times 750^3}{12} = 12.3 \times 10^9 \text{ mm}^4$$

Fig. 8.53 Strain and stress distributions for Example 8.12: (a) section; (b) strain due to prestress; (c) total strain; (d) total stress

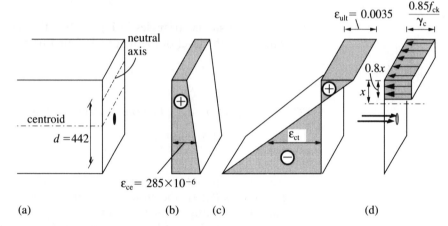

Assuming no secondary effects, equation (8.55) gives the compressive strain in the concrete due to prestress, ε_{ce}, at the level of the prestressing tendon as:

$$\varepsilon_{ce} = \frac{1}{E_c}\left(\frac{\rho P}{A_g} + \frac{(\rho Pe)e}{I_g}\right)$$

$$= \frac{1}{35\,000}\left(\frac{0.79 \times 3023 \times 10^3}{350 \times 750} + \frac{0.79 \times 3023 \times 10^3 \times (-67)^2}{12.3 \times 10^9}\right)$$

$$= 285 \times 10^{-6}$$

From equation (8.61), the total strain in the concrete at the level of the tendon at ULS is equal to:

$$\varepsilon_{ct} = -\frac{0.0035(d - x)}{x} = -\frac{0.0035(442 - x)}{x}$$

Hence, from equation (8.58), the total strain in the tendon at the ultimate limit state is:

$$\varepsilon_{pu} = \frac{-\rho P}{A_p E_p} + \varepsilon_{ct} - \varepsilon_{ce}$$

$$= \frac{-0.79 \times 3023 \times 10^3}{2400 \times 190\,000} - \frac{0.0035(442 - x)}{x} - 285 \times 10^{-6}$$

$$= -5522 \times 10^{-6} - \frac{0.0035(442 - x)}{x}$$

Fig. 8.54 Typical stress/strain relationship for prestressing steel (from BS8110)

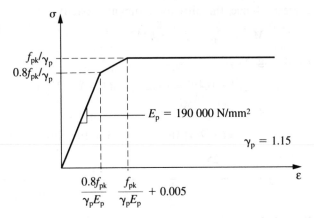

EC2 specifies that producers of prestressing strand supply stress–strain diagrams for their products. A typical such diagram is illustrated in Fig. 8.54 (adapted from BS8110). Initially, assume the steel has not yielded (i.e. $\varepsilon_{pu} < 0.8 f_{pk}/\gamma_p E_p$). Hence, the total force in the tendon, F_p, is:

$$F_p = A_p(E_p \varepsilon_{pu})$$

$$= 2400(190\,000)\left(-5522 \times 10^{-6} - \frac{0.0035(442 - x)}{x}\right)$$

The compressive force, F_c, acting on the concrete in compression is (Fig. 8.53(d)):

$$F_c = 0.8x(350)\frac{0.85 f_{ck}}{\gamma_c'} = 0.8x(350)\frac{0.85(40)}{1.5}$$

$$= 6347x$$

By equilibrium, $F_p + F_c = 0$. Thus:

$$2400(190\,000)\left(-5522 \times 10^{-6} - \frac{0.0035(442 - x)}{x}\right) + 6347x = 0$$

Multiplying by x and rearranging gives:

$$6.347x^2 - 922x - 705\,432 = 0$$

The only positive root of this quadratic equation is:

$$x = 414 \text{ mm}$$

To check that the steel has not yielded, substitute $x = 414$ mm into the expression for ε_{pu} to give:

$$\varepsilon_{pu} = \left(-5522 \times 10^{-6} - \frac{0.0035(442 - x)}{x}\right)$$

$$= -0.005\,76$$

The initial yield strain (see Fig. 8.54) is $-0.8 f_{pk}/\gamma_p E_p = -(0.8 \times 1770)/(1.15 \times 190\,000) = -0.006\,48$. Thus, the assumption that the steel had not yielded is

351

correct. Hence, the ultimate moment capacity is:

$$M_{ult} = F_p z = F_c z = 6347xz$$

From Fig. 8.53:

$$z = (d - 0.4x) = (442 - 0.4(414)) = 276 \text{ mm}$$

Thus:

$$M_{ult} = 6347(414)(276) = 725 \times 10^6 \text{ Nmm}$$
$$= 725 \text{ kNm}$$

8.9 Partially prestressed members

Traditionally in the UK, only full prestressing has been practised, that is transfer and service stresses have limited to the extent that no cracking whatsoever can occur. Partial prestressing, in which some flexural cracking is allowed, has been applied in European and North American practice for many years and, with the advent of EC2, is now also allowed in the UK. The relaxation of the strict stress limits applied in the past is acceptable for most residential and office buildings and similar occupancies with a dry environment. Although practices vary, members in such environments are often designed to have no tensile stress under quasi-permanent loads. The EC2 recommendation for Exposure Class 1 (Table 6.2) is that crack widths should not exceed 0.2 mm under the frequent load combinations (see Table 3.10 for frequent load combination factor, ψ_1). Prestressed concrete exposed to a humid environment may also be partially prestressed but is designed to have no tensile stress within 25 mm of the duct under the frequent load combination.

There are two major potential benefits from using partial prestressing:

1. Cost savings, because less prestressing steel is required or a smaller concrete section can be used.
2. The upward deflection (camber) caused by prestressing can be significantly reduced by the reduction in prestress force that is possible with partial prestressing, as illustrated in Fig. 8.55(a).

Fig. 8.55 Deflections in simply supported member: (a) upward camber due to prestress and permanent gravity load; (b) total deflection due to prestress and permanent and variable gravity loads at SLS

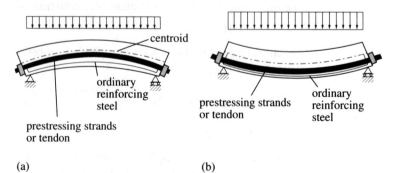

centroid

ordinary reinforcing steel

prestressing strands or tendon

prestressing strands or tendon

ordinary reinforcing steel

(a)

(b)

Partial prestressing is achieved by decreasing the initial stress in the prestressing steel or by reducing the area of prestressing steel provided. In the latter case, ordinary reinforcing steel may be needed to supplement the prestressing steel in order to obtain the required ultimate moment capacity. As this reinforcing steel is in compression under low loads (see Fig. 8.55(a)) it contributes substantially towards reducing camber.

Changes to the Magnel diagram

The potential for reducing the prestressing force is evident from a Magnel diagram such as that for Example 8.5 which is illustrated in Fig. 8.25. The inequality which dictates the minimum prestress force (maximum $1/P$) is inequality (8.28) which is:

$$\frac{1}{P}\sigma_{bmin} \leq \frac{1}{A} + \frac{e}{Z_b}$$

$$\Rightarrow \qquad \frac{1}{P} \leq \frac{1/A + e/Z_b}{\sigma_{bmin}} \qquad\qquad (8.62)$$

If cracking is allowed at the bottom fibre of the section, a greater tensile stress will be allowed and σ_{bmin} will be reduced (tension is negative). Thus, $(1/P)$ will be required to be less than a larger number. The result is that $(1/P)$ can be larger and P can be accordingly smaller. Therefore, the effect of relaxing this limit is greatly to expand the feasible zone and to reduce the required prestressing force.

Consider a design for decompression of the bottom fibre under permanent plus one-half of variable gravity load, that is the bottom fibre stress is allowed to become tensile only for load in excess of this. The moment due to permanent plus one half of the variable load is (refer to Example 8.4):

$$M_d = [(3.75 + 5 + 10)(9)^2]/8 = 190 \text{ kNm}$$

Thus, σ_{bmin} becomes:

$$\sigma_{bmin} = \text{greater of} \left(p_{0min} - \frac{M_0}{Z_b} \right) \text{ and } \frac{1}{\rho}\left(p_{Smin} - \frac{M_d}{Z_b} \right)$$

$$= \text{greater of} \left(-2.5 - \frac{41.1 \times 10^6}{-17.6 \times 10^6} \right)$$

$$\text{and } \frac{1}{0.75}\left(-2.0 - \frac{190 \times 10^6}{-17.6 \times 10^6} \right)$$

$$= 11.73 \text{ N/mm}^2$$

and inequality (8.62) becomes:

$$\frac{1}{P} \le \frac{1/162\,500 + e/(-17.6 \times 10^6)}{11.73}$$

$$\Rightarrow \quad \frac{1}{P} \le (524.6 - 4.844e) \times 10^{-9} \tag{8.63}$$

Cracking is likely to be sensitive to variations in prestress. EC2 therefore requires that computations for the prediction of cracking and decompression moments be based on characteristic values for prestress which are taken to be 0.9 and 1.1 times the mean. A force of 0.9P should therefore be used which changes inequality (8.63) to:

$$\frac{1}{P} \le 0.9(524.6 - 4.844e) \times 10^{-9} \tag{8.64}$$

This revised version of inequality (8.28) is plotted, together with all the previous inequalities, in Fig. 8.56. It is clear from the figure that partial prestressing substantially increases the size of the feasible zone. For Example 8.5, the change to partial prestressing allows the prestressing force to be reduced from 1300 kN to about 750 kN.

Fig. 8.56 Magnel diagram for partial prestressing in Example 8.5

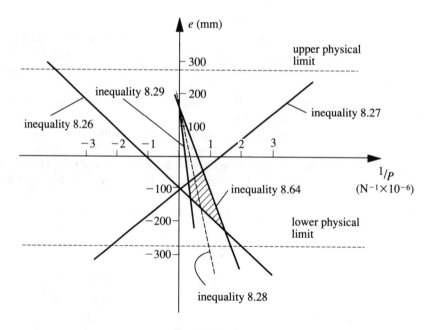

Clearly, with partial prestressing, a smaller section can be used than would otherwise have been possible. Variations on inequalities (8.12), (8.13), (8.16) and (8.17) can be used to determine the minimum required section moduli for preliminary design purposes. When the applied moment is sagging, inequalities (8.12) and (8.17) govern the design. Of these, inequality (8.12) ensures that there is capacity to provide a prestress which is sufficiently large to prevent tension at the top fibre at transfer but which does not result in excessive compressive stresses there at service. As it does not pertain to tension at service, this inequality is unchanged for the case of partial prestressing. Inequality (8.17), on the other hand, ensures a capacity to provide a prestress sufficiently large to prevent tension at the bottom fibre at service while not causing excessive compression there at transfer. Thus, for the case of partial prestressing, M_s should be replaced with the decompression moment M_d in inequality (8.17). Similarly, when the applied moment is hogging, inequalities (8.13) and (8.16) govern and M_s should be replaced with M_d in inequality (8.13).

In the above the design criterion has been to prevent decompression of the bottom fibre at service. The EC2 rule is in fact less stringent than this in that it specifies that decompression be prevented at a point above the bottom fibre (25 mm below the duct). The procedure in this case is identical to the above except that Z_b is replaced by I_g/x_d, where x_d is the distance from the centroid to the point where decompression is to be prevented (negative below centroid).

Crack control

Because tensile stresses are allowed to exceed the tensile strength of the concrete, partially prestressed concrete is expected to crack if the full design service load is reached during its design life. As for reinforced concrete, crack widths need to be kept within desired limits to inhibit the corrosion of the reinforcement. At a crack, the full tensile force is taken by the reinforcement. Between cracks, the bond between the reinforcement and the concrete gradually transfers the tension into the concrete as illustrated in Fig. 8.57(b). This transfer of stress builds up with distance from the crack as illustrated in Fig. 8.57(c). If the load is increased, a new crack occurs about half-way between existing cracks when the tensile stress reaches the tensile strength of the concrete. If the bond is poor, the build-up of tensile stress is slow and the distance between cracks is large. This results in a small number of large cracks. Thus, crack widths can be kept smaller by increasing bond through the use of a larger number of smaller-diameter bars (and hence a larger perimeter length per unit area).

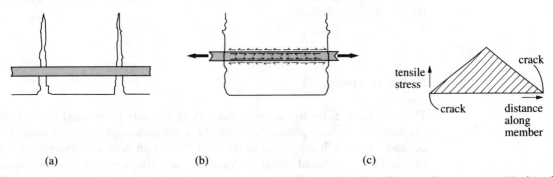

(a)   (b)   (c)

Fig. 8.57 Crack spacing: (a) reinforcement crossing cracks; (b) transfer of tensile stress into concrete; (c) plot of tensile stress in concrete versus distance from crack

In practice, concrete crack width is controlled in partially prestressed concrete by the area, spacing and level of stress in the ordinary reinforcement and by the level of prestressing. EC2 specifies an equation for the minimum area of ordinary reinforcement required to control cracking.

The selection of appropriate limits on crack width is a very subjective process and limits vary greatly between codes. The need for crack width computations can generally be avoided by designing members to remain in compression under frequent loads, that is allow decompression for load in excess of the frequent level only. When tension is allowed under frequent loads acceptable crack widths for the various classes of exposure are specified in EC2.

Deflections, fatigue and vibrations

Detailed discussion of deflections, fatigue and vibrations is beyond the scope of this book. Because of cracking, the deflection of partially prestressed concrete under moments higher than the decompression moment cannot be predicted using gross section properties. For realistic deflection estimates, the extent of cracking must be estimated at sections along the beam; the slope and deflection is then based on the curvature at these sections.

If moments due to cyclic load produce decompression at the level of the reinforcement, the stress range in the steel is substantially increased so that the fatigue life of the member is reduced and may need to be reviewed. This is not generally a concern for building structures because the cyclic loads usually do not produce tension in the concrete at the level of the tendon. Members should be designed so that they remain in compression at the level of the steel under those cyclic loads which can be expected to occur frequently.

Vibratory loads are generally low enough that decompression does not occur under these loads and the analysis can continue to be based on the gross section with no special attention needed to account for partial prestressing. When the decompression moment is exceeded so that cracking occurs, partially prestressed members are not as stiff as fully prestressed members. Thus the amplitude of vibrations may be greater. Cracks provide vibration damping and change the natural frequency of the member so that under some circumstances they could be beneficial.

Ultimate strength

The requirements for the ultimate strength of partially prestressed concrete are the same as for fully prestressed sections. Because less prestressing steel is, in general, required, it may have to be augmented with ordinary reinforcing steel to satisfy the required ultimate moment capacity. Some of this additional strength can be provided by the ordinary reinforcement provided for crack control.

Problems

Case study

A four-span continuous bridge section is to be designed. Because of symmetry, only half of the four-span bridge needs to be considered. This leaves ten sections for consideration, as illustrated in Fig. 8.58, each of which has the same geometric properties. The concrete to be used has a 28-day characteristic cylinder strength of 45 N/mm^2 and a transfer strength of 35 N/mm^2. The strands have an area of 165 mm^2, an ultimate strength of 1820 N/mm^2 and a modulus of elasticity of 190 000 N/mm^2. The applied bending moments are as follows:

Section No.	Minimum moment, M_s (greatest hog or least sag in kNm)	Maximum moment, M_s (greatest sag or least hog in kNm)
1	0	0
2	51	741
3	−37	831
4	−584	846
5	−1368	−86
6	−771	343
7	−43	1247
8	62	1167
9	−548	719
10	−1551	−328

The allowable tension at transfer = 1.0 N/mm^2, and at SLS = 0.0.

The self weight and permanent moments are as follows:

Section No.	Self weight and permanent moments, M_o
1	0
2	159
3	182
4	23
5	−387
6	−154
7	224
8	275
9	6
10	−486

The friction loss coefficients are $k = 0.008$ and $\mu = 0.19$ and the draw-in is 10 mm. The 1000 hour relaxation loss is 2.5 per cent and the shrinkage strain has been specified as 0.35×10^{-3}. The creep coefficient, φ, is 2.0. The tendons in the edge section illustrated are stressed last, and these are stressed in turn using multistrand jacks, starting with tendons no. 1.

Section 8.4

8.1 (a) Calculate the limits on the stress due to prestress for sections 1 and 10 of Fig. 8.58.

(b) Verify that the section moduli at these sections are in excess of the minimum required.

Fig. 8.58 Case study for Problems 8.1, 8.4–8.7: (a) longitudinal section; (b) transverse section through the edge of the bridge

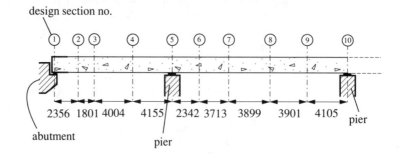

design section no.

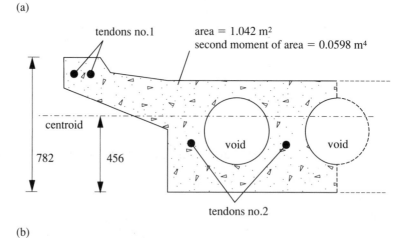

(a)

tendons no.1

area = 1.042 m²

second moment of area = 0.0598 m⁴

centroid

void   void

tendons no.2

(b)

8.2 In the unsymmetrical I-beam illustrated in Fig. 8.59, the permissible stresses are:

Fig. 8.59 Beam of Problem 8.2

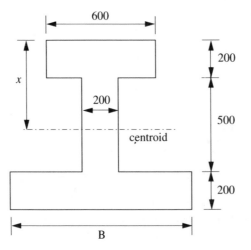

Transfer: minimum $= -1 \, \text{N/mm}^2$ (i.e. tension not to exceed $1 \, \text{N/mm}^2$)
maximum $= 20 \, \text{N/mm}^2$

SLS: minimum $= 0 \, \text{N/mm}^2$
maximum $= 20 \, \text{N/mm}^2$

Find the breadth of the bottom flange, B, for which both section moduli equal their minimum values at an applied service moment of M_s and an applied transfer moment of M_0. Assume a loss of 20 per cent between transfer and service.

8.3 Show from first principles that the prestressed beam section illustrated in Fig. 8.60 is adequate given that the permissible stresses are:

Transfer: minimum $= -1 \, \text{N/mm}^2$
maximum $= 15 \, \text{N/mm}^2$

SLS: minimum $= 0 \, \text{N/mm}^2$
maximum $= 20 \, \text{N/mm}^2$

The section is subjected to an applied transfer moment, $M_o = 112 \, \text{kNm}$ and an applied service moment, $M_s = 500 \, \text{kNm}$. A loss of 15 per cent may be assumed between transfer and service.

Fig. 8.60 Beam of Problem 8.3

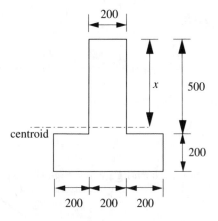

Section 8.5

8.4 With reference to the bridge illustrated in Fig. 8.58:

(a) construct the Magnel diagrams for sections 1 and 10;
(b) given that the Magnel diagrams for all other sections are as illustrated in Fig. 8.61, decide on a prestress force for preliminary design.

8.5 Using the Magnel diagrams of Problem 8.4, calculate the limits on tendon eccentricity for sections 1 and 10 given that the prestress force has been estimated at 10 000 kN.

8.6 For the bridge illustrated in Fig. 8.58, the feasible zone for tendons no. 2 is as illustrated in Fig. 8.62. Determine a preliminary profile.

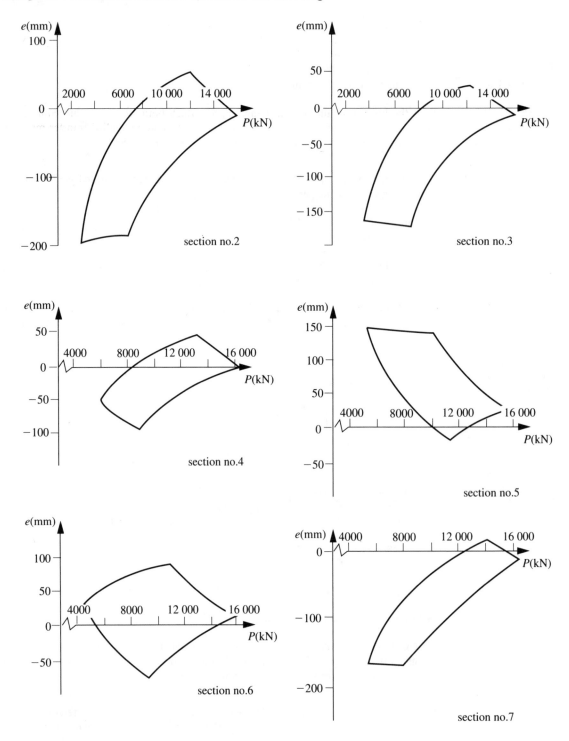

Fig. 8.61 Magnel
diagrams for
Problem 8.4(b)

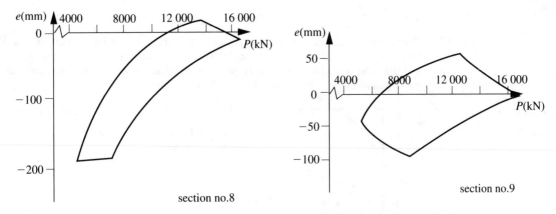

section no.8

section no.9

Fig. 8.61
(*continued*)

Fig. 8.62 Feasible
zone of bridge of
Fig. 8.58 for
Problem 8.6

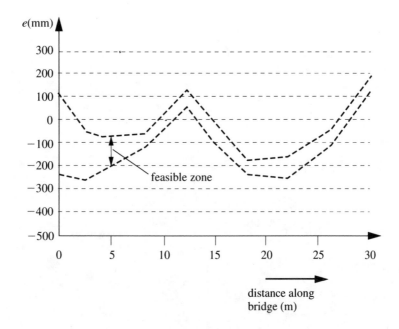

Section 8.6

8.7 For tendons 1 and 2 in the bridge of Fig. 8.58:

(a) calculate the wobble losses, the draw-in losses and the maximum internal force after
draw-in;

(b) calculate the elastic shortening losses and the creep losses for sections 1 and 5.

Section 8.8

8.8 Determine the ultimate moment capacity of the pretensioned beam illustrated in Fig. 8.63 given that the strands are stressed to 70 per cent of their ultimate strength at transfer and the total post-transfer loss is 18 per cent. Assume no secondary effects and use the stress/strain diagram for strand illustrated in Fig. 8.54.

Fig. 8.63 Beam of Problem 8.8

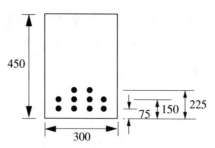

Area of strand = 150 mm²
E_p = 190 000 N/mm²
f_{pk} = 1770 N/mm²
f_{ck} = 45 N/mm²

9

Combined Axial Force and Bending of Reinforced Concrete Members

9.1 Introduction

Axially loaded members are classified as those which carry their load primarily in tension or compression. Since tension members are not commonly used in concrete structures (remember concrete is weak in tension), this chapter deals for the most part with the design of compression members.

The majority of compression/tension members carry a portion of their load in bending. This may be due to the load not being applied at the centroid of the member (i.e. load is applied eccentrically), as illustrated in Fig. 9.1(a). Alternatively, bending moments in a compression member may result from unbalanced moments in the members connected to its ends, as illustrated in Figs 9.1(b) and (c). The result of such bending moments in axially loaded members is to reduce the range of axial force which the member can safely carry. For this reason, it is essential that the effects of bending in axially loaded members are considered.

In this chapter, one-dimensional compression members are often referred to as columns and two-dimensional members as walls. Strictly speaking, a column is a particular type of compression member which is vertical. However, the term 'column' is used loosely to describe all one-dimensional compression members, irrespective of their orientation.

9.2 Classification of compression members (columns)

Braced/unbraced columns

As was stated above, the axial load-carrying capacity of a member depends on the magnitude of bending moment in it. An unbraced structure is one in which frame action is used to resist horizontal (wind) loads. In such a structure, the horizontal loads are transmitted to the foundations through bending action in the beams and columns. The moments in the columns due to this bending can substantially reduce their axial (vertical) load-carrying capacity. Unbraced structures are generally quite flexible and allow horizontal displacement as illustrated in Fig. 9.2. When this displacement is sufficiently large to influence

Fig. 9.1 Columns subjected to axial load and moment: (a) eccentric load; (b) frame; (c) balanced and out-of-balance moments

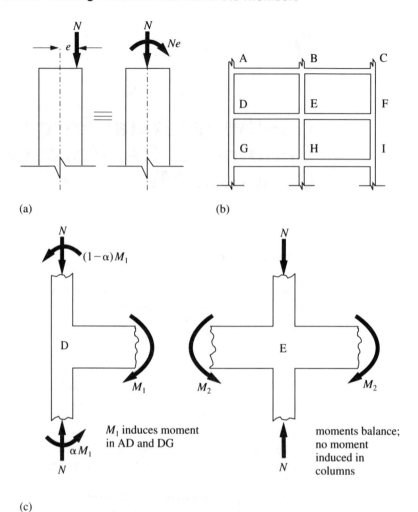

(a)

(b)

M_1 induces moment in AD and DG

moments balance; no moment induced in columns

(c)

Fig. 9.2 Sway frame

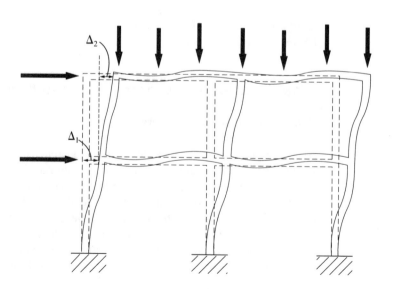

significantly the column moments, the structure is termed a **sway frame**. Columns in a sway structure undergo lateral displacement between their ends, as illustrated in Fig. 9.2, which reduces their resistance to buckling. For this reason, such columns, termed **unbraced** columns, are treated differently from a design viewpoint than other columns. In an unbraced column, the bending moment is increased by an additional amount $N\Delta$, where N is the axial force and Δ is the relative horizontal displacement of the ends of the column. Thus, to maximize the axial load capacity of columns, non-sway structures should be used whenever possible. Fully non-sway structures are difficult to achieve in practice but EC2 allows a structure to be classified as non-sway if it is braced against lateral loads using substantial bracing members such as cores and/or shear walls (specific values for the minimum stiffness of bracing elements for non-sway structures are also given in the code). A column within such a non-sway structure is considered to be **braced** and the second-order moment on such columns, $N\Delta$, is negligible.

Slender/short columns

When an unbalanced moment or a moment due to eccentric loading (Fig. 9.1) is applied to a column, the member responds by bending, as illustrated in Fig. 9.3. If, as shown, the deflection at the centre of the member is δ, then at the centre there is a force N and a total moment of $M + N\delta$. The second-order bending component, $N\delta$, is due to the extra eccentricity of the axial force which results from the deflection. If the column is short and squat, δ is small and this second-order moment is negligible. If, on the other hand, the column is long and slender, δ is large and $N\delta$ must be calculated and added to the applied moment, M. A column in which the moment $N\delta$ is negligible is commonly known as a **short column**, while a column in which $N\delta$ is not negligible is known as a **slender column**. It should be noted that in the case of tension members, there is no second-order bending component.

Slenderness

The significance of the $N\delta$ term (i.e. whether a column is short or slender) is normally defined by a **slenderness ratio**, which is a function of the parameters which determine the lateral deflection of the column. In EC2, the slenderness ratio is defined by:

$$\lambda = l_e/r \tag{9.1}$$

where l_e is the effective length (explained below) of the member and r is the radius of gyration. The radius of gyration is equal to:

$$r = \sqrt{I/A} \tag{9.2}$$

where I is the second moment of area of the section (usually taken as the gross value, I_g) and A is its cross-sectional area (usually taken as the gross area, A_g). By definition, second-order moments in a column can be ignored if:

$$\lambda \leq \text{greater of 25 and } 15/\sqrt{v_u} \tag{9.3}$$

Fig. 9.3 Additional moment due to deflection: (a) geometry and loading; (b) deflected shape; (c) additional moment, $N\delta$

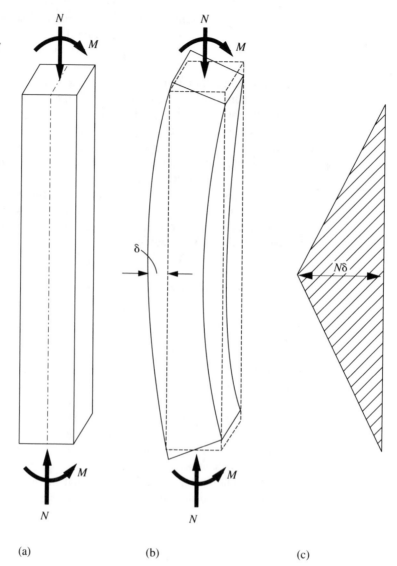

(a)   (b)   (c)

where v_u is the longitudinal force coefficient which is equal to $(N\gamma_c)/(A_g f_{ck})$. In other words, columns which satisfy inequality (9.3) are classified as **short columns** while those which do not satisfy inequality (9.3) are classed as **slender columns**.

Effective length of columns

The effective length of a column can be looked on as the length which is effective against buckling. The greater the effective length, the more likely the column is to buckle. The effective length is dependent on the support conditions at the ends of the column, whether or not the column is braced and, of course, the actual length (height) of the member. Theoretically, the effective length of a column is equal to the distance between the points of contraflexure in the member.

Fig. 9.4 Effective lengths of columns in non-sway structures: (a) pinned–pinned column; (b) fixed–fixed column

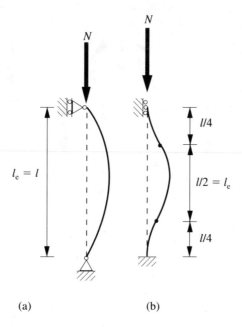

(a)   (b)

A braced column pinned at both ends (pinned–pinned) in a non-sway frame buckles in a straightforward manner and its effective length, l_e, is equal to its actual length, l, as illustrated in Fig. 9.4(a). However, a fixed–fixed column (no rotation at either end) in a non-sway frame is much less likely to buckle and its effective length is substantially less. The points of contraflexure in a braced fixed–fixed column occur at distances of $l/4$ from the fixed ends, as illustrated in Fig. 9.4(b). Thus, the effective length of a fixed–fixed braced column is half its actual length.

For a fixed–fixed column in a sway frame (i.e. unbraced), such as that illustrated in Fig. 9.5(a), there is only one point of contraflexure which occurs at mid-span. However, another point of contraflexure exists on an imaginary line produced through the member as illustrated in the figure. Thus, the effective length of the unbraced column of Fig. 9.5(a) is equal to its actual length. For an unbraced column, fixed against rotation at one end and pinned at the other (Fig. 9.5(b)), the effective length is equal to twice the actual length. Notice that the effective length in an unbraced frame is always greater than that in a braced frame.

In real structures, the ends of columns are neither pinned nor fixed but are actually something in between. The actual rotational restraint provided by adjoining members at the end of a column can be expressed by a function of their stiffnesses (refer to Appendix A):

$$k = \frac{\sum (E_c I_{col}/l)}{\sum (\alpha E_c I_b/l_{eff})} \tag{9.4}$$

where the summation is for all beams or columns meeting at the joint and where:

I_{col} = second moment of area of columns
I_b = second moment of area of beams

Combined Axial Force and Bending of Reinforced Concrete Members

Fig. 9.5 Effective
lengths in sway frames:
(a) fixed–fixed column;
(b) pinned–fixed column

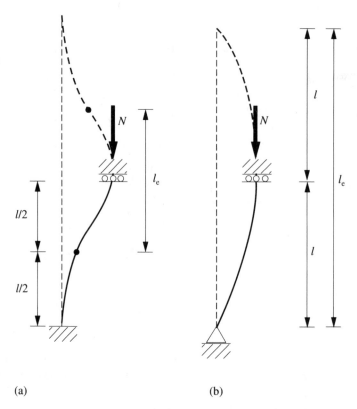

(a)   (b)

l = length of column measured between centres of connections
l_{eff} = effective length of beams (typically measured from centre to centre of columns)
α = factor taking into account the conditions of restraint of the beam at the opposite end: $\alpha = 1.0$ for opposite end rigidly restrained; $\alpha = 0.5$ for opposite end free to rotate; $\alpha = 0$ for cantilever

Once the rotational stiffness, k, for each end of the column is known, the effective length can be computed as:

$$l_e = \beta l \tag{9.5}$$

The value of β is most readily calculated using alignment charts or nomograms. The nomogram recommended for use by EC2 is illustrated in Fig. 9.6. Its use is illustrated in the following example.

Example 9.1 Classification of column

Problem For the frame of Fig. 9.7, determine whether the column is short or slender. It may be assumed that the same grade of concrete is used for all members (i.e. E_c constant throughout) and that $f_{ck} = 25 \text{ N/mm}^2$. The applied axial force is 525 kN.

Solution For members AB and BC, the second moment of area is:

$$I_b = bh^3/12 = 0.3(0.6)^3/12 = 5.4 \times 10^{-3} \text{ m}^4$$

Fig. 9.6 Nomograms for the calculation of effective length (from EC2): (a) non-sway frame; (b) sway frame. k_A and k_B are the rotational restraints at ends A and B (equation 9.4)

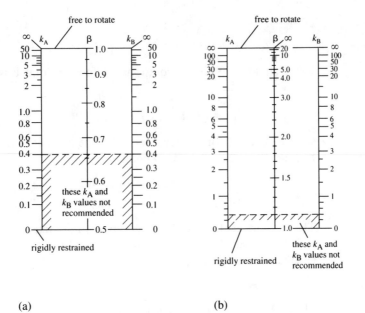

(a)   (b)

Fig. 9.7 Column of Example 9.1: (a) geometry; (b) cross sectional dimensions

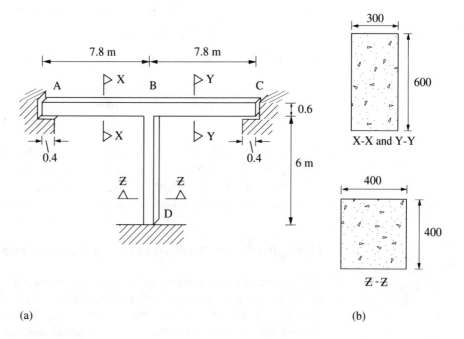

(a)   (b)

For member BD:

$$I_{col} = bh^3/12 = (0.4)^4/12 = 2.13 \times 10^{-3} \text{ m}^4$$

The effective length of the beams, l_{eff}, is equal to:

$$l_{eff} = 7.8 - 0.2 - 0.2 = 7.4 \text{ m}$$

The length of the column, l, measured between the centres of connections is given by:

$$l = 6.0 + 0.3 = 6.3 \text{ m}$$

Using equation (9.4), the rotational stiffness at joint B is:

$$k_B = \frac{\sum (E_c I_{col}/l)}{\sum (\alpha E_c I_b/l_{eff})} = \frac{\sum (I_{col}/l)}{\sum (\alpha I_b/l_{eff})}$$

since E_c is equal for all members. Since the opposite ends of the beams are free to rotate, $\alpha = 0.5$. Thus, k_B becomes:

$$k_B = \frac{(2.13 \times 10^{-3})/6.3}{0.5[(5.4 \times 10^{-3})/7.4 + (5.4 \times 10^{-3})/7.4]} = 0.46$$

The support at the base of the column is fully fixed. Therefore, the rotational stiffness of joint D is $k_D = 0$. However, EC2 recommends that k not be taken less than 0.4. The effective length of the column is given by equation (9.5), that is:

$$l_e = \beta l$$

In Fig. 9.6(a), a straight line from 0.46 on the left ($k_B = 0.46$) and 0.4 on the right ($k_D = 0.4$) gives β as 0.67 (for non-sway frame). Therefore, the effective length is:

$$l_e = 0.67l = 0.67(6.3) = 4.22 \text{ m}$$

The slenderness ratio, λ, is given by equation (9.1) as:

$$\lambda = \frac{l_e}{r} = \frac{l_e}{\sqrt{I/A}} = \frac{4.22}{\sqrt{(2.13 \times 10^{-3})/0.4^2}}$$

$$\Rightarrow \qquad \lambda = 37$$

From equation (9.3), a column is short if λ is less than the greater of 25 and $15/\sqrt{v_u}$, where $v_u = 525 \times 10^3 \times 1.5/(400^2 \times 25) = 0.197$, that is $\lambda < 34$. As this is not the case, the column is classified as 'slender'.

9.3 Design of short members for axial force

The reinforcement detail in a typical reinforced concrete column is illustrated in Fig. 9.8. In compression, both the longitudinal steel and the concrete contribute to the resistance of the applied axial force. The stirrups serve to confine the concrete and prevent buckling of the longitudinal reinforcement in compression. For the design of short columns in pure compression, EC2 limits the strain in the concrete to 0.002, since generally this is the strain at which the stress in the concrete is maximum (refer to Fig. 7.17). Based on the stress/strain curve of Fig. 7.22(a) or (b), reinforcement with strength $f_y = 460 \text{ N/mm}^2$ will also have yielded at this strain (assuming $E_s = 200\,000 \text{ N/mm}^2$). The capacity to resist compressive force, N_{ult}, is approximately equal to:

$$N_{ult} = \alpha f_{ck}(A_g - A_s) + f_y A_s \tag{9.6}$$

Fig. 9.8 Typical
connection
reinforcement: (a) part
elevation; (b) section
X–X

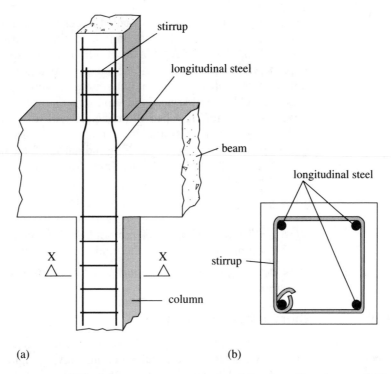

(a)   (b)

where α = coefficient defined in section 7.5, generally taken as 0.85
$\quad\quad f_{ck}$ = characteristic compressive cylinder strength of the concrete
$\quad\quad A_g$ = gross cross-sectional area (bh)
$\quad\quad A_s$ = area of longitudinal reinforcement
$\quad\quad f_y$ = yield strength of reinforcement

Once N_{ult} is reached, failure occurs by crushing of the concrete or local buckling
of the longitudinal reinforcement between the links. Adjusting equation (9.6)
to allow for variability in the material strengths, it becomes:

$$N_{ult} = \frac{0.85 f_{ck}}{\gamma_c}(A_g - A_s) + \frac{f_y}{\gamma_s}A_s \tag{9.7}$$

where γ_c and γ_s are the partial factors of safety for concrete and steel,
respectively.

In tension, it is common practice to ignore the small tensile strength of
concrete and to assume that resistance is only provided by the reinforcement,
that is $N_{ult} = (f_y/\gamma_s)A_s$. Although tension does not often arise in reinforced
concrete members, it does occur in some circumstances. For example, pure
tension does occur in deep beams as described in section 9.7.

Detailing requirements

Most codes of practice place restrictions on the maximum and minimum areas
of reinforcement provided in columns. As for beams, a maximum percentage,

371

$A_{s,max}$, of longitudinal steel is specified to ensure that the steel yields before the concrete crushes at the ultimate limit state, thus ensuring ductile behaviour. It is also important to limit the area of reinforcement so that there will be room for concrete to pass between the bars, particularly at laps. Limits on the minimum percentage of longitudinal reinforcement, $A_{s,min}$, ensure that the steel does not yield under service loads and that there is sufficient reinforcement to prevent sudden failure when the concrete reaches its tensile strength (under ULS loads). Upper limits are also commonly specified on the spacing of links, s_{max}, to prevent buckling of the longitudinal reinforcement. For an applied ultimate compressive force N, the limits for reinforcement in columns recommended by EC2 are as follows:

(a) $A_{s,min}$ = greater of $0.15N/(f_y/\gamma_s)$ and $0.003A_g$
(b) $A_{s,max} = 0.08A_g$ (even where bars are lapped)
(c) s_{max} = lesser of $12\phi_{min}$, 300 mm and the least cross-sectional dimension of the column, where ϕ_{min} is the minimum diameter of the longitudinal bars

These detailing requirements apply to all columns, short or slender, braced or unbraced.

Example 9.2 Short circular column

Problem Calculate the ultimate capacity to resist compressive and tensile force for a 300 mm diameter, short circular column having eight 16 mm diameter bars, where $f_y = 460$ N/mm² and $f_{ck} = 35$ N/mm².

Solution The gross area of the section is:

$$A_g = \pi(300)^2/4 = 70\ 686 \text{ mm}^2$$

The total area of reinforcement is:

$$A_s = 8\pi(16)^2/4 = 1608 \text{ mm}^2$$

From equation (9.7), the ultimate compressive force is:

$$N_{ult} = \frac{0.85f_{ck}}{\gamma_c}(A_g - A_s) + \frac{f_y}{\gamma_s}A_s$$

Taking $\gamma_c = 1.5$ and $\gamma_s = 1.15$ gives:

$$N_{ult} = \frac{0.85(35)}{1.5}(70\ 686 - 1608) + \frac{460}{1.15}1608$$

$$= 2\ 013\ 000 \text{ N}$$

$$= 2013 \text{ kN}$$

In tension, the ultimate axial load is simply given by:

$$N_{ult} = -\frac{f_y}{\gamma_s}A_s = -\frac{460}{1.15}1608 \text{ N} = -643 \text{ kN}$$

Example 9.3 Short rectangular column

Problem Design a short rectangular column to carry a design ultimate compressive force of 2500 kN given $f_{ck} = 40$ N/mm² and $f_y = 460$ N/mm².

Solution From equation (9.7):

$$N_{ult} = \frac{0.85 f_{ck}}{\gamma_c}(A_g - A_s) + \frac{f_y}{\gamma_s}A_s$$

Assume an area of longitudinal reinforcement, A_s, equal to $0.02A_g$ (see Chapter 6 for preliminary design recommendations). For a design load of 2500 kN, equation (9.7) then becomes:

$$2500 \times 10^3 \leq \frac{0.85 f_{ck}}{\gamma_c}(1 - 0.02)A_g + \frac{f_y}{\gamma_s}0.02A_g$$

$$\Rightarrow \quad 2500 \times 10^3 \leq \frac{0.85(40)}{1.5}(0.98A_g) + \frac{460}{1.15}0.02A_g$$

$$\Rightarrow \quad A_g \geq 82\ 745 \text{ mm}^2$$

$$\Rightarrow \quad \sqrt{A_g} \geq 288 \text{ mm}$$

Therefore, take a square section of side length 300 mm. For this section, $A_g = 90\ 000$ mm². The exact area of reinforcement required is found using equation (9.7), that is:

$$2500 \times 10^3 \leq \frac{0.85(40)}{1.5}(90\ 000 - A_s) + \frac{460}{1.15}A_s$$

$$\Rightarrow \quad 2500 \times 10^3 \leq 2.04 \times 10^6 + 377A_s$$

$$\Rightarrow \quad A_s \geq 1220 \text{ mm}^2$$

Thus, adopt four 20 mm diameter bars of total area $A_s = 1257$ mm².

9.4 Design of short members for axial force and uniaxial bending

When the load applied to a column is eccentric about one axis only, as illustrated in Fig. 9.9, a bending moment results about that one axis and the column is said to be subjected to uniaxial bending. Uniaxial bending also occurs when unbalanced moments are transferred from beams or slabs to the column and are confined to a single axis of the column. As stated in section 9.1, the presence of this form of bending in axially loaded members can reduce the axial load capacity of the member. At the serviceability limit state, the applied axial compression in columns tends to close up any cracks caused by bending. Thus, the presence of compression with bending is actually beneficial at SLS and crack widths are unlikely to be excessive. It is therefore the combined effect of axial compression and bending at the ultimate limit state that tends to govern the design. In the case of tension members, cracking at SLS may cause problems and does need to be checked.

Fig. 9.9 Column with eccentric load: (a) geometry and loading; (b) section A–A

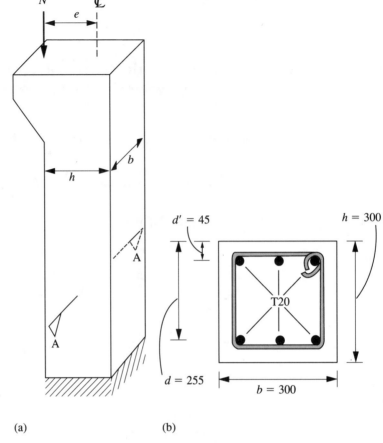

(a)   (b)

Imperfections

To cover dimensional inaccuracies and any uncertainty as to the location of the applied axial load, an eccentricity, e_a, should always be assumed. In the case of eccentrically applied loads, e_a should be added to the known eccentricity of the load in the least favourable direction. Thus, the design moment acting on a short member due to an applied force N at an eccentricity e becomes:

$$M = N(e + e_a) \tag{9.8}$$

The value for e_a recommended in EC2 is given by:

$$e_a = \nu l_e / 2 \tag{9.9}$$

where l_e is the effective length of the member and ν is the inclination of the member from the vertical. In most cases, the value of ν for a non-sway structure of total height l_{tot} can be taken as:

$$\nu = \text{greater of } \frac{1}{100\sqrt{l_{tot}}} \text{ and } \frac{1}{400} \tag{9.10}$$

Ultimate strength

The ultimate strength of short members subjected to combined axial force and bending is evaluated using the same techniques as those described in Chapter 7 for members in pure bending. As for members in pure bending, the ultimate limit state of failure is reached when the strain in the concrete reaches a specified ultimate value, ε_{ult}. The value for ε_{ult} is taken as 0.0035 by EC2. To simplify design, EC2 also allows the use of the simplified stress diagrams of Fig. 7.21 in place of the actual stress distribution.

As an example, consider the member of Fig. 9.9 which is subjected to a compressive force N at an eccentricity e about the member centre line. To highlight the similarity with beam design, the column with its eccentric load is plotted on its side in Fig. 9.10(a) and the equivalent loading of N and $M = N(e + e_a)$ is illustrated in Fig. 9.10(b). The strain and (equivalent rectangular)

Fig. 9.10 Equivalent loading on eccentrically loaded column: (a) column with original loading; (b) equivalent loading

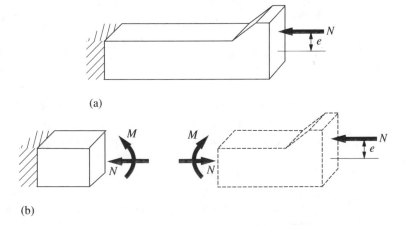

(a)

(b)

stress distributions due to the combined effects of M and N are illustrated in Figs 9.11(a) and (b). As the section is on the point of failure, the maximum compressive strain has reached its ultimate value and the distribution of stress is also that for a section on the point of failure. The precise location of the neutral axis depends on the relative magnitude of the compressive force N and the moment M. In broad terms, two design cases can be distinguished:

1. The neutral axis lies between the rows of reinforcement as normally occurs in beams (Fig. 9.11(a)). In this case, the reinforcement is in tension on one side of the neutral axis and in compression on the other side.
2. The neutral axis lies outside the section and the entire section is in compression (Fig. 9.11(b)). In this case, all of the reinforcement (both layers) is in compression.

Each location of the neutral axis corresponds to a unique combination of M and N. The stresses in the reinforcement and the concrete can be determined for any given location of the neutral axis and, from this, the respective values of moment and axial force that would cause this distribution of stress can be calculated.

Fig. 9.11 Alternative
strain and stress
distributions: (a) case
(1); (b) case (2)

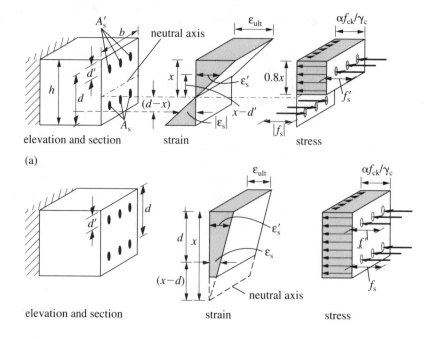

(a)

Consider the case where the neutral axis lies between the rows of reinforcement, at a depth x from the extreme fibre in compression as illustrated in Fig. 9.11(a). The compressive force in the concrete, F_c, is given by:

$$F_c = 0.8xb \frac{\alpha f_{ck}}{\gamma_c} \qquad (9.11)$$

By similar triangles of the strain diagram, the strain in the 'top' layer of reinforcement is:

$$\varepsilon'_s = \frac{\varepsilon_{ult}(x - d')}{x} \qquad (9.12)$$

Similarly, the strain in the 'bottom' layer of reinforcement is:

$$\varepsilon_s = -\frac{\varepsilon_{ult}(d - x)}{x} \qquad (9.13)$$

When the neutral axis lies outside the section, equations (9.12) and (9.13) still hold true but, as $(d - x)$ will be negative, equation (9.13) gives a compressive strain.

The corresponding steel stresses, f'_s and f_s, are found from the stress/strain diagram of Fig. 7.22(a) or (b). The compressive force in the top reinforcement is then:

$$f'_s A'_s \qquad (9.14)$$

and in the bottom reinforcement is:

$$f_s A_s \qquad (9.15)$$

When the neutral axis lies between the rows of reinforcement, f_s is negative giving a negative compressive force, that is a tensile force. By equilibrium of all axial forces, the applied axial force which has caused this distribution of stress must be:

$$N_{ult} = F_c + f'_s A'_s + f_s A_s \tag{9.16}$$

where f'_s and f_s are both considered positive when compressive. Similarly, by taking moments about a point on the section, the ultimate moment capacity in the presence of this axial force can be found. Moments can be taken about any point and it is conventional, for pure bending, to take moments about the neutral axis. However, in the case of combined axial force and bending, this would involve a component due to N_{ult} which is conventionally applied at the centre line (an axial force applied at any point is equivalent to an axial force at the centre line plus a moment). Accordingly, it is more convenient to take moments about the **centre line**. Hence:

$$M_{ult} = F_c\left(\frac{h}{2} - 0.4x\right) + f'_s A'_s\left(\frac{h}{2} - d'\right) - f_s A_s\left(d - \frac{h}{2}\right) \tag{9.17}$$

The third term is subtracted from the others because a compressive force in the 'bottom' reinforcement acts against the other components of moment. When the neutral axis lies between the layers of reinforcement, f_s is negative and this term becomes positive.

In order to design a section to carry an ultimate design moment and axial force, the procedure outlined above must be repeated using different values of x until the value for N_{ult} equals the applied ultimate force. Then M_{ult} is calculated and compared with the applied ultimate moment. This procedure can be very laborious and as a result it is common practice to use a design aid known as an **interaction diagram**. The following example illustrates the construction of one such interaction diagram.

Example 9.4 Construction of interaction diagram

Problem Construct the interaction diagram for the column of Fig. 9.9, assuming that all six bars are high tensile reinforcement ($f_y = 460$ N/mm²) of 20 mm diameter (total area equals 1885 mm²) and that the characteristic strength of the concrete is $f_{ck} = 35$ N/mm². Use the EC2 equivalent rectangular stress block for concrete in compression.

Solution An interaction diagram is a chart or graph illustrating the capacity of a column to resist a range of combinations of force and moment. It is found by assuming a number of strain distributions (i.e. assume a number of values for x) and calculating in each case the combination of force, N_{ult}, and moment, M_{ult}, that would cause that strain distribution. These results form isolated points on a plot of N_{ult} versus M_{ult} and, once sufficient points have been determined, the points can be joined to form a design curve representing the general solution.

For the first point on the interaction diagram, assume that the column is in pure compression under the action of a force at the centre line (i.e. $M_{ult} = 0$ and x is infinite). As stated in section 9.3, the strain at failure is then uniform

377

Combined Axial Force and Bending of Reinforced Concrete Members

Fig. 9.12 Strain distributions for Example 9.4: (a) elevation and section; (b) pure compression (A); (c) balanced failure (B); (d) $x = 45$ (C); (e) $x = 100$ (D); (f) $x = 300$ (E); (g) pure tension (F)

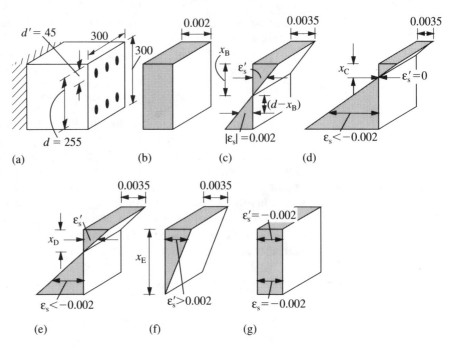

across the section and equal to 0.002, as illustrated in Fig. 9.12(b). From equation (9.7), the ultimate force capacity is:

$$
\begin{aligned}
N_{\text{ult}} &= \frac{0.85 f_{\text{ck}}}{\gamma_{\text{c}}} (A_{\text{g}} - A_{\text{s}}) + \frac{f_{\text{y}}}{\gamma_{\text{s}}} A_{\text{s}} \\
&= \frac{0.85(35)}{1.5} [(300)^2 - 1885] + \frac{460}{1.15} 1885 \text{ N} \\
&= 2502 \text{ kN}
\end{aligned}
$$

This result forms the first point, point A, on the interaction diagram for the column as illustrated in Fig. 9.13.

Next, consider the case when the strain in the tension reinforcement is on the point of yielding at the same time as concrete is failing in compression. This is referred to as balanced design and the corresponding force capacity is denoted N_{bal}. However, it differs slightly from the definition of balanced design given in section 7.6 in that material factors of safety, γ_{s} and γ_{c}, are included here. Thus, the strain in the bottom reinforcement is:

$$
\varepsilon_{\text{s}} = -\frac{f_{\text{y}}}{\gamma_{\text{s}} E_{\text{s}}} = -0.002
$$

The complete strain distribution for this case is illustrated in Fig. 9.12(c). By similar triangles:

$$
\frac{0.002}{d - x_{\text{B}}} = \frac{0.0035}{x_{\text{B}}}
$$

$$
\Rightarrow \quad x_{\text{B}} = \frac{0.0035(255)}{(0.002 + 0.0035)} = 162 \text{ mm}
$$

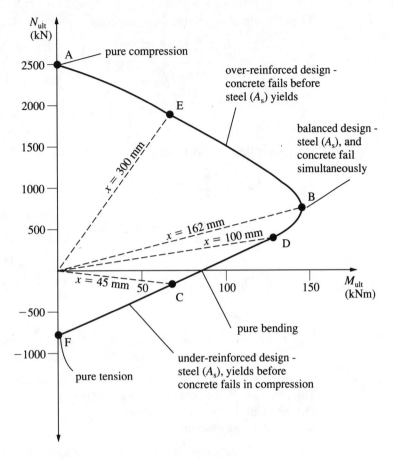

Fig. 9.13 Interaction diagram of Example 9.4

The strain in the top reinforcement, A_s', is found by similar triangles:

$$\frac{\varepsilon_s'}{x_B - d'} = \frac{0.0035}{x}$$

$$\Rightarrow \quad \varepsilon_s' = \frac{0.0035(162 - 45)}{162} = 0.002\,53$$

As this exceeds the yield strain of 0.002, the top reinforcement has yielded in compression.

The compressive force in the concrete, F_c, is given by equation (9.11):

$$F_c = 0.8 x_B b \left(\frac{\alpha f_{ck}}{\gamma_c} \right)$$

$$= 0.8(162)(300)\left(\frac{0.85 \times 35}{1.5} \right)$$

$$= 771\,120 \text{ N}$$

The ultimate compressive force capacity is given by equation (9.16), that is:

$$N_{ult} = F_c + f'_s A'_s + f_s A_s$$

$$= 771\,120 + (f_y/\gamma_s)\left(\frac{1885}{2}\right) - (f_y/\gamma_s)\left(\frac{1885}{2}\right)$$

$$= 771\,120 \text{ N}$$

$$= 771 \text{ kN}$$

Corresponding to this force we have, from equation (9.17):

$$M_{ult} = F_c\left(\frac{h}{2} - 0.4x_B\right) + f'_s A'_s\left(\frac{h}{2} - d'\right) - f_s A_s\left(d - \frac{h}{2}\right)$$

$$= 771\,120[150 - (0.4)(162)] + \left(\frac{460}{1.15}\right)\left(\frac{1885}{2}\right)(150 - 45)$$

$$- \left(\frac{-460}{1.15}\right)\left(\frac{1885}{2}\right)(255 - 150)$$

$$= 145 \times 10^6 \text{ Nmm}$$

$$= 145 \text{ kNm}$$

This result is represented by point B on the interaction diagram of Fig. 9.13. A further four design cases have been considered as follows:

(i) point C: $x = 45$ mm (zero strain in top reinforcement)
(ii) point D: $45 < x < 162$ mm (say, 100 mm)
(iii) point E: $162 < x$ (say, 300 mm)
(iv) point F: member in pure tension

The strain distributions for these cases are illustrated in Figs 9.12(d), (e), (f) and (g), respectively. The solution in each case is derived in the same way as above and each solution is illustrated on the interaction diagram of Fig. 9.13. In the case of pure tension, the ultimate tensile force is simply given by $-(f_y/\gamma_s)(A_s + A'_s)$ and the moment is zero.

The complete interaction diagram of Fig. 9.13 represents all combinations of N and M that can cause failure of the column of Fig. 9.9 where the reinforcement is 2.1 per cent of gross area. All combinations of axial force and moment between points A and B on the diagram will cause the concrete to fail in compression before the bottom reinforcement, A_s, yields. On the other hand, all combinations between points B and F will result in the tensile yielding of A_s before the concrete fails in compression.

For a complete design chart, interaction diagrams are plotted for other percentages of reinforcement in Fig. 9.14 up to the maximum of 8 per cent allowed in EC2. If the depth, breadth, location of reinforcement (d or d') or concrete strength of the member is altered, the interaction diagram must be redrawn, that is, other design charts must be used. Interaction diagrams are published for a wide range of cross-sectional geometries and concrete strengths and are used extensively in design offices for the design of columns.

Fig. 9.14 Interaction diagram for different percentages of reinforcement

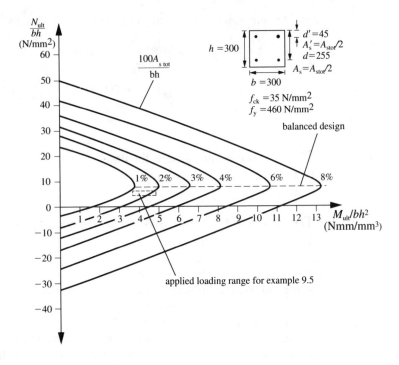

Example 9.5 Application of design chart

Problem   Using an appropriate design chart, calculate the area of reinforcement required to resist an ultimate compressive force which varies in the range 440–600 kN combined with a uniaxial ultimate moment in the range 100–130 kNm. The column is short and has the dimensions and material properties given for Example 9.4.

Solution   The ratio N/bh ranges between:

$$\frac{440 \times 10^3}{(300)^2} = 4.89 \quad \text{and} \quad \frac{600 \times 10^3}{(300)^2} = 6.67$$

Similarly, the ratio M/bh^2 ranges between:

$$\frac{100 \times 10^6}{(300)^3} = 3.70 \quad \text{and} \quad \frac{130 \times 10^6}{(300)^3} = 4.82$$

This range of combinations of N/bh and M/bh^2 is indicated in Fig. 9.14. It can be seen that for this example (and indeed throughout the interaction diagram) increasing the moment increases the required area of reinforcement. However, this is not true of applied force, for, as can be seen in the figure, the force/moment combination for which the required area of reinforcement is maximum consists of the largest moment with the **smallest** force in the specified range. The total area of reinforcement required to provide a capacity of $N/bh = 4.89$ and $M/bh^2 = 4.82$ is about 2.3 per cent or $(0.023(300)^2 =) 2070$ mm^2, of which half should be placed at each face.

The minimum area of reinforcement required by EC2 is:

$$A_{s,min} = \text{greater of } \frac{0.15N}{f_y/\gamma_s} \text{ and } 0.003A_g$$

$$= \text{greater of } \frac{0.15 \times 600 \times 10^3}{460/1.15} \text{ and } 0.003(300)^2$$

$$= 270 \text{ mm}^2$$

Clearly this does not govern the design.

9.5 Design of short members for axial force and biaxial bending

Up to this point in the chapter, we have only considered short columns subjected to axial force and bending about a single axis. In fact, many columns are simultaneously subjected to bending about two (usually perpendicular) axes. Such bending, when it arises, is known as biaxial bending. Like uniaxial bending, biaxial bending is caused by eccentric loading of the column or by unbalanced moments transferred to the column from the members joined to its ends. In the case of the column in Fig. 9.15, the axial force N is applied at eccentricities e_y and e_z to the centre line and the member is subjected to a moment $M_y = Ne_z$ about the Y-axis and to a moment $M_z = Ne_y$ about the

Fig. 9.15 Column in biaxial bending: (a) top of column; (b) section A–A

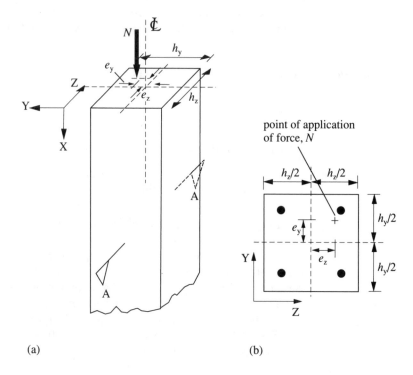

(a)   (b)

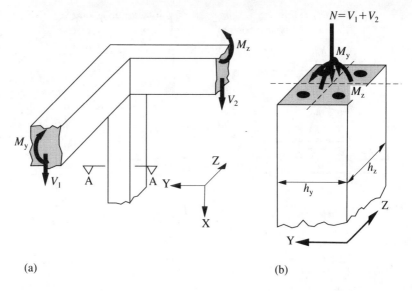

Fig. 9.16 Corner column: (a) geometry and loading; (b) section A–A

(a)   (b)

Z-axis. In a similar manner, biaxial bending can result from unbalanced moments in both axes of the column. This form of biaxial bending commonly arises in corner columns, as illustrated in Fig. 9.16.

Substantially uniaxial bending

For many members with biaxial bending, the applied bending moment (or the eccentricity of the axial load) about one of the axes is often much smaller than the applied bending moment (or eccentricity) about the other axis. Where such cases arise, sufficient accuracy can be attained by carrying out two separate designs in which only bending about one axis is considered in turn. Each design is carried out in accordance with section 9.4.

For members with rectangular cross-sections in which moments are applied about the two principal planes, *Y* and *Z*, EC2 allows separate (uniaxial) checks to be made provided one of the following two conditions is met:

$$\left|\frac{e_y}{e_z}\right| \geq 5\frac{h_y}{h_z} \tag{9.18}$$

or

$$\left|\frac{e_y}{e_z}\right| \leq \frac{1}{5}\frac{h_y}{h_z} \tag{9.19}$$

These conditions are satisfied if the point of application of the axial load is located within the hatched area of Fig. 9.17(a). In the case of biaxial moments M_y and M_z, resulting from the transfer of unbalanced moments, the equivalent eccentricities are: $e_y = M_z/N$, $e_z = M_y/N$. The values for e_y and e_z should not include the additional eccentricity, e_a, as given by equation (9.9), when checking equations (9.18) and (9.19).

When the eccentricity e_z exceeds one-fifth of the depth h_z, a further restriction

Combined Axial Force and Bending of Reinforced Concrete Members

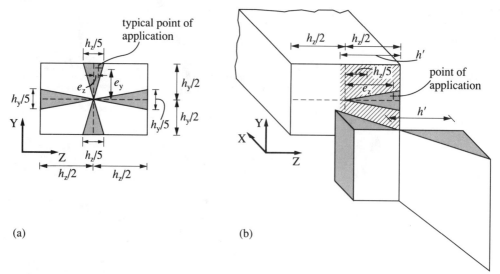

(a)   (b)

Fig. 9.17 Substantially uniaxial bending: (a) zone in which bending is substantially uniaxial; (b) strain distribution due to M_y

applies in addition to inequalities (9.18) and (9.19). Separate uniaxial checks can be made but, for the moment about the minor axis (Z-axis in this case), the calculation is based on a reduced breadth, h', as illustrated in Fig. 9.17(b). The distance h' is equal to the depth of the compression zone for bending about the major axis (Y-axis in this case).

For members where both applied moments (or eccentricities) are substantial, however, the biaxial (combined) effect of the moments must be considered. The method of design of such members is described in the remainder of this section.

Ultimate capacity

The ultimate capacity of short members subjected to combined axial load and biaxial bending is evaluated using the same principles as those used for members with uniaxial bending. Failure at any given section is reached when the strain in the concrete reaches its ultimate value of $\varepsilon_{ult} = 0.0035$. Consider the case of the eccentrically loaded column of Fig. 9.18(a). By resolution of the moment vectors, the bending moments M_y and M_z at any section can be replaced by a single resultant moment, M_r, as illustrated in Fig. 9.18(b). The magnitude and direction of the resultant moment are respectively given by:

$$M_r = \sqrt{(M_y)^2 + (M_z)^2} \tag{9.20}$$

$$\theta = \tan^{-1}(M_y/M_z) \tag{9.21}$$

When moments are due to eccentric loading, equation (9.20) can be written as:

$$M_r = \sqrt{(Ne_z)^2 + (Ne_y)^2} = N\sqrt{(e_z)^2 + (e_y)^2}$$

and equation (9.21) becomes:

$$\theta = \tan^{-1}(Ne_z/Ne_y) = \tan^{-1}(e_z/e_y)$$

Fig. 9.18 Resolution of biaxial bending moments: (a) applied force and moments; (b) resolution of moment vectors

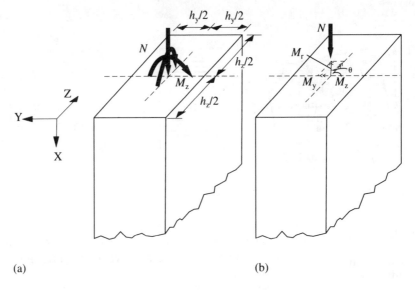

(a)   (b)

The section is then designed to resist the **uniaxial** moment, M_r, and the axial force, N. However, the calculation of the ultimate axial force and moment capacities is complicated by the fact that the neutral axis is inclined at an angle θ to the Z-axis of the section.

A typical distribution of strain and stress for the section of Fig. 9.18(b) is illustrated in Fig. 9.19. As for uniaxial bending, strain increases linearly with

Fig. 9.19 Distributions of strain and stress due to N and M_r: (a) elevation and section; (b) strain distribution; (c) stress distribution

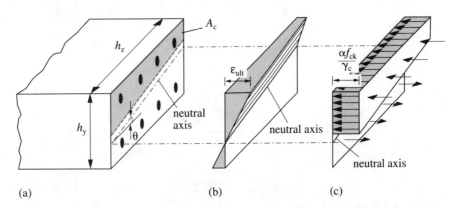

(a)   (b)   (c)

distance from the neutral axis. Thus, in an elevation **perpendicular to the plane of the neutral axis**, the distribution of strain is triangular. This is illustrated in Fig. 9.20. The strains in the reinforcing bars, now all at different distances from the neutral axis, are found from the strain diagram of Fig. 9.20(c). The corresponding stresses are then determined from the stress/strain relationship and, from these, the forces can be calculated. The compressive force in the concrete, F_c, is equal to $(\alpha f_{ck}/\gamma_c)A_c$, where A_c is the area of concrete in compression (the shaded area of Figs 9.19(a) and 9.20(a)). It should be remembered when calculating F_c that the shape of the compression zone depends on the inclination

385

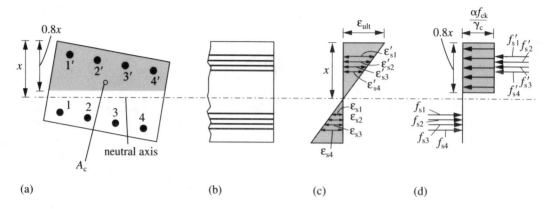

Fig. 9.20 Elevation perpendicular to plane of neutral axis: (a) section; (b) elevation; (c) strain; (d) stress

of the neutral axis as is illustrated in Fig. 9.21. When the shape of the zone is as illustrated in Fig. 9.20(a), we have, by equilibrium:

$$N_{\text{ult}} = F_{\text{c}} + \sum_{i=1}^{4} f'_{\text{s}i} A'_{\text{s}i} + \sum_{j=1}^{4} f_{\text{s}j} A_{\text{s}j} \tag{9.22}$$

Taking moments of the internal forces about the centre then gives the ultimate moment capacity for bending about this axis.

Fig. 9.21 Shapes of compression zone

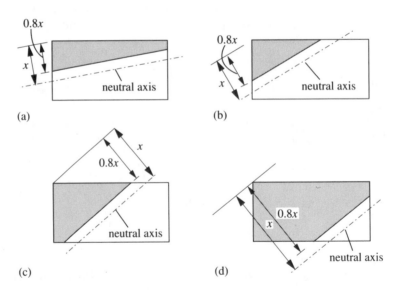

As for columns with uniaxial bending, the design of a section with biaxial bending and axial force involves iteration of the procedure outlined above for various values of x until the value of N_{ult} equals the applied axial force. Then the capacity to resist the resultant moment is compared with the applied moment. For this reason, this method of designing short members with combined axial force and biaxial bending is impractical unless applied on a computer. In practice, computer programs are used or the design of such members is carried out using three-dimensional interaction diagrams known as **interaction surfaces**.

Interaction surfaces

In section 9.4, interaction diagrams were introduced as the most convenient method for checking the force/moment capacity of short columns with uniaxial bending. Recall that such interaction diagrams are derived by assuming a range of strain distributions (values for x), deriving the force/moment capacity for each distribution and using the results to draw the design curve. The more strain distributions chosen, the more accurate the design curve.

In a manner similar to the construction of interaction diagrams, the general solution for the moment/force capacity of short columns with biaxial bending can be represented by a three-dimensional design surface, known as an **interaction surface**. Like an interaction diagram, an interaction surface is found by assuming a number of strain distributions for the design section. However, for a member with biaxial bending, this means assuming a number of inclinations of the neutral axis in addition to a number of values for x.

Consider a member subjected to an axial force N and moments M_y and M_z. The resultant moment, M_r, is given by equation (9.20) and its inclination to the Z-axis, θ, is given by equation (9.21). By definition, the inclination of the neutral axis is also equal to θ. Say, first of all, that $M_z = 0$, that is $M_r = M_y$ and $\theta = 90°$. The interaction diagram corresponding to this condition is found as before (by assuming different values for x) and is illustrated in Fig. 9.22(a). Next, assume $M_y = 0$, that is $M_r = M_z$ and $\theta = 0°$. The interaction diagram in this case is illustrated in Fig. 9.22(b). Next, consider a case where both M_y and M_z are non-zero, say $M_y = M_z$. In this case, $M_r = \sqrt{2}M_y$ and $\theta = 45°$. The interaction diagram in this case is illustrated in Fig. 9.22(c). By taking other values for θ, deriving the interaction diagram for each by assuming a range of values for x and plotting the diagrams together, the complete interaction surface for the section is found to be as illustrated in Fig. 9.22(d). Each point on this surface represents a unique set of N, M_y and M_z for the particular section which can (just) be resisted. All points within the interaction surface represent 'safe' designs.

A horizontal section through the interaction surface of Fig. 9.22(d) is considered where the axial force equals N_1. The interaction curve obtained, that of Fig. 9.23, represents all combinations of M_y and M_z which will cause failure in the presence of the axial load, N_1. The precise shape of this interaction curve varies for different values of N_1. For small N_1, the curve is approximately linear while for large N_1, the curve is almost circular. Therefore, an expression defining the precise shape of the interaction surface for the general case is not easily found. A number of approximate expressions have been developed for the equation of the curve including the following from CP110, the predecessor to BS 8110:

$$\left(\frac{M_{y,\,\text{ult}}}{M_{y,\,\text{max}}}\right)^\alpha + \left(\frac{M_{z,\,\text{ult}}}{M_{z,\,\text{max}}}\right)^\alpha = 1 \tag{9.23}$$

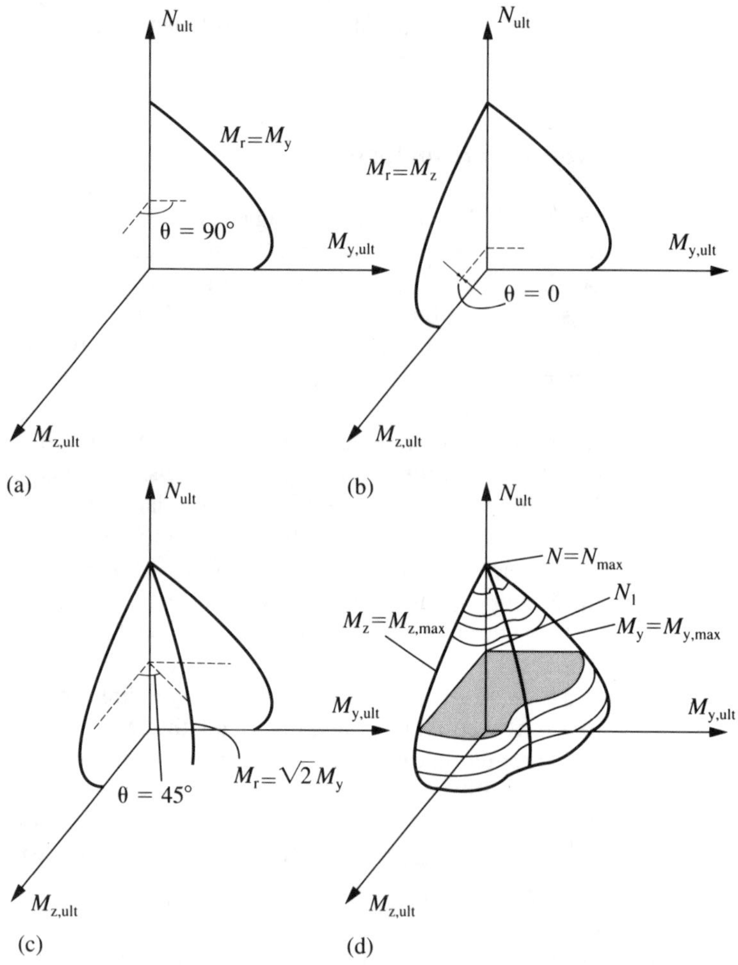

Fig. 9.22 Construction
of interaction surface:
(a) uniaxial bending,
$M_n = M_y$; (b) uniaxial
bending, $M_n = M_z$;
(c) $M_n = \sqrt{2}M_y$;
(d) interaction surface

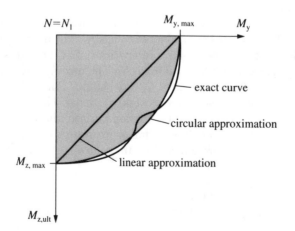

Fig. 9.23 Section
through interaction
surface at $N = N_1$

In equation (9.23) $M_{y,\max}$ is the ultimate capacity required to resist M_y in the presence of N (only) and $M_{z,\max}$ is the ultimate capacity required to resist M_z in the presence of N (only) (see Fig. 9.22(d)). The coefficient α is a function of the ratio, $N/N_{\max}$. It varies linearly from a value of unity at $N/N_{\max} = 0.2$ to 2 at $N/N_{\max} = 0.8$. No formula corresponding to equation (9.23) is given in EC2.

9.6 Design of slender members for axial force and uniaxial bending

Recall from section 9.2 that a slender member is defined by EC2 as one for which the second-order moments due to deflection of the member must be accounted for in design. Equation (9.3) gives the slenderness ratio, λ, above which a column is classed as slender. The principal effect of the second-order moment in a slender column is to reduce significantly its load-carrying capacity.

Consider a typical column subject to a compressive force N and a moment M as illustrated in Fig. 9.24(a). This is equivalent to the column illustrated in Fig. 9.24(b) where the force is applied at an eccentricity $e = M/N$. In a short column, the distribution of internal moment due to the eccentric force is uniform over the member as illustrated in Fig. 9.24(c). Allowing for inaccuracies in the placement of the load, the magnitude of the so-called 'first-order' moment is

Fig. 9.24 Slender column: (a) geometry and loading; (b) equivalent geometry and loading; (c) distribution of first order moment; (d) deflection due to applied moment; (e) distribution of total moment

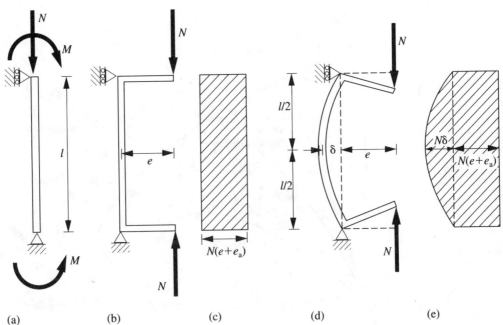

given by equation (9.8), that is:

$$\text{moment} = M + Ne_a = N(e + e_a)$$

Now, if the member is slender it will deflect as illustrated in Fig. 9.24(d) with the maximum deflection, δ, occurring at mid-span. This lateral deflection increases the eccentricity of the axial force and consequently increases the bending moments in the member. Hence, the total design moment for this column at mid-height becomes:

$$M = N(e + e_a + \delta) \tag{9.24}$$

where δ is the maximum lateral deflection in the member. The total bending moment diagram for the member of Fig. 9.24(a) is then as illustrated in Fig. 9.24(e).

The effect of slenderness on the capacity of a member to resist load can be visualized using interaction diagrams. Figure 9.25 shows the interaction diagram for a typical short member. The line OP in Fig. 9.25 represents the path followed when the force is increased while the eccentricity of that force is kept constant. When there is no lateral deflection, δ, the behaviour of a column under increasing axial force can be represented by the linear force/moment path OP. The corresponding path for a short column, in which δ is small, is represented by OA. The slope of the line OA remains nearly constant

Fig. 9.25 Effect of slenderness on force/moment path

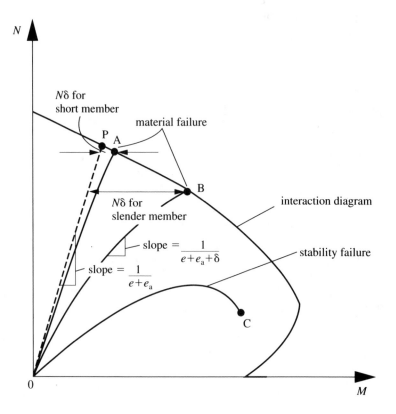

at $1/(e + e_a)$ up to the point of failure because the moment, $N\delta$, due to deflection is negligible. This form of failure is commonly known as **material failure** since the member remains stable until its constituent material reaches its ultimate capacity.

For a moderately slender braced member, the lateral deflection is not negligible and the total moment is given by equation (9.24). The behaviour of such a member can be represented by the force/moment path OB of Fig. 9.25. For small N, the lateral deflection is small and the contribution of δ is negligible. As N is increased, however, the lateral deflection, and hence δ, increases more rapidly and causes the force/moment path to become significantly non-linear. Material failure of the member occurs when the interaction diagram is reached at B. At failure, the horizontal distance between the force/ moment paths, OB and OP, is equal to $N\delta$. Also, it can be seen from Fig. 9.25 that the ultimate load at point B is less than at point A, indicating that slenderness has reduced the load-carrying capacity of the column.

The behaviour of a very slender member is represented by the force/moment path OC. In this case, the additional eccentricity, δ, increases so rapidly that it reaches a peak load before a material failure can occur. When the load reaches the peak value, the member becomes unstable and buckles, since any increase in deflection results in a reduction in the force capacity. This type of failure, commonly known as **stability failure**, is rare in reinforced concrete members.

Quantifying δ

In order to calculate the maximum lateral deflection, δ, in a slender member caused by the applied loading, we must first consider the curvature of the deflected member. Recall that for a section in bending, the elastic moment–curvature relationship is given by:

$$\frac{M}{I} = \frac{E}{R} = E\kappa \tag{9.25}$$

where $\kappa = 1/R$ is the curvature. It has been shown by Cranston (1972) that, for a reinforced concrete column under the action of ultimate loads, the maximum curvature, κ_{max}, which occurs at the critical section, depends only on the effective length of the member and the depth of the section.

The precise curvature distribution in slender columns is not known as the cross-section may be cracked to a greater or lesser extent. However, it is reasonable to assume that the shape is somewhere between a triangular and a rectangular distribution as illustrated in Fig. 9.26. The triangular distribution assumes only one critical section (at the apex of the triangle as illustrated in Fig. 9.26(c)) while the rectangular distribution assumes that the member is critical along its entire length. For these two curvature distributions, the maximum deflection can readily be calculated using corollaries to the moment–area theorems known as the curvature–area theorems. The corollaries are found by simply replacing $M/(EI)$ in the moment–area theorems by κ. Hence equation

Fig. 9.26 Curvature distributions for slender moment: (a) geometry and loading; (b) actual curvature distribution; (c) triangular curvature distribution; (d) rectangular curvature distribution

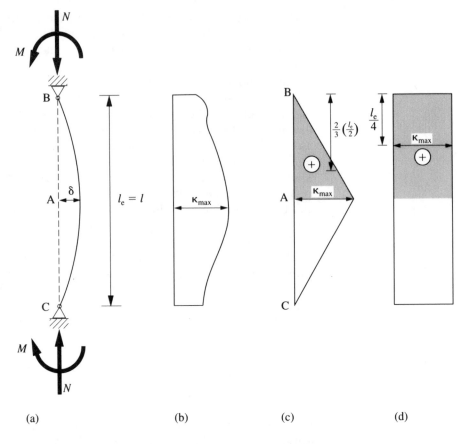

(a)   (b)   (c)   (d)

(4.28) gives the deviation at B, t_{AB}, of the tangent to point A, as illustrated in Fig. 4.48:

$$t_{AB} = \bar{x}_B \int_A^B \kappa \, dx \qquad (9.26)$$

where $\bar{x}_B$ is the distance from point B to the centroid of the portion of the κ diagram between A and B. As point A, by definition, is the point of maximum deflection, the tangent at A is vertical. Thus:

$$t_{AB} = \delta$$

For the triangular curvature diagram of Fig. 9.26(c), equation (9.26) then yields:

δ = moment of κ diagram between A and B about point B

$$= \left(\frac{1}{2} \kappa_{max} \frac{l_e}{2} \right) \left(\frac{2}{3} \frac{l_e}{2} \right)$$

$$\Rightarrow \qquad \delta = \frac{\kappa_{max} l_e^2}{12} \qquad (9.27)$$

Alternatively, for the rectangular curvature diagram of Fig. 9.26(d), equation

(9.26) yields:

$$\delta = \left(\kappa_{\max} \frac{l_e}{2} \right) \left(\frac{l_e}{4} \right)$$

$$\Rightarrow \qquad \delta = \frac{\kappa_{\max} l_e^2}{8} \tag{9.28}$$

From equations (9.27) and (9.28), a logical estimate of the actual maximum deflection of the member is therefore given by:

$$\delta = \frac{\kappa_{\max} l_e^2}{10} \tag{9.29}$$

Consequently, the expression for the maximum deflection δ, recommended for design by EC2, is given by:

$$\delta = K_1 \frac{(1/R_{\min}) l_e^2}{10} \tag{9.30}$$

where $R_{\min}$ is the minimum radius of curvature (i.e. $1/R_{\min}$ is the maximum curvature) and K_1 is a reduction factor which depends on the slenderness of the member. The magnitude of K_1 is taken from:

$$\left. \begin{aligned} K_1 &= \frac{\lambda}{20} - 0.75 \quad \text{for } 15 \le \lambda \le 35 \\ K_1 &= 1 \qquad\qquad \text{for } \lambda > 35 \end{aligned} \right\} \tag{9.31}$$

Where great accuracy is not required, EC2 recommends that $1/R_{\min}$ in equation (9.30) be approximated by:

$$\frac{1}{R_{\min}} = \frac{2K_2 \varepsilon_y}{0.9d} \tag{9.32}$$

where K_2 = reduction factor which takes account of the decrease of the curvature due to the presence of axial force
ε_y = design yield strain of the reinforcement, $\varepsilon_y = f_y/(\gamma_s E_s)$
d = effective depth in the direction in which δ is measured

The factor K_2 is defined by:

$$K_2 = \text{lesser of } \frac{N_{\max} - N}{N_{\max} - N_{\text{bal}}} \text{ and } 1 \tag{9.33}$$

where $N_{\max}$ = ultimate capacity of section subjected to axial force only; $N_{\max}$ is calculated using equation (9.7)
N = design applied axial force
N_{bal} = axial load which maximizes the moment capacity, that is balanced design; for a symmetrical arrangement of reinforcement, N_{bal} may be taken as:

$$N_{\text{bal}} = \frac{0.4 f_{ck}}{\gamma_c} A_g \tag{9.34}$$

393

Note that it is always conservative to assume that $K_2 = 1$. The application of equations (9.30) to (9.34) is illustrated by the following example.

Example 9.6 Slender braced column

Problem Determine the quantity of reinforcement required in a slender braced column of dimensions $l = 6$ m, $b = h = 300$ mm to resist a compressive force of 1650 kN and first order end moments of 80 kN m. The cover is 25 mm, $l_e = 0.7l$, $f_{ck} = 35$ N/mm^2 and $f_y = 460$ N/mm^2.

Solution The basic procedure for the design of slender braced columns with equal end moments consists of the following six steps.

Step 1

Determine the interaction diagram to be used. From this diagram, read off N_{bal}.

Assuming, as in Example 9.5, that the reinforcement consists of 20 mm diameter longitudinal bars and 10 mm diameter links, we have:

$$d = 300 - 25 - 10 - \tfrac{1}{2}(20) = 255 \text{ mm}$$

Therefore, the design chart of Fig. 9.14 can be used. The value for N_{bal} is the axial force corresponding to the maximum moment. From the design chart of Fig. 9.14:

$$N_{bal}/bh = 8.5$$

$$\Rightarrow \qquad N_{bal} = 8.5(300)^2 \text{ N}$$

$$\Rightarrow \qquad N_{bal} = 765 \text{ kN}$$

Alternatively, equation (9.34), which is based on an estimated d/h ratio, can be used to determine an approximate (and more conservative) value for N_{bal}.

Step 2

Assume initially that $K_2 = 1$ when calculating δ and derive the total design moment.

Taking $K_2 = 1$, equation (9.32) yields:

$$\frac{1}{R_{min}} = \frac{2\varepsilon_y}{0.9d} = \frac{2(f_y/(\gamma_s E_s))}{0.9d}$$

$$= \frac{2(460/(1.15 \times 200\,000))}{0.9 \times 255}$$

$$= 17.43 \times 10^{-6} \text{ mm}^{-1}$$

The magnitude of K_1 depends on the slenderness ratio, λ, of the member. From equations (9.1) and (9.2), we have:

$$\lambda = \frac{l_e}{\sqrt{I_g/A_g}} = \frac{l_e}{\sqrt{(bh^3/12)/(bh)}} = \frac{l_e}{\sqrt{(h^2/12)}}$$

$$= \frac{0.7(6000)}{\sqrt{300^2/12}} = 48.5$$

From equation (9.31):

$$K_1 = 1$$

The second-order eccentricity, δ, is given by equation (9.30), that is:

$$\delta = K_1 \frac{(1/R_{min})l_e^2}{10}$$

$$\delta = (1) \frac{(17.43 \times 10^{-6})(0.7 \times 6000)^2}{10}$$

$$= 31 \text{ mm}$$

The second-order moment corresponding to this additional eccentricity is:

$$M_2 = (1650 \times 10^3)(31)$$

$$= 51 \times 10^6 \text{ Nmm}$$

$$= 51 \text{ kNm}$$

Thus, the total design moment is equal to:

$$M = 80 + 51 = 131 \text{ kN m}$$

Step 3

Determine A_s from the interaction diagram:

$$\frac{N}{bh} = \frac{1650 \times 10^3}{(300)^2} = 18.3$$

$$\frac{M}{bh^2} = \frac{131 \times 10^6}{(300)^3} = 4.85$$

From Fig. 9.14, $100A_{s,tot}/(bh) \approx 3.3$ (i.e. 3.3 per cent). Therefore:

$$A_{s,tot} = 3.3(300)^2/100 = 2970 \text{ mm}^2$$

Step 4

Based on this value for the area of reinforcement, recalculate K_2.
    Recalculate K_2 using equation (9.33):

$$K_2 = \frac{N_{max} - N}{N_{max} - N_{bal}}$$

From the design chart of Fig. 9.14, the axial force capacity in the absence of moment, N_{max}, for 3.3 per cent reinforcement is:

$$N_{max}/(bh) = 33$$

$$\Rightarrow \quad N_{max} = 33(300)^2 \text{ N}$$

$$= 2970 \text{ kN}$$

(Alternatively, N_{max} can be calculated exactly using equation (9.7).) Thus:

$$K_2 = \frac{2970 - 1650}{2970 - 765} = 0.60$$

Step 5

If the new K_2 is almost equal to the old value, the design is complete. If not, return to step 2, using the new K_2 value to recalculate δ.

Since the new K_2 does not equal the old value (of unity), we must return to step 2 and repeat the design process. Starting again with $K_2 = 0.60$, the maximum curvature becomes $1/R_{min} = 10.46 \times 10^{-6}$, the deflection becomes 18.5 mm and the second-order moment becomes 31 kNm. Accordingly, the required area of reinforcement reduces to 2.6 per cent or $A_{s,tot} = 2340 \text{ mm}^2$. Recalculation of K_2 (with $N_{max} = 2655$ kN) then yields $K_2 = 0.53$. The process can be repeated in this way until there is convergence of K_2. As six T25 bars provide an area in excess of 2340 mm^2 and it seems unlikely that the required area will reduce to a level for which 4T25 or 6T20 will be adequate, no further iteration is performed in this example.

Note: As T25 bars are required, the original calculation of d is now incorrect by 3 mm. However, as the area provided is well in excess of that required, this seems unlikely to affect the end result.

Step 6

Ensure that the completed design satisfies the detailing requirements of section 9.3.

The maximum area of reinforcement allowed is 8 per cent of the gross area. The area provided by 6T25 is:

$$\frac{(6\pi 25^2/4)}{300^2} 100 \text{ per cent} = 3.3 \text{ per cent}$$

Even when the bars are lapped, this is less than the maximum allowed. The minimum allowable reinforcement is:

$$A_{s,min} = \text{greater of } \frac{0.15N}{f_y/\gamma_s} \text{ and } 0.003A_g$$

$$\Rightarrow \quad A_{s,min} = \frac{0.15 \times 1650 \times 10^3}{460/1.15} = 619 \text{ mm}^2$$

It can be seen that the solution is well within the required limits.

Design of slender braced members with unequal end moments

Up to this point, we have assumed that the first-order end moments acting on a slender column are of equal magnitude (i.e. that the eccentricity of axial force is equal at both ends of the member as illustrated in Fig. 9.24(b)). Consequently, the distribution of first-order moment is constant throughout the member as illustrated in Fig. 9.24(c). However, in most practical design situations the applied end moments are unequal and, in some cases, of opposite eccentricity. These two general cases are illustrated in Figs 9.27 and 9.28. From these figures, it can be seen that a member with unequal end moments in which the end eccentricities are of the same sign undergoes single curvature (as illustrated in Fig. 9.27(c)) while a member having eccentricities of opposite sign undergoes double curvature (as illustrated in Fig. 9.28(c)).

For a member bent in single curvature, the distributions of first- and second-order moment combine to give the total moment distribution of Fig. 9.27(e). The design moment (i.e. the maximum moment in the member) no longer occurs at the centre. Let M_1 equal the first-order moment at the point of maximum total moment. The deflection at this point will be less than the **maximum** deflection in the member. Therefore, it is conservative to take the

Fig. 9.27 Eccentric loading with eccentricities of like sign: (a) geometry and loading; (b) first order moment; (c) deflected shape; (d) second order moment; (e) total moment

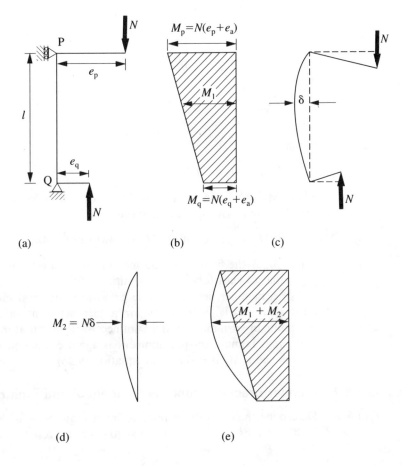

(a) (b) (c)

(d) (e)

Fig. 9.28 Eccentric loading with eccentricites of opposite sign: (a) geometry and loading; (b) first order moment; (c) deflected shape; (d) second order moment; (e) total moment

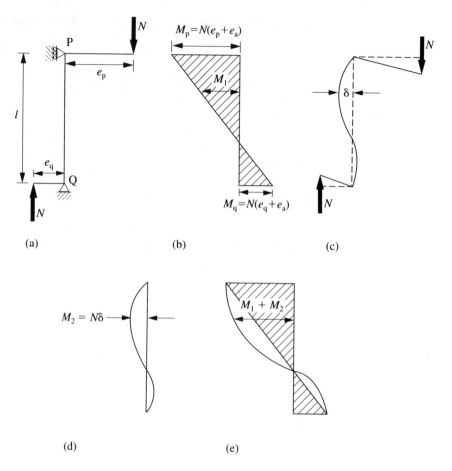

design moment as:

$$M = M_1 + M_2 \tag{9.35}$$

where $M_2 = N\delta$ and where δ is the maximum deflection in the member. EC2 specifies the following approximation for M_1:

$$M_1 = \text{greater of } (0.6M_p + 0.4M_q) \text{ and } 0.4M_p \tag{9.36}$$

where M_p is the first-order end moment of greatest (absolute) magnitude and M_q is that of least (absolute) magnitude, that is, $|M_p| > |M_q|$.

For a member bent in double curvature, the total moment distribution is illustrated in Fig. 9.28(e). As for a member bent in single curvature, the maximum in-span moment does not necessarily occur at mid-span. EC2 suggests that the maximum in-span moment is again calculated using equation (9.35) where, again, M_1 is taken from equation (9.36).

Example 9.7 Slender braced column with unequal end moments

Problem Determine the quantity of reinforcement required in a slender braced column of dimensions $l = 5.5 \, \text{m}$, $b = h = 300 \, \text{mm}$ to resist a compressive force of

1400 kN and first order end moments of 145 kNm and -120 kNm. The effective length is $l_e = 0.66l$, $f_{ck} = 35$ N/mm² and $f_y = 460$ N/mm².

Solution To design this member, and indeed all members with unequal end moments, the procedure of Example 9.6 is followed. For step 2, however, the total design moment is calculated using equations (9.35) and (9.36).

Step 1

Assume, as for Example 9.6, that the reinforcement for this member consists of 20 mm diameter longitudinal bars and 10 mm diameter links with 25 mm cover to the links. The effective depth is then, as before, $d = 255$ mm and the design chart of Fig. 9.14 applies. As before, $N_{bal} = 765$ kN.

Step 2

Assuming initially that $K_2 = 1$, equation (9.32) gives (as for Example 9.6):

$$1/R_{min} = 17.43 \times 10^{-6} \text{ mm}^{-1}$$

From equations (9.1) and (9.2), we have:

$$\lambda = \frac{l_e}{\sqrt{I_g/A_g}} = \frac{l_e}{\sqrt{h^2/12}}$$

$$= \frac{0.66(5500)}{\sqrt{300^2/12}} = 42$$

From equation (9.31):

$$K_1 = 1$$

Therefore, from equation (9.30):

$$\delta = K_1 \frac{(1/R_{min})l_e^2}{10}$$

$$\delta = (1) \frac{17.43 \times 10^{-6}(0.66 \times 5500)^2}{10}$$

$$= 23 \text{ mm}$$

The second-order moment corresponding to this additional eccentricity is:

$$M_2 = (1400 \times 10^3)(23) = 32 \times 10^6 \text{ Nmm}$$

$$= 32 \text{ kNm}$$

From equation (9.36), the first-order moment, M_1, at the critical point is:

$$M_1 = \text{greater of } (0.6M_p + 0.4M_q) \text{ and } 0.4M_p$$

$$= \text{greater of } [0.6(145) + 0.4(-120)] \text{ and } 0.4(145)$$

$$= \text{greater of } 39 \text{ and } 58$$

$$= 58 \text{ kNm}$$

From equation (9.35), the total design moment is given by:

$$M = M_1 + M_2 = 58 + 32$$

$$= 90 \text{ kNm}$$

As this is less than the end moment of 145 kNm, the latter governs the design.

Step 3

$$\frac{N}{bh} = \frac{1400 \times 10^3}{(300)^2} = 15.6$$

$$\frac{M}{bh^2} = \frac{145 \times 10^6}{(300)^3} = 5.37$$

From the design chart of Fig. 9.14, $100A_{s,tot}/(bh) = 3.2$ (i.e. 3.2 per cent). Therefore:

$$A_{s,tot} = 0.032(300)^2 = 2880 \text{ mm}^2$$

Step 4

Recalculate K_2 using equation (9.33):

$$K_2 = \frac{N_{max} - N}{N_{max} - N_{bal}}$$

From the design chart of Fig. 9.14, $N_{max} = 2925$ kN. Hence:

$$K_2 = \frac{2925 - 1400}{2925 - 765} = 0.71$$

Step 5

The new K_2 does not equal the old K_2 and so we must return to step 2 and repeat the design process. Starting again with $K_2 = 0.71$ does not alter the design moment, that is M_p is still maximum. Therefore, the value of $A_{s,tot} = 2880 \text{ mm}^2$ is taken as the final design requirement. Six T25 bars provide an area of 2945 mm^2 which, as can be seen from Example 9.6, is within the allowable limits.

Design of slender unbraced members

In a frame structure, the stability of the frame against lateral loads depends on the moment capacity of the individual columns. As stated in section 9.2, columns in such sway frames are termed 'unbraced' because, unlike braced columns, the ends are not fixed against lateral displacement. In a sway frame all the columns at a particular level must displace by the same amount. For this reason it is necessary to consider the entire structure when designing the columns.

Fig. 9.29 Sway frame: (a) geometry and loading; (b) first order moment; (c) deflected shape; (d) second order moment

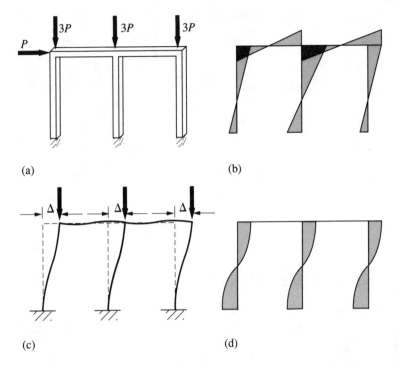

(a)

(b)

(c)

(d)

A simple sway frame is illustrated in Fig. 9.29. The first-order moments due to lateral load and the resulting deflections of the members are illustrated in Figs 9.29(b) and (c), respectively. The additional second-order moments due to the deflection are illustrated in Fig. 9.29(d). The maximum second-order moment in each column occurs at the end of greater stiffness, say, in this case, at the bottom of each column. The second-order moment at the other end of each column can, in the general case, be reduced by an amount in proportion to the ratio of the joint stiffnesses at either end as defined by equation (9.4). The magnitude of second-order moment in a column at a particular floor level is given by:

$$M_2 = N\Delta \tag{9.37}$$

where N is the vertical force and Δ is the average lateral deflection at that level (see Fig. 9.29(c)).

It can be seen from Fig. 9.29 that the total bending moment will be maximum at the ends of the columns rather than within the span, as is often the case with braced members. Thus, to design an unbraced slender member, the design procedure of Example 9.6 is followed, taking the design moment as the greater of the two total end moments.

9.7 Design of reinforced concrete deep beams

Recall that a deep beam was defined as one in which the span is less than twice the depth. In concrete buildings, deep beams are most commonly found in shear

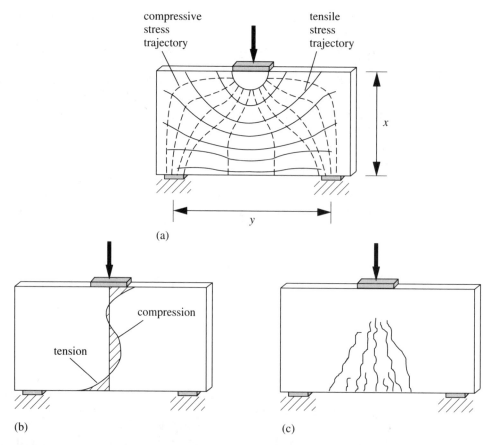

Fig. 9.30 Simply supported deep beam (from McGregor (1992)): (a) stress trajectories; (b) elastic stress distribution; (c) crack pattern

wall systems (deep cantilevers) and are often located on the perimeter of framed structures where they provide stiffness against horizontal loads (see section 2.3). The applied loads in deep beams are carried by membrane action, which effectively means the load is carried through tension and compression zones. Figure 9.30(a) illustrates the stress trajectories (load paths) by which load is carried to the supports of a simply supported deep beam subject to a central point load. It can be seen from Fig. 9.30(a) that the compressive stress trajectories are mainly confined to direct paths between the applied force and the supports. Also, the tensile stress trajectories are roughly horizontal and are concentrated near the bottom face of the member.

The intensity and trajectory of the internal stresses in a deep beam can only be determined using sophisticated linear elastic methods of analysis (such as the finite element method). Even when such programs are available, they do not usually take account of the redistribution of stress which occurs when concrete cracks in tension. For the member of Fig. 9.30(a), the distribution of internal stress at a section directly below the concentrated load is illustrated in Fig. 9.30(b). In contrast to a simply supported beam in flexure (i.e. where the span/depth ratio exceeds two), it can be shown that the tensile stress at the bottom of a deep beam is approximately constant along the entire length of the member and that the compressive stress is also constant along the inclined load paths.

The results of a linear elastic analysis are only valid while the deep beam remains uncracked. In practice, however, tensile cracks develop in most deep beams at between one-third and one-half of the ultimate load. The crack pattern for the simply supported member of Fig. 9.30(a) is illustrated in Fig. 9.30(c). It can be seen that the tensile cracks form along lines which are perpendicular to the tensile stress trajectories (and hence parallel to the compressive stress trajectories). Thus, to prevent failure of the beam, steel reinforcement is required along the bottom face of the member. If sufficient tension reinforcement is provided, failure of the beam can only occur when the compressive strength of the concrete is exceeded on the diagonals between the applied load and the supports. However, it is usually the tension reinforcement requirement which governs the design of deep beams.

Deep beam models for design

Linear elastic methods of analysis are not commonly used in the design of deep beams owing to their inherent complexity and because they are only valid while the member remains uncracked. In practice, most deep beams are idealized as statically determinate strut and tie (truss) models. EC2 does not place any restriction on the type of truss models which are used. In any model, the force in each strut/tie is established from considerations of equilibrium. Reinforcement is then provided to carry the tension in the ties. In addition, the compressive stress in the struts is compared with the compressive strength of the concrete in those zones. The models presented here are based on the work of Kotsovos (1988).

For the simply supported deep beam of Fig. 9.30(a), the model proposed by Kotsovos is illustrated in Fig. 9.31(a). This simplified model consists of two inclined compressive struts (of concrete) and a single horizontal tie (of reinforcement). The width of the concrete struts is assumed to be constant at $a_v/3$, where a_v is the shear span (distance from the support to the first load). Figure 9.31(b) illustrates the proposed model for a simply supported deep beam subject to two concentrated loads. The model of Fig. 9.31(b) may also be used for a simply supported member subject to a uniform load if (and only if) the equivalent two point loads are applied at the third points, that is a distance of $\frac{1}{3}L$ from each end. Similar models can be derived for continuous deep beams (Fig. 9.31(c)) and for shear walls (Fig. 9.31(d)). The following example illustrates the procedure by which the models of Fig. 9.31 are used in design.

Example 9.8 Simply supported deep beam

Problem The simply supported deep beam of Fig. 9.32(a) has a characteristic compressive cylinder strength, f_{ck}, of 30 N/mm². If the beam is subject to the following ultimate factored uniform loads:

$$\text{permanent gravity (incl. self-weight)} = 25 \text{ kN/m}$$
$$\text{variable gravity} = 20 \text{ kN/m}$$

Fig. 9.31 Deep beam strut/tie models: (a) simply supported beam with concentrated load; (b) simply supported beam with two concentrated loads; (c) model for continuous beam; (d) model for shear wall

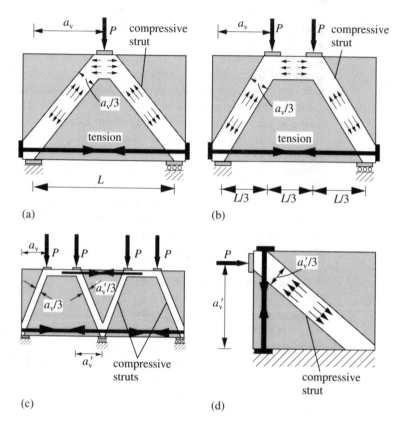

(a)

(b)

(c)

(d)

Fig. 9.32 Deep beam of Example 9.8: (a) geometry; (b) strut/tie model

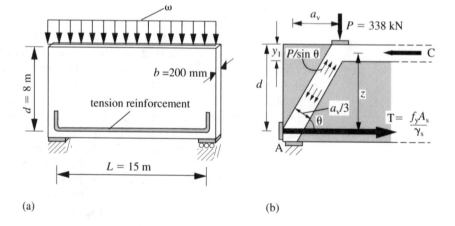

(a)

(b)

check the compressive stress in the diagonal struts and calculate the area of reinforcement required in the tie.

Solution   The total uniform loading, ω, acting on the beam is 45 kN/m. This uniform load is represented by two point loads, P, of magnitude:

$$P = \omega L/2 = (45)(15)/2 = 338 \text{ kN}$$

Figure 9.32(b) illustrates one-half on the strut and tie model at failure. Let T equal the tensile force in the tie and let C equal the compressive force in the horizontal portion of strut between the two concentrated loads. The allowable compressive force in the horizontal strut is:

$$C = \left(\frac{vf_{ck}}{\gamma_c}\right) by_1 \tag{9.38}$$

where b is the breadth, y_1 is the depth of the strut and v is the effectiveness factor, which allows for the difference between a cylinder tested in the laboratory and the actual compressive strength of a strut. For shear design, EC2 specifies an effectiveness factor of:

$$v = \text{greater of}\left(0.7 - \frac{f_{ck}}{200}\right) \text{ and } 0.5 \tag{9.39}$$

Hence:

$$v = 0.55$$

Taking moments about the centre of the. tension force at A gives:

$$Pa_v = Cz \tag{9.40}$$

where z is the lever arm:

$$z = d - y_1/2 \tag{9.41}$$

$$\Rightarrow \qquad Pa_v = C(d - y_1/2)$$

$$\Rightarrow \qquad C = \frac{Pa_v}{(d - y_1/2)}$$

Equating this expression for C with equation (9.38) gives:

$$\left(\frac{vf_{ck}}{\gamma_c}\right) by_1 = \frac{Pa_v}{(d - y_1/2)}$$

which reduces to the following quadratic equation in y_1:

$$y_1^2 - (2d)y_1 + \frac{2Pa_v}{bvf_{ck}/\gamma_c} = 0$$

$$\Rightarrow \qquad y_1^2 - 2(8000)y_1 + \frac{2(338 \times 10^3)(5000)}{(200)(0.55)(30)/1.5} = 0$$

$$\Rightarrow \qquad y_1 = 97 \text{ mm}$$

Thus, from equation (9.38), we have:

$$C = \left(\frac{vf_{ck}}{\gamma_c}\right) by_1 = \left(\frac{(0.55)(30)}{1.5}\right)(200)(97) \text{ N}$$

$$= 213 \text{ kN}$$

By horizontal equilibrium $C = T$ and, assuming the tensile reinforcement to have yielded, this gives:

$$T = \frac{f_y A_s}{\gamma_s} = 213 \times 10^3 \text{ N}$$

$$\Rightarrow \quad A_s = \frac{213 \times 10^3}{460/1.15} = 533 \text{ mm}^2$$

This can be provided using six bars (three on each face) of 12 mm diameter at a spacing of 150 mm, that is, in a zone 300 mm deep.

The force in the diagonal struts, P_s, is equal to $P/\sin \theta$ (refer to Fig. 9.32(b)). From the geometry of the figure:

$$\theta = \tan^{-1}(8000/a_v) = 58°$$

$$\Rightarrow \quad P_s = 338/\sin 58° = 399 \text{ kN}$$

The strength of the diagonal struts is:

$$\left(\frac{v f_{ck}}{\gamma_c}\right)(b)\left(\frac{a_v}{3}\right) = \frac{0.55(30)}{1.5}(200)\left(\frac{5000}{3}\right) \text{ N}$$

$$= 3666 \text{ kN}$$

Thus, the concrete struts are more than strong enough to carry the applied force of 399 kN. If it was found that the struts could not sustain the applied load, their strength could be enhanced by either (a) increasing the wall thickness b or (b) designing the struts as columns in which the compressive strength is increased by the provision of reinforcement.

Although the dimensions of the deep beam of Fig. 9.32(a) are sufficient to keep the stresses in the struts within acceptable limits, it is entirely possible that the deep beam may fail under a smaller applied load due to buckling of slender column struts. Thus, in addition to the calculations described above, the slenderness of each strut should be checked and it should, if necessary, be designed as a column in accordance with section 9.6.

Continuous deep beams are also sensitive to differential settlement. For this reason, EC2 suggests that a range of support reactions, corresponding to possible settlements, be considered in the design of continuous members.

Detailing of deep beams

Tensile reinforcing bars in deep beams, corresponding to the ties in the design model, should be fully anchored at their ends by using hooks or some form of anchoring device. In addition, EC2 recommends that deep beams are provided with an orthogonal mesh of reinforcement having a minimum area of 0.15 per cent of the gross area in both directions. If the struts are particularly slender, the area of reinforcement required in these regions may, of course, be even greater.

Problems

Section 9.2

9.1 The roof of a building is supported on vertical cantilevers with clear heights of 8 m and rectangular cross-sections of dimensions 500 mm × 1000 mm. Determine the slenderness ratios given that the structure is braced.

9.2 Repeat Problem 9.1 assuming an unbraced structure.

9.3 The column of Problem 9.1 is subjected to a factored axial force of 7000 kN. Determine if the column is short or slender as defined in EC2.

Section 9.4

9.4 A rectangular column of breadth 350 mm, depth 700 mm and effective depth 650 mm is constructed from concrete with $f_{ck} = 40$ N/mm². Eight T25 bars placed in the mid-sides and corners of the column give a total area of reinforcement of 3 927 mm². Determine if the column has the capacity to resist a combination of $N = 2000$ kN and $M = 750$ kNm (these are factored ultimate values).

9.5 Check the capacity of the section of Problem 9.4 for an applied ultimate axial force of 2500 kN at an eccentricity (including e_a) of 300 mm.

9.6 Find four key points on the interaction diagram for the section illustrated in Fig. 9.33 given that $f_{ck} = 35$ N/mm². The external loads are applied at the centres of the adjoining beams as illustrated.

Fig. 9.33. Section of Problem 9.6

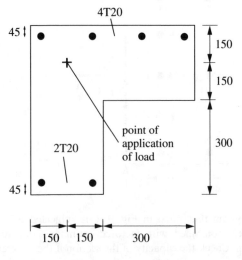

9.7 Construct the interaction diagram for the octagonal section illustrated in Fig. 9.34, given that $f_y = 460$ N/mm² and $f_{ck} = 35$ N/mm².

Section 9.5

9.8 The column whose section is illustrated in Fig. 9.35 is subjected to factored ULS moments (excluding additional eccentricity, e_a) of $M_y = 400$ kNm and $M_z = 30$ kNm about the major and minor axes respectively, together with a factored axial compressive force of 1000 kN. Determine if the section has sufficient capacity to resist this combination given that $f_{ck} = 35$ N/mm² and $f_y = 460$ N/mm².

407

Fig. 9.34 Section of Problem 9.7

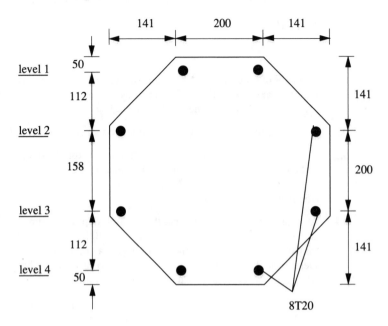

Fig. 9.35 Section of Problem 9.8

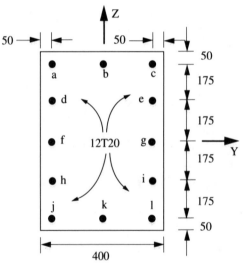

9.9 The square column illustrated in Fig. 9.36 is subjected to a factored ultimate **tensile** force of 400 kN combined with factored ultimate moments of $M_y = 500$ kNm and $M_z = 375$ kNm. Check the capacity of the section if the concrete has a cylinder strength of 40 N/mm².

9.10 Construct an interaction surface for the column whose section is illustrated in Fig. 9.37 given that $f_{ck} = 30$ N/mm².

Section 9.6

9.11 Determine the quantity of reinforcement required in a slender braced column of dimensions, $l = 5.5$ m, $b = h = 300$ mm, to resist a compressive force of 1400 kN and

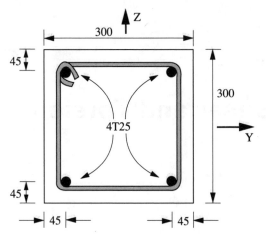

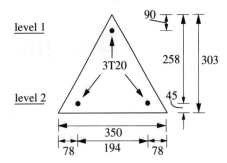

Fig. 9.36 Column of Problem 9.9

Fig. 9.37 Section of Problem 9.10

Fig. 9.38 Cantilever
of Problem 9.12

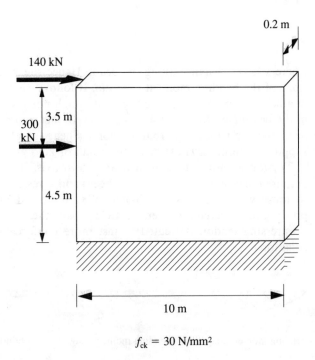

$f_{ck} = 30 \text{ N/mm}^2$

first order end moments of 145 kNm and 100 kNm. The effective length is $l_e = 0.66l$ and the characteristic strengths are, $f_{ck} = 35 \text{ N/mm}^2$ and $f_y = 460 \text{ N/mm}^2$.

9.12 Design the deep cantilever illustrated in Fig. 9.38 to resist the factored ultimate loads given. The wall may be assumed to be braced against lateral buckling.

10

Design for Shear and Torsion

10.1 Introduction

It can readily be shown that shear force in a beam, V, is related to moment, M, by:

$$V = \frac{dM}{dx}$$

where x is the distance along the beam. Hence shear equals in magnitude the slope of the bending moment diagram. It follows that, where there is moment in a beam, there must also be shear, although not necessarily in the same part of the beam (see, for example, Fig. 10.1). In fact, moment tends to be greatest near mid-span or at internal supports. Shear force tends to be large at all supports, internal and external, and small or zero at mid-span.

Torsion is moment about a beam's own axis. It is treated in this chapter because it causes shear stresses in a beam and is resisted by mechanisms similar to those which resist shear. While BS8110 treated ordinary reinforced and prestressed concrete differently, EC2 treats them in the same manner. A prestressing tendon is treated as just more reinforcement combined with an axial force.

Fig. 10.1 Bending moment and corresponding shear force diagrams:
(a) simply supported, uniformly loaded;
(b) simply supported with point load;
(c) three-span uniformly loaded

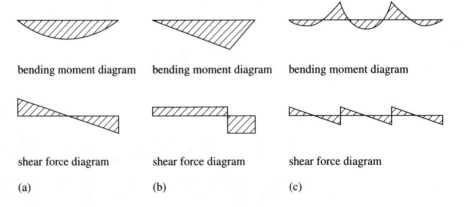

bending moment diagram   bending moment diagram   bending moment diagram

shear force diagram   shear force diagram   shear force diagram

(a)   (b)   (c)

Fig. 10.2 Shear and torsion reinforcement: (a) stirrups for resistance of shear; (b) stirrups and longitudinal bars for resistance of torsion; (c) bent up longitudinal bars for resistance of shear

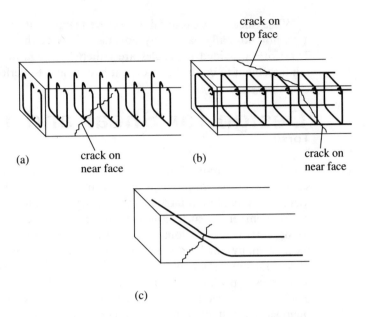

Both shear and torsion are resisted by the concrete itself and by shear/torsion reinforcement where this is present. Reinforcement generally takes the form of stirrups (also called links) such as illustrated in Fig. 10.2(a). Shear tends to cause roughly diagonal cracking and stirrups are provided to carry shear force across such cracks. While stirrups can be inclined, it is far more usual for them to be vertical. Torsion tends to cause spiral cracking. Thus, closed stirrups, as illustrated in Fig. 10.2(b), are used to control torsional cracking on all surfaces of the beam including the top surface. Longitudinal reinforcement is also required for the resistance of torsion. Ordinary longitudinal reinforcement can be used to resist shear if it is 'bent up' as illustrated in Fig. 10.2(c). However, this is no longer widespread practice in Western Europe and North America.

Shear

As shear failure in unreinforced members is sudden, shear reinforcement in some form must always be provided in beams (even when not theoretically required) except in beams of minor structural importance. Fortunately shear does not generally govern the design of reinforced and prestressed concrete members, that is, the section dimensions are not normally dictated by the requirements for shear. Exceptions to this are beams which are deep relative to their span and flanged beams where the web is not very wide. In contrast to beams, slabs do **not** normally have shear reinforcement and EC2 does not require reinforcement when it is not theoretically required. This is because slabs generally have a higher degree of redundancy than beams; that is alternative load paths are available should part of the slab be defective. This may also reflect the fact that it is difficult to provide shear reinforcement in floor slabs owing to their small depth. In some circumstances, particularly for flat slab construction, shear can govern the depth of slabs.

While shear can cause failure of the member at the ultimate limit state, it does not normally cause any adverse effects at the serviceability limit state. Therefore, provided that rules are adhered to regarding the spacing of shear reinforcement to control cracking, it is sufficient to perform a ULS shear safety check only.

Torsion

Torsion occurs quite frequently in structures but generally its importance is secondary to that of moment and shear. Two types of torsion are commonly identified. **Equilibrium torsion** is that which is required to maintain equilibrium of the member. In such situations, the external load has no other option but to be carried by torsion; that is, there exists no alternative load path or mechanism by which the load can be transferred through the member. Examples of equilibrium torsion, illustrated in Fig. 10.3, are an eccentrically loaded beam and a beam in which there is a change in the direction of its longitudinal axis. Equilibrium torsion is of primary interest in design because failure of the member is inevitable if it has insufficient torsional strength.

Fig. 10.3 Examples of equilibrium torsion: (a) eccentrically loaded member; (b) member with change in direction of longitudinal axis

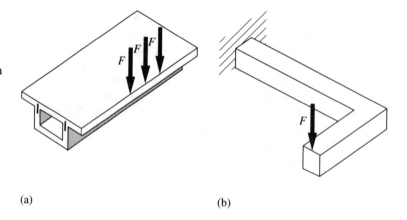

(a)　　　　　　　　　　　　　　　　　　(b)

The other type of torsion, known as **compatibility torsion**, arises in indeterminate structures having rigidly connected members. It results from the compatibility of deformations of members meeting at a joint. Owing to the monolithic nature of their construction, most members in concrete structures undergo a certain degree of compatibility torsion. Perhaps the most common example is an external drop beam supporting a floor slab, as illustrated in Fig. 10.4(a). Bending due to applied loads causes rotation at the edges of the slab which in turn causes rotation of the edge beam (Fig. 10.4(b)). If the edge beam were free to rotate, no torsion would result. However, if (because it is attached to columns at its ends) it resists rotation, torsion does result. Although such torsion may result in the formation of large cracks at the joint, more serious consequences are unlikely if the member possesses adequate ductility to redistribute the torsional moments. Further, torsional stiffness tends to be less

Fig. 10.4 Example of compatibility torsion: (a) floor slab and drop beams; (b) deflected shape of part of structure

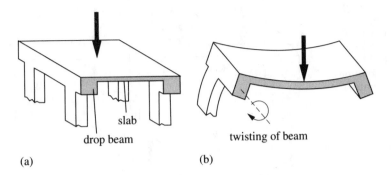

slab

drop beam

twisting of beam

(a)　　　　　　　　　　　　(b)

than bending stiffness. This causes an edge beam in such a situation to offer less resistance to rotation with the result that the bulk of the load is transferred from the edge beam to its supporting columns in the form of bending. For these reasons, compatibility torsion is generally of secondary interest in design and can often be ignored.

EC2 and other codes of practice allow compatibility torsion to be ignored if sufficient steel is provided to ensure ductile behaviour and the spacing of stirrups is sufficiently small to control cracking at the serviceability limit state. Specifically for EC2, it is unnecessary to design for compatibility torsion if the crack control requirements of section 7.3 are satisfied and the amount of suitably spaced shear reinforcement (stirrups also act as torsional reinforcement as will be shown in subsequent sections) equals or exceeds the minimum.

10.2　Types of cracking

Types of cracking are often identified by reference to the arrangement of loading illustrated in Fig. 10.5. At the neutral axis of this beam, the axial stress due to bending is zero and a small segment of concrete is subjected to shear stress only (Fig. 10.5(b)). By vertical equilibrium, there exist two equal and opposite vertical stresses, v, acting on either side of the segment illustrated in the figure. However, these two stresses tend to cause clockwise rotation of the element and so two further balancing stresses, as illustrated in Fig. 10.5(c), are required for overall equilibrium of the element. The forces corresponding to these four shear stresses can be resolved into components at 45° to the horizontal as illustrated in Fig. 10.5(d). Combining components, it can be seen that there is a total compressive force (arrows out) of $2v\,\delta x/\sqrt{2}$ which equals $\sqrt{2}v\,\delta x$ in one direction and a total tensile force of $\sqrt{2}\,v\,\delta x$ in the other. As these act on a length $\sqrt{2}\,\delta x$ (diagonal of the square) the corresponding stresses are $\sqrt{2}\,v\,\delta x/\sqrt{2}\,\delta x$ which reduces to v. These stresses, known as principal stresses, are illustrated in Fig. 10.5(e) where it can be seen that the tensile stresses tend to cause cracking at 45° to the horizontal. This cracking occurs when the principal tensile stress exceeds the tensile strength of the concrete. Once a diagonal crack has developed in a section at the point of maximum stress, it will propagate through the section and may eventually cause failure of the member.

The segment of concrete considered above was at the neutral axis of the

Design for Shear and Torsion

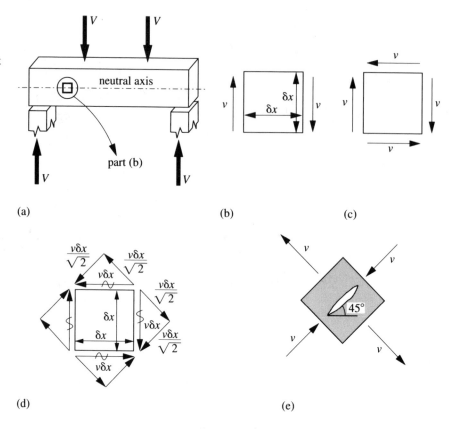

Fig. 10.5 Stresses of segment of concrete: (a) simply supported beam; (b) small segment of concrete; (c) shear stresses; (d) resolution of forces; (e) principal stresses

(a)

(b)

(c)

(d)

(e)

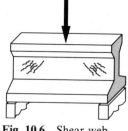

Fig. 10.6 Shear web cracking

section where axial stress is, by definition, of zero magnitude. For segments at other points, axial stress exists as a result of moment and, in the case of prestressed members, the prestress force. Sections where shear stress predominates throughout the depth are not common but do occur in some cases where the bending moment is negligible or where the web width of the member is small. In such situations, cracking, known as **shear–web cracking**, occurs, as illustrated in Fig. 10.6.

For a more comprehensive understanding of the behaviour of members with shear, we must consider the combined effects of stresses due to shear, moment and prestress force. For the uncracked rectangular member of Fig. 10.7(a), the distribution of axial stress due to bending is illustrated in Fig. 10.7(b). For such a rectangular section, the distribution of shear stress is parabolic, varying from zero at the top and bottom to a maximum at the centre as illustrated in 10.7(c). Consider a small segment, A, at the bottom of the member (Fig. 10.7(a)). From Figs 10.7(b) and (c), the shear stress, v, in this segment is zero and the bending stress, σ, is at its maximum value. The equivalent principal stresses, illustrated in Fig. 10.7(d), act on a vertical plane. Thus, a vertical flexural crack will form at this point when the tensile stress due to moment exceeds the tensile strength of the concrete. A small segment at the neutral axis of the member (such as B in Fig. 10.7(a)) was considered above. At this point, the shear stresses are maximum and the bending stress is zero. The corresponding principal stresses,

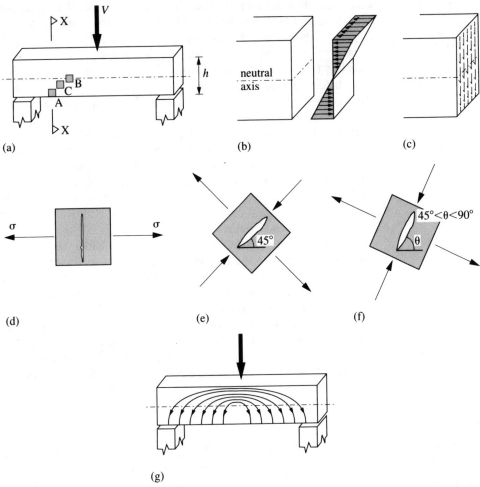

Fig. 10.7 Orientation of principal stresses: (a) geometry and loading; (b) axial stress distribution at section X–X; (c) shear stress distribution at section X–X; (d) segment A of part (a); (e) segment B of part (a); (f) segment C of part (a); (g) stress paths

illustrated in Fig. 10.7(e), act on planes inclined at 45° to the horizontal. Therefore, a diagonal crack inclined at 45° will develop at this point under the principal tensile stresses. For a segment, C, between the neutral axis and the bottom of the member, a combination of bending and shear stresses is acting on the element. The principal tensile stresses for this element act on planes inclined at an angle between 45° and 90° to the horizontal, as illustrated in Fig. 10.7(f). Therefore, diagonal cracks with an inclination of between 45° and 90° will develop in this element. Clearly, the inclination of the cracks will decrease towards the neutral axis as the shear stress becomes larger and the axial stress due to bending approaches zero.

By considering small segments at other points within the member, the orientation of the principal stress planes can be determined. These planes can be represented by stress paths through the member, as illustrated in Fig. 10.7(g). The stress paths represent the direction of the principal compressive stresses through an uncracked member. Since the principal tensile stresses act normal to the principal compressive stresses, these lines also represent the lines along

415

which tension cracks will tend to develop in the member. Thus, for a member with shear and bending, the cracks tend to be curved and to vary in slope from 90° at the extreme fibre (point of pure bending) to 45° at the neutral axis (point of pure shear). This form of cracking is commonly known as **shear–flexure cracking**. In most reinforced and prestressed concrete members, flexural cracks form before the principal stresses at the neutral axis are large enough for shear–web cracking to occur. Thus, shear–flexure cracking is generally the more common type of failure.

In prestressed concrete members, compressive prestress force is applied which reduces the total tensile stress at the serviceability limit state to little or nothing. However, at the ultimate limit state, the applied loads are significantly larger and, hence, tensile stress does occur. Thus, prestressing has the effect of increasing the applied load at which flexural cracks form. Therefore sections which are prestressed tend to have a greater effective shear strength than ordinary reinforced concrete members. In addition, the prestress force also affects the orientation of inclined cracks. Consider the introduction of a prestressed tendon to the member of Fig. 10.7(a), as illustrated in Fig. 10.8(a). Segment B is located at the axis where direct stress due to bending is zero; that is, it is located at what was the neutral axis in the absence of prestress. Now, if the prestress force in the tendon at service is ρP, then the compressive stress due to prestress at B is $\sigma = \rho P / A_g$, where A_g is the (gross) cross-sectional area. Thus, the segment is now subjected to horizontal compressive stress in addition to the shear stress, as illustrated in Fig. 10.8(b). The corresponding principal stresses acting on the element are given in Fig. 10.8(c). The compressive stress

Fig. 10.8 ULS cracking in prestressed beam: (a) geometry and loading; (b) stresses on segment B; (c) orientation of crack at segment B

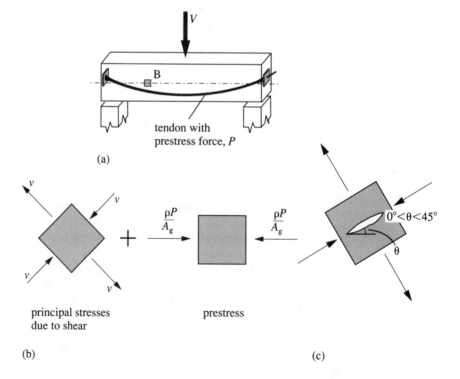

has the effect of reducing the tensile shear stress and, as a consequence, cracks tend to form at an angle of between 0° and 45° making them flatter than the cracks which form at this point in an ordinary reinforced beam. It will be seen in subsequent sections that these flatter inclined cracks improve the shear strength of members because a greater number of stirrups (shear reinforcement) cross the cracks.

Although the stress paths described above do roughly model the pattern by which shear–flexure cracks develop in practice, they are by no means exact. For instance, the model does not account for the redistribution of the stresses which occurs when cracks are formed and does not account for the relative magnitudes of the shear, bending and compressive stresses which determine the precise inclination of the cracks. Thus, the shear strength of a concrete member cannot be predicted solely by calculating its principal tensile stresses in its uncracked state. In practice, the prediction of shear strength relies on the application of empirical formulae based on observed experimental results.

Average shear stress

In order to determine the capacity of a member to resist shear force, it is necessary to be able to relate the shear stress within the cracked member to the corresponding applied shear force, V. For an uncracked linear elastic material, the distribution of shear stress over a cross-section is given by the formula:

$$v_y = \frac{VQ}{Ib} \tag{10.1}$$

where v_y = shear stress at a point located a distance y from the centroid of the section

V = shear force acting on the section

Q = first moment of area of the portion of the section, with cross-sectional area A_y, lying beyond the point where the shear stress is measured (see Fig. 10.9); $Q = A_y \bar{y}$, where $\bar{y}$ is the distance from the centroid of the section to the centroid of the area, A_y

I = second moment of area of the section

b = breadth of the member at the point where the shear stress is calculated

Examples of the respective stress distributions due to a shear force, V, for a rectangular section and an I-section, as derived from equation (10.1), are illustrated in Fig. 10.10. In each case, the maximum shear stress, v_{max}, which occurs at the centroid, and the average shear stress, v, across the section are indicated in the figure. Since the proportion of shear carried by the flanges of the I-section is small, the average shear stress for the section is approximated by dividing V by the cross-sectional area of the web. The same approximation can be made for all other forms of flanged section.

As stated above, equation (10.1) is only valid for uncracked linear elastic sections. Concrete does not fall within this category as it generally cracks under

Fig. 10.9 Interpretation of equation 10.1

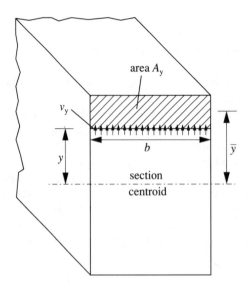

Fig. 10.10 Distributions of shear stress:
(a) rectangular section;
(b) I-section

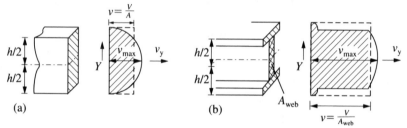

the action of bending and shear. Further, at the ultimate limit state, the distribution of stress through the section is highly non-linear. Consequently, the distribution of shear stress in a cracked concrete section is somewhat different from that in an uncracked section. A complete understanding of distributions of shear stress within cracked concrete members has yet to be attained. However, a good approximation of the **average** shear stress in a concrete section is given by:

$$v = \frac{V}{b_w z} \tag{10.2}$$

where b_w = breadth of the section or, in the case of flanged sections, b_w is the minimum web breadth

z = lever arm between the tensile force in the reinforcement and the compressive force in the concrete

The average shear stress, as derived from equation (10.2), closely resembles the average shear stress for an uncracked section. The lever arm, z, is approximately equal to $0.9d$ in most cases, where d is the effective depth of the section. EC2 recommends the use of a simplified version of equation (10.2) when considering the shear applied to reinforced and prestressed concrete members. It is given by:

$$v = \frac{V}{b_w d} \tag{10.3}$$

Shear strength is a term commonly used to refer to the average shear stress at which failure of a member occurs. Thus, shear strength is the capacity of a member to resist shear force, divided by the area, $b_w d$.

10.3 Types of shear failure

Inclined cracks must develop in a member before complete shear failure can occur. As described above, inclined cracks can form by shear–web cracking or, more commonly, by shear–flexure cracking. It is possible to identify several different types of shear failure in reinforced and prestressed members, each of which exhibits particular failure characteristics. The type of shear failure which occurs in a particular member depends on various factors including its geometry, the load configuration and the quantity of longitudinal reinforcement. One of the most significant factors is the shear span/effective depth ratio (a_v/d), where the shear span, a_v, is defined as the distance between points of zero and maximum moment, as illustrated in Fig. 10.11. Much of the research

Fig. 10.11 Shear span in experimental beam: (a) geometry and loading; (b) bending moment diagram

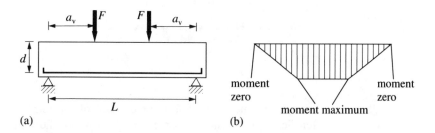

into the shear behaviour of concrete beams and slabs, from which empirical formulae have been derived, has been carried out using the load arrangement illustrated in the figure. It has been found that shear span/effective depth ratios can be divided into the following four general categories:

Category I: $0 < a_v/d \leq 1$
Category II: $1 < a_v/d \leq 2.5$
Category III: $2.5 < a_v/d \leq 6$
Category IV: $6 < a_v/d$

For different members falling into the same category, the sequence of events and the nature of the failure are (approximately) the same.

Members with very short shear spans or which have a large effective depth, that is, deep beams, fall into category I. The type of failure associated with such members is commonly known as **deep beam failure**, illustrated in Fig. 10.12. Diagonal shear–web cracks form almost in a direct line between the applied load and the support owing to the splitting action of the compressive force between the two points. Since the shear force cannot be transmitted across the cracks, the member exhibits a truss-like behaviour with the longitudinal reinforcement (or prestressing tendon) acting as a tie and the compression zones acting as struts (Fig. 10.13). Most commonly, the mode of failure in such a

Fig. 10.12 Deep beam failure (category I)

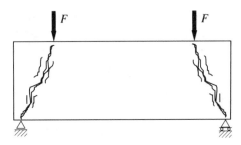

Fig. 10.13 Truss model for deep beams

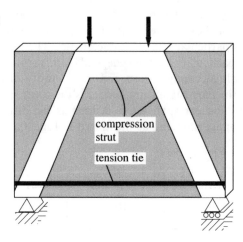

member is anchorage failure at the end of the reinforcement due to the large tensile force. If the reinforcement is bent at the ends, as illustrated in Fig. 10.14, failure can be delayed until the concrete in front of the hook fails in compression.

Members in category II, where a_v/d is larger, behave in a similar manner to category I members initially, that is, shear–web cracks develop in the region between the loaded sections and the supports. However, unlike deep beam failure, the crack often then propagates along the tension reinforcement destroying the bond between the reinforcement and the surrounding concrete, as illustrated in Fig. 10.15. This form of failure is known as **shear bond failure**. Alternatively, category II members can fail owing to dowel failure of the reinforcement at the point of the inclined crack, as illustrated in Fig. 10.16 or by crushing failure of the concrete at the points of application of the loads (Fig. 10.17). The latter type of failure is known as **shear compression failure**, which often occurs explosively and at a load substantially less than that for deep beam failure.

Members with an a_v/d ratio in category III are most likely to develop flexural cracks before the compressive force is great enough to develop shear–web cracks. The flexural cracks nearest the support (where the shear force is greatest) develop into inclined shear–flexure cracks and propagate towards the applied load as illustrated in Fig. 10.18, splitting the member. This type of failure is most usually known as **diagonal tension failure**. The load at which diagonal tension failure occurs is approximately half that for shear compression/shear

Fig. 10.14 Failure of bond in category I members

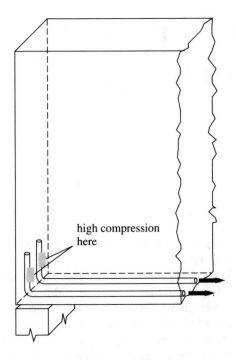

high compression here

Fig. 10.15 Shear bond failure (category II)

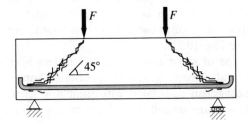

F F

45°

Fig. 10.16 Dowel failure (category II)

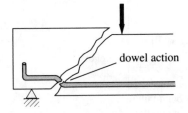

dowel action

Fig. 10.17 Shear compression failure (category II)

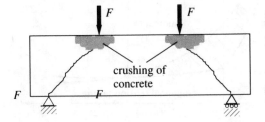

F F

crushing of concrete

F F

Fig. 10.18 Diagonal tension failure (category III)

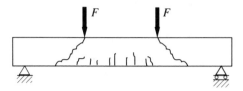

Fig. 10.19 Flexural failure (category IV)

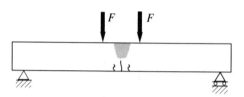

bond failure. Since the majority of beams fall within this category it is one of the most common types of failure.

Category IV members are so slender that they tend to fail in pure flexure, as illustrated in Fig. 10.19; that is, the longitudinal reinforcement yields and the concrete above it crushes before shear cracking occurs. A summary of the four categories described above is presented in Fig. 10.20.

Up to this point, we have only considered the shear failure of members which exhibit beam-type behaviour. This is applicable to beams and to one-way spanning slabs. It is also applicable to two-way spanning slabs supported on beams or walls such as in beam and slab construction. For such two-way spanning slabs, the shear in each of the two coordinate directions is considered separately (see Fig. 10.21).

Another type of shear failure, known as **punching shear failure**, can arise under concentrated loads or at isolated column supports in flat slab construction. The mechanism associated with punching shear failure is illustrated in Fig. 10.22. It is characterized by inclined cracks propagating around the edges

Fig. 10.20 Summary of types of failure for beam of Fig. 10.11 (after McGregor 1992)

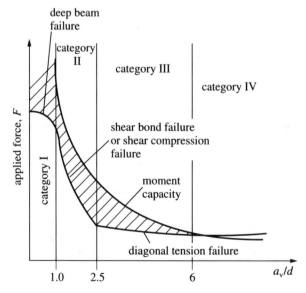

Fig. 10.21 Two-way spanning slab: (a) plan; (b) failure on face parallel to Y-axis; (c) failure on face parallel to X-axis

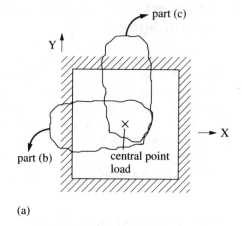

(a)

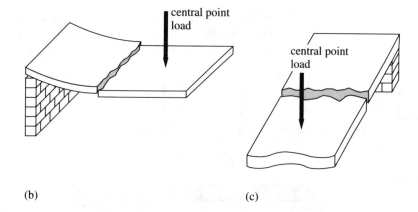

(b)   (c)

Fig. 10.22 Punching shear failure

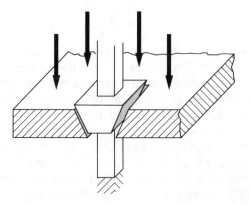

of the concentrated load or support to form a pyramidal or conically shaped wedge of concrete. Thus, the connection between the slab and the load or support is effectively destroyed. This type of failure can often occur very suddenly with little or no warning. In general, flat slabs have a punching shear capacity less than their capacity for regular shear but it is necessary to check for both forms of failure.

10.4   Shear strength of members without shear reinforcement

For the design of reinforced and prestressed concrete members with shear, a knowledge of the natural shear resistance of the concrete is essential. To date, most estimates of the shear strength of members without shear reinforcement are based on empirical equations derived from extensive experimental work. These account for the different influencing parameters which affect the shear capacity. Before such equations are presented, however, it is first necessary to illustrate the mechanisms of shear transfer in a cracked section and to review briefly the parameters which influence the capacity to resist shear.

Mechanisms of shear transfer

Consider a typical member in bending and shear, which is reinforced with longitudinal steel against bending. Under its applied loads, the member will crack in one of the characteristic manners, such as that illustrated in Fig. 10.23.

Fig. 10.23 Member with longitudinal reinforcement

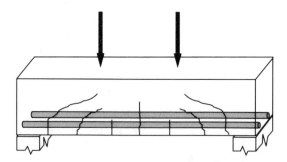

The shear force is transmitted through the cracked member by a combination of three mechanisms. The first is **dowel action** of the reinforcement (ordinary reinforcement or tendons/strands in the case of a prestressed concrete member). This results from the resistance of the reinforcement to local bending as illustrated in Fig. 10.24(a) and the resistance of the concrete to localized crushing near the reinforcement. The second mechanism by which cracked members resist shear is known as **aggregate interlock** and results from the forces transmitted across the crack by interlocking pieces of aggregate (Fig. 10.24(b)). The third mechanism by which cracked members resist shear is through the resistance of the concrete in the uncracked portion of the beam where the axial stress is compressive. The free-body diagram for the portion of the member between the left-hand support and the first inclined crack is illustrated in Fig. 10.25. The total capacity to resist shear force, V_c, is the result of the combined effects of the three mechanisms illustrated in that figure.

It has been found that dowel action is generally the first to reach its capacity, followed by failure of the aggregate interlock mechanism followed by shear failure of the concrete in the compression zone. However, the precise proportion

Fig. 10.24 Mechanisms of shear transfer: (a) dowel action; (b) aggregate interlock; (c) shear stresses in uncracked concrete

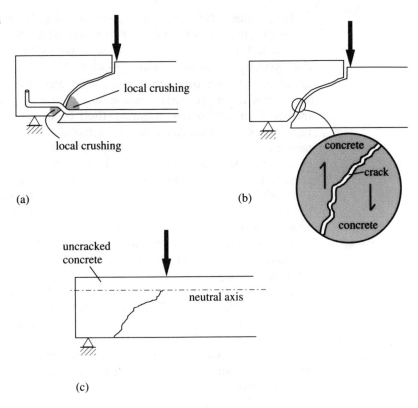

(a)   (b)

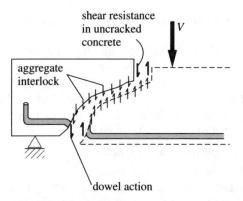

(c)

Fig. 10.25 Transfer of shear force through members

of the total shear force carried by each mechanism is difficult to establish and, consequently, the shear strength of the concrete is most often represented by a single expression which accounts for all three mechanisms acting together.

Factors affecting shear capacity

The shear capacity of a member is strongly dependent on the shear span/effective depth ratio, with the capacity decreasing with increasing a_v/d as was seen in the previous section.

425

The strength (grade) of the concrete and the amount of longitudinal reinforcement are also dominant factors affecting the shear capacity of members. Increasing the strength of the concrete increases the capacity through each of the three mechanisms of shear resistance. The total shear capacity increases approximately in proportion to $\sqrt{f_{ck}}$. The area of longitudinal reinforcement determines the dowel shear resistance at a cracked section. In addition, an increase in the area of reinforcement effectively increases the aggregate interlock resistance by reducing the width of the inclined crack. However, the reinforcement ratio, A_s/bd, has a relatively small influence on the overall shear strength of the member. Increasing the reinforcement by 1 per cent of the gross cross-sectional area enhances the shear strength by approximately 30 per cent. For high reinforcement ratios (2 to 4 per cent), the increase is less and, for particularly heavily reinforced members, the increase in shear resistance resulting from an increase in reinforcement is quite small.

The introduction of compressive axial forces, including prestressing, usually enhances the shear capacity of a member while tensile axial forces have the effect of reducing the capacity.

The size of the member is also a major factor affecting its shear capacity. Larger members are, in fact, less effective at resisting shear than smaller members; that is, the shear strength (average shear stress required to cause failure) is significantly less in larger members.

The influences of all the parameters described above on the shear strength of a concrete member are incorporated into one empirically derived formula in the next section.

Recommended design formula for shear strength

EC2 uses a parameter, k, to allow for the scale effect (i.e. the fact that smaller members are more effective). Commonly in beams, longitudinal bottom reinforcement, required at mid-span to resist sag moment, is not required near the support and some of the bars are curtailed; that is, they do not extend all of the way into the support. Where 50 per cent or more of such reinforcement is curtailed, $k = 1$. Otherwise, k is given by:

$$k = \text{greater of } (1.6 - d) \text{ and } 1 \tag{10.4}$$

where d is the effective depth in metres.

The expression for the design shear strength of both prestressed and reinforced concrete members without shear reinforcement, recommended by EC2, is:

$$v_c = \tau_{rd}k(1.2 + 40\rho_l) + 0.15\sigma_{cp} \tag{10.5}$$

where τ_{rd} = basic design shear strength. The value of τ_{rd} should be taken as 25 per cent of a conservatively low value for the tensile strength of the concrete, that is $\tau_{rd} = 0.25f_{ct,0.05}/\gamma_c$ where $\gamma_c = 1.5$ and $f_{ct,0.05}$ is the tensile strength with a 5 per cent probability of not being exceeded. This is given in EC2 as:

$$f_{ct,0.05} = 0.21f_{ck}^{2/3} \tag{10.6}$$

Values for τ_{rd} are given in Table 10.1.

Table 10.1 Values for basic shear strength, τ_{rd} (from EC2)

Characteristic cylinder strength, f_{ck} (N/mm^2)	Basic shear strength, τ_{rd} (N/mm^2)
12	0.18
16	0.22
20	0.26
25	0.30
30	0.34
35	0.37
40	0.41
45	0.44
50	0.48

ρ_l = lesser of the longitudinal tension reinforcement ratio, $A_s/b_w d$, and 0.02. The value for A_s should be taken as the smallest value in a region within a length of (d + required anchorage length) on both sides of the section considered.

$\sigma_{cp} = N/A_g$, where N is the factored design axial force either from an external force or from a prestress force (compression positive) and A_g is the gross area of the cross-section.

Recall from section 10.2 that the average shear stress at a section is given by equation (10.3). Therefore, the maximum capacity to resist shear force, V_c, which can be applied to members without shear reinforcement, is given by:

$$V_c = v_c b_w d \tag{10.7}$$

$$\Rightarrow \quad V_c = [\tau_{rd} k(1.2 + 40\rho_l) + 0.15\sigma_{cp}]b_w d \tag{10.8}$$

Variable depth members

Up to now, only members of prismatic (constant) cross-section have been considered. For members of variable depth, such as illustrated in Fig. 10.26,

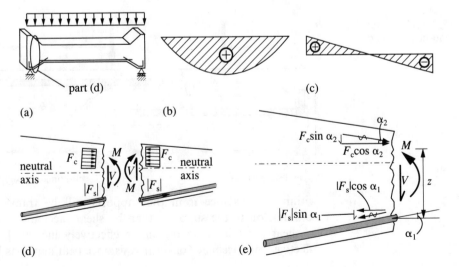

Fig. 10.26 Beam of variable depth: (a) elevation; (b) bending moment diagram; (c) shear force diagram; (d) part elevation of beam; (e) resolution of forces

the capacity to resist shear is generally different from the shear capacity of prismatic members. This is due to the direction of the forces which resist the applied moment. The force in the tension reinforcement, F_s, clearly acts parallel to that reinforcement. The force in the concrete acts at some angle intermediate between that of the top of the member and that of the neutral axis. These forces are resolved into horizontal and vertical components in Fig. 10.26(e). It can be seen from this figure that the horizontal components, $F_s \cos \alpha_1$ and $F_c \cos \alpha_2$, resist the applied moment. However, the vertical components act in sympathy with the applied shear, V. Thus, the total shear to be resisted by the member is:

$$\text{force to be resisted} = V + |F_s| \sin \alpha_1 + F_c \sin \alpha_2 \qquad (10.9)$$

The moment is related to the horizontal components by:

$$M = (|F_s| \cos \alpha_1)z = (F_c \cos \alpha_2)z \qquad (10.10)$$

where z is the lever arm. Equation (10.10) can be used to get expressions for $|F_s|$ and F_c as functions of M. Substituting for $|F_s|$ and F_c in equation (10.9) then gives:

$$\text{force to be resisted} = V + \frac{M \tan \alpha_1}{z} + \frac{M \tan \alpha_2}{z} \qquad (10.11)$$

For the example in Fig. 10.26, the variation in depth effectively increased the applied shear. This happens when the effective depth of the member is decreasing in the same direction as moment is numerically increasing. In other cases, the effect can be beneficial.

Shear enhancement

In a member where inclined shear–flexure cracks form, the cracks closest to the supports extend outwards from those supports at an inclination of between 30° and 45°, extending a maximum distance of approximately $2d$, as illustrated in Fig. 10.27. In such circumstances, concentrated loads which are located

Fig. 10.27 Applied load near supports

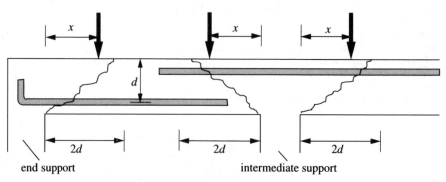

end support   intermediate support

within this distance from the supports will be transferred more by direct compression to the supports than by shear mechanisms. Thus, the resistance to shear force in these regions is effectively increased. EC2 allows for this effective 'enhancement' of shear resistance near supports by permitting the use of an increased value for τ_{rd} when concentrated loads are situated within a distance of 2.5d from the face of the support. In such circumstances, the value for τ_{rd} may be increased by multiplying by a factor, β, given by:

$$\beta = \text{lesser of } 2.5d/x \text{ and } 5.0 \qquad (10.12)$$

where x is the distance from the face of the support to the point of application of the load. However, shear enhancement at supports using equation (10.12) is only allowed when the following conditions are satisfied:

(a) The loading and support reactions are such that they cause only diagonal **compression** in the member. Thus, the support must be provided at the bottom of the beam when the loading is applied from the top.

(b) At an end support, all of the longitudinal tension reinforcement required within a distance of 2.5d from the support must be anchored into the support.

(c) At an intermediate support, the tension reinforcement required at the face of the support should continue for at least $2.5d + l_{b,\,net}$ into the span, where $l_{b,\,net}$ is the required anchorage length.

Uniformly loaded members do not involve a concentrated load near the support. However, enhancement does apply for that portion of the load which is applied near the support. In recognition of this, EC2 specifies that it will normally be conservative to compare the (unenhanced) shear capacity with the applied shear at a distance d from the support (see Example 10.2).

In some situations, shear enhancement is sufficient to reduce the shear reinforcement requirement to the minimum level. In such situations EC2 recommends caution and specifies that 'the designer may wish to base the resistance on the unenhanced' shear capacity (see Example 10.4).

Example 10.1 Shear enhancement

Problem Check the capacity of the member, AB, illustrated in Fig. 10.28, to resist the given distribution of shear force. The characteristic cylinder strength of the concrete is $f_{ck} = 35 \text{ N/mm}^2$.

Solution The shear force at point A is 200 kN and this extends to point C. Clearly, enhancement is least at the point furthest from the column face, so the factor, β, is calculated at C:

$$\beta = \text{lesser of } 2.5d/x \text{ and } 5.0$$

$$= \text{lesser of } (2.5 \times 500/750) \text{ and } 5.0$$

$$= 1.67$$

From Table 10.1, $\tau_{rd} = 0.37 \text{ N/mm}^2$ (for $f_{ck} = 35 \text{ N/mm}^2$). The enhanced value for τ_{rd} is $1.67 \times 0.37 = 0.62 \text{ N/mm}^2$. The scale effect factor, k, is given by:

$$k = \text{greater of } (1.6 - d) \text{ and } 1$$

$$= \text{greater of } (1.6 - 0.5) \text{ and } 1$$

$$= 1.1$$

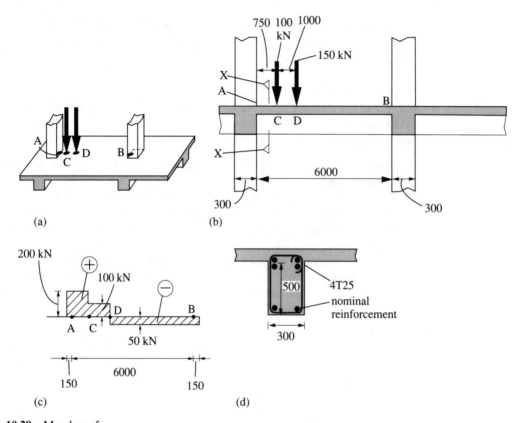

Fig. 10.28 Member of
Example 10.1:
(a) pictorial view;
(b) elevation
(dimensions in mm);
(c) shear force diagram;
(d) section X–X

The area of four 25 mm diameter bars is $4(\pi 25^2/4) = 1963$ mm². Hence, the reinforcement ratio for the (longitudinal) top reinforcement is:

$$\rho_l = \frac{1963}{300 \times 500}$$

$$= 0.013$$

(This reinforcement is assumed to extend the required distance on both sides of the point under consideration.) Thus, the shear force capacity at this point is, from equation (10.8):

$$V_c = [\tau_{rd}k(1.2 + 40\rho_l) + 0.15\sigma_{cp}]b_w d$$

$$= [0.62 \times 1.1(1.2 + 40 \times 0.013) + 0]300 \times 500$$

$$= 175\,956 \text{ N}$$

$$= 176 \text{ kN}$$

Thus, shear reinforcement will be required to resist the difference between the applied shear, $V = 200$ kN, and the shear capacity of the concrete section, $V_c = 176$ kN (Example 10.4 deals with this).

At the point of application of the second load, D, the maximum shear is 100 kN. This point is beyond 2.5d from the face of the support, so no enhancement factor applies. Hence, the shear capacity is:

$$V_c = [\tau_{rd}k(1.2 + 40\rho_l) + 0.15\sigma_{cp}]b_w d$$
$$= [0.37 \times 1.1(1.2 + 40 \times 0.013) + 0]300 \times 500$$
$$= 105\,006 \text{ N}$$
$$= 105 \text{ kN}$$

As the capacity exceeds the applied shear of 100 kN at this point, minimum shear reinforcement is sufficient here. Between D and B, the shear capacity is also 105 kN and the applied shear is 50 kN. Thus, minimum shear reinforcement is also sufficient for this part of the beam.

Maximum compressive stress

As illustrated in Fig. 10.5(e), tensile stress is generated at the neutral axis of members subjected to shear which can cause diagonal shear–web cracking. If the associated compressive stress, v, is sufficiently large and/or the breadth of the web of the member is small, this stress can lead to crushing of the web. Equations are given in EC2 for the maximum shear force that can be carried without such crushing. This phenomenon is more relevant for members with high levels of shear reinforcement and is dealt with in greater detail in section 10.5.

Members with inclined prestressing tendons

In prestressed members with inclined tendons, the vertical component of the prestress force is $\rho P(\sin \theta_p)$, where ρ is the loss ratio and θ_p is the inclination of the tendon (Fig. 10.29(e)). As mentioned in Chapter 8, the profile of a tendon often mimics the shape of the bending moment diagram (see Fig. 10.29). When such is the case, the vertical component of prestress frequently acts in the opposite direction to the applied shear force, V. When this occurs, the magnitude of the applied shear force is reduced by the prestress and the net magnitude is given by:

$$V_{net} = V - \gamma_p\rho P(\sin \theta_p) \tag{10.13}$$

where γ_p is the partial factor of safety for prestress (EC2 specifies $\gamma_p = 0.9$). Thus, the introduction of a profiled prestressing tendon can be seen effectively to increase the shear strength of a member. It should, however, be noted that there can be situations where prestress acts in sympathy with the applied shear and increases rather than reduces the shear to be allowed for in design.

When post-tensioned tendons are placed in ungrouted ducts, the effective web width for the calculation of shear capacity should be reduced by the diameter of the duct. When ducts are grouted (which is normally the case), EC2 specifies an effective web width equal to the actual web width minus half the duct diameter. This allows for the fact that the grout does offer some resistance to

Fig. 10.29 Post-tensioned prestressed beam of Example 10.2: (a) geometry and loading; (b) elevation showing tendon and ordinary reinforcement; (c) bending moment diagram; (d) shear force diagram; (e) free body diagram near left support (elevation)

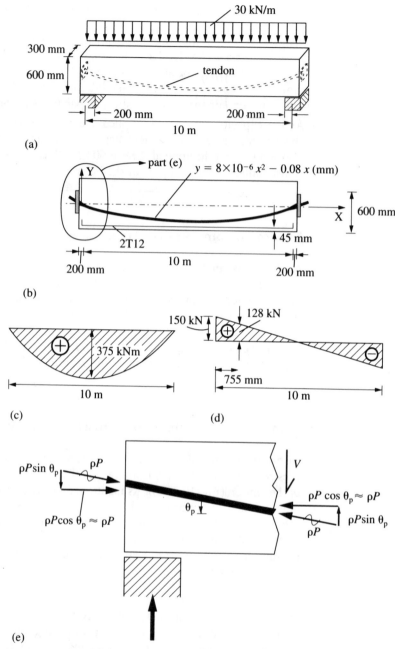

(a)

(b)

(c)

(d)

(e)

shear. The reduction in the web width due to the presence of post-tensioning ducts can significantly affect the requirements for shear, particularly in flanged members where the web may already not be very wide.

Example 10.2   Post-tensioned beam with uniform loading

Problem   Determine the shear capacity of the beam of Fig. 10.29 where the area of prestressing strand is 1800 mm² and the prestress force after all losses is

$\rho P = 1400$ kN. The factored ULS loading is 30 kN/m and the resulting distributions of moment and shear are given in Figs 10.29(c) and (d). The duct diameter is 75 mm and the characteristic cylinder strength is 40 N/mm^2.

Solution For the calculation of effective depth, EC2 specifies that prestressing strand should be ignored. Hence, the effective depth is, from Fig. 10.29(b), $d = 555$ mm. Allowing for enhancement of shear capacity near the support in uniformly loaded members, it is necessary only to check the shear at a distance of $d = 555$ mm from the face of the support or, in this case, 755 mm from the centre of the support. At that point, the applied shear force (by linear interpolation in Fig. 10.29(d)) is 128 kN.

The slope of the tendon profile at this point is found by differentiating the equation given in Fig. 10.29(b):

$$y = 8 \times 10^{-6}x^2 - 0.08x$$

$$\Rightarrow \quad \frac{dy}{dx} = 16 \times 10^{-6}x - 0.08$$

At $x = 755$, the slope is:

$$\frac{dy}{dx}(x = 755) = -0.0679$$

Hence the tendon is inclined at an angle of $\tan^{-1}(-0.0679) = 3.88°$ and the vertical component of prestress is $1400 \times \sin(3.88°) = 95$ kN. The net applied shear force is therefore:

$$V_{net} = 128 - 0.9 \times 95$$

$$= 43 \text{ kN}$$

The basic design shear strength is, from Table 10.1:

$$\tau_{rd} = 0.41 \text{ N/mm}^2$$

The scale factor is:

$$k = \text{greater of } (1.6 - d) \text{ and } 1$$

$$= 1.6 - 0.555$$

$$= 1.045$$

Allowing for the presence of a grouted duct of 75 mm diameter, the net web width is:

$$b_w = 300 - 0.5 \times 75$$

$$= 262 \text{ mm}$$

As the ordinary reinforcement consists of two 12 mm diameter bars, the

433

longitudinal reinforcement ratio is:

$$\rho_l = \text{lesser of}\left(\frac{2(\pi 12^2/4)}{262 \times 555}\right) \text{ and } 0.02$$

$$= 0.001\,55$$

The prestress in the member is*:

$$\sigma_{cp} = \frac{\gamma_p \rho P}{A_g}$$

$$= \frac{0.9 \times 1400 \times 10^3}{300 \times 600}$$

$$= 7.0 \text{ N/mm}^2$$

Hence, the shear capacity is, from equation (10.8):

$$V_c = [\tau_{rd}k(1.2 + 40\rho_l) + 0.15\sigma_{cp}]b_w d$$

$$= [0.41 \times 1.045(1.2 + 40 \times 0.001\,55) + 0.15 \times 7]262 \times 555$$

$$= 231\,304 \text{ N}$$

$$= 231 \text{ kN}$$

As V_c greatly exceeds the net applied force of 43 kN, there is ample capacity to resist shear and minimum shear reinforcement throughout the beam will be sufficient.

10.5 Shear strength of members with shear reinforcement

When the applied shear force, V, exceeds the shear capacity of the concrete, V_c, shear reinforcement must be provided to resist the difference, $V - V_c$. Even when the applied shear in beams is less than the shear capacity, a minimum quantity of shear reinforcement must be provided.

Stirrups (also known as links), such as those illustrated in Fig. 10.2(a), are the most common form of shear reinforcement. Stirrups are generally placed vertically in the web of the member as illustrated and should be fixed around the longitudinal reinforcement. For variable depth members, vertical stirrups are no longer perpendicular to the longitudinal axis of the beam and special care must be taken in their design. Another form of shear reinforcement is bent-up bars (Fig. 10.2(c)), which are bars of bottom longitudinal reinforcement bent up at their ends so as to cross the cracks developed by shear stresses. For beams, however, EC2 recommends that bent-up bars should not be used for

---

* While EC2 does not specifically say so, it seems sensibly conservative to assume that prestress losses will have occurred and that prestress is factored by γ_p.

Table 10.2 Minimum values for shear ratio, ρ_w, for beams, as recommended by EC2

Concrete strength, f_{ck} (N/mm²)	Yield strength of steel (N/mm²)	
	250*	460*
12 and 20	0.0015	0.0008
25 to 35	0.0022	0.0012
40 to 50	0.0028	0.0014

* Values found by interpolation of values specified in EC2.

shear reinforcement except in combination with stirrups and, furthermore, at least 50 per cent of the shear reinforcement should be in the form of stirrups. More importantly perhaps, bent-up bars, while effective in terms of material, have a high associated labour cost and tend to add to the overall cost of construction.

The minimum quantity of shear reinforcement for beams, recommended by EC2, is given in Table 10.2. These minimum values may be disregarded in minor structural elements such as lintels of less than 2 m span. For slabs, it is not necessary to provide any shear reinforcement provided $V > V_c$. However, where links are to be provided, a minimum quantity of 60 per cent of the values in Table 10.2 should be used. The minimum quantities of reinforcement are specified in terms of the shear reinforcement ratio, ρ_w. For vertical stirrups in prismatic members, this is the ratio of the cross-sectional area of stirrup to the area of concrete in plan (see Fig. 10.30), that is:

$$\rho_w = \frac{A_{sw}}{sb_w} \tag{10.14}$$

Fig. 10.30 Definition of ρ_w: (a) pictorial view; (b) section X–X (plan view)

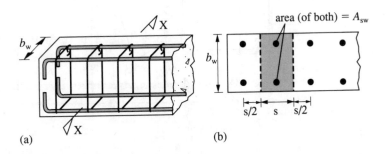

where A_{sw} = area of shear reinforcement within length, s; for stirrups with two legs each, the combined area from both legs is used in the calculation of A_{sw}

s = spacing of shear reinforcement

b_w = web breadth or, in the case of rectangular sections, actual breadth

435

Mechanism of shear transfer

Shear reinforcement, like flexural reinforcement, does not prevent cracks from forming in a member. Its purpose is to ensure that the member will not undergo shear failure before the full bending capacity of the member is reached. When inclined cracks form in a member with shear reinforcement, the bars which cross the cracks contribute to the shear resistance of the member, as illustrated in Fig. 10.31. The total shear capacity provided by the section can be considered as a combination of the capacity of the reinforcement and that of the concrete as illustrated in Fig. 10.31. Thus:

$$V_{cap} = V_s + V_c \qquad (10.15)$$

where V_s is the contribution of the shear reinforcement and V_c is the contribution from dowel action, aggregate interlock and the shear stresses in the uncracked concrete (V_c is given by equation (10.8)).

Fig. 10.31 Shear resistance of member with stirrups: (a) elevation showing reinforcement; (b) transfer of shear force through member

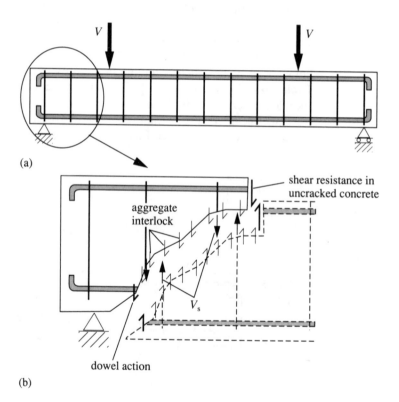

If the applied shear force is sufficiently large, the shear reinforcement will reach its yield strength. Beyond this point, the reinforcement behaves plastically and the cracks open more rapidly. As the cracks widen, the proportion of the shear resisted by aggregate interlock is reduced forcing an increase in dowel action and shear stress in the uncracked portion of the section. Failure finally occurs by dowel splitting or by crushing of the concrete in the compression zone.

Truss model for members with shear reinforcement

In order to quantify the behaviour and strength of members with shear reinforcement, an equivalent truss model is frequently used to represent the behaviour of members in shear. The simplest and earliest of the available models is described here as it clearly demonstrates the basic principles.

When subjected to combined moment and shear, a member with shear reinforcement, such as that illustrated in Fig. 10.32(a), develops inclined cracks in the same way as a member without shear reinforcement. For the cracked member illustrated, compressive stress develops in the concrete at the top of the section and tensile stress develops in the longitudinal reinforcement at the bottom of the section as illustrated in Fig. 10.32(b). In addition, compressive forces are developed along inclined paths between the cracks and tensile forces are developed in each of the stirrups crossing the cracks. All of these internal forces are illustrated in Fig. 10.32(b), and together they are similar in pattern to the forces generated in a truss such as that illustrated in Fig. 10.32(c).

Fig. 10.32 Truss model for reinforced beam: (a) cracking in reinforced beam; (b) zones of compression between cracks; (c) truss model

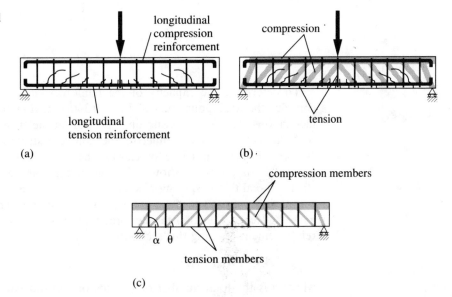

longitudinal compression reinforcement

compression

longitudinal tension reinforcement

tension

(a)

(b)

compression members

tension members

α θ

(c)

For the general truss analogy, the tension members corresponding to the shear reinforcement are inclined at an angle, α, to the horizontal equal to the inclination of the shear reinforcement. For vertical stirrups, this angle, α, is 90° as illustrated in Fig. 10.32(c). In a similar manner, the compression members are inclined at an angle, θ, to the horizontal where $0° < \theta < 90°$. In the simplest of the truss models, the inclination of the concrete 'struts' is assumed to be constant, generally with $\theta = 45°$ as illustrated in Fig. 10.32(c). In more complex models, the inclination of the concrete struts is assumed to vary along the length of the member. This latter is considered to be more accurate and its use can result in significant savings in shear reinforcement. An example of such a model is illustrated in Fig. 10.33.

437

Design for Shear and Torsion

Fig. 10.33 Truss with struts at various angles

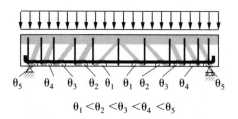

$$\theta_1 < \theta_2 < \theta_3 < \theta_4 < \theta_5$$

Fig. 10.34 Details of truss model:
(a) elevation;
(b) elevation of beam left of X–X

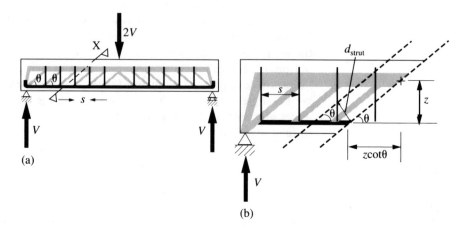

(a)

(b)

To determine the capacity of the reinforcement to resist shear force, V_s, consider the analogous truss of Fig. 10.34(a), having constant concrete strut inclinations. The forces acting on a portion of the truss to the left of section X–X (a section taken parallel to the concrete strut) are illustrated in Fig. 10.34(b). For the purpose of clarity, the truss illustrated in Figs 10.34(a) and (b) is simplified to show only one stirrup joining adjacent compression struts. In reality, the spacing of stirrups is typically less than the spacing of the compression struts and there are a number of stirrups crossing a section such as X–X. Hence, the total tensile force that can safely be carried by the shear reinforcement crossing this section is:

$$V_s = n A_{sw}(f_y/\gamma_s) \tag{10.16}$$

where n is the total number of stirrups or bent-up bars crossing the section, A_{sw} is the cross-sectional area of each stirrup, f_y is the characteristic stirrup yield strength and γ_s the material factor of safety for steel ($\gamma_s = 1.15$). If the spacing of the stirrups is given by s, the total number of bars crossing section X–X and the total number of compression struts is:

$$n = z(\cot \theta)/s \tag{10.17}$$

where z is the lever arm between the tensile force in the longitudinal bottom reinforcement and the horizontal compression force in the concrete. Thus, equation (10.16) becomes:

$$V_s = \frac{A_{sw} f_y z(\cot \theta)}{\gamma_s s} \tag{10.18}$$

438

The compressive force acting on the concrete struts is assumed to generate a uniform stress distribution in each strut. This will cause compression failure in the strut when the applied stress equals the compressive strength of the concrete. The maximum force that can be taken by each strut is:

$$C_{\text{strut}} = v \frac{f_{\text{ck}}}{\gamma_{\text{c}}} b_{\text{w}} d_{\text{strut}} \qquad (10.19)$$

where $v(f_{\text{ck}}/\gamma_{\text{c}})$ is the compressive strength of the concrete struts and d_{strut} is the depth, as illustrated in Fig. 10.34(b). The **effectiveness factor, v,** reflects the difference between the compressive strength of a cylinder tested in the laboratory and the actual compressive strength of a strut. The shear force required to cause failure of the compressive struts, V_{w}, is the vertical component of C_{strut} multiplied by the number of struts and is given by:

$$V_{\text{w}} = C_{\text{strut}}(\sin \theta)n = v \frac{f_{\text{ck}}}{\gamma_{\text{c}}} b_{\text{w}} d_{\text{strut}}(\sin \theta)n \qquad (10.20)$$

From Fig. 10.34(b), the effective depth of the struts, d_{strut}, is given by:

$$d_{\text{strut}} = s(\sin \theta) \qquad (10.21)$$

Thus, equation (10.20) becomes:

$$V_{\text{w}} = v \frac{f_{\text{ck}}}{\gamma_{\text{c}}} b_{\text{w}} s(\sin^2 \theta)n$$

$$= v \frac{f_{\text{ck}}}{\gamma_{\text{c}}} b_{\text{w}} s(\sin^2 \theta) \frac{z(\cot \theta)}{s}$$

$$\Rightarrow \qquad V_{\text{w}} = v \frac{f_{\text{ck}}}{\gamma_{\text{c}}} b_{\text{w}} z(\cos \theta)(\sin \theta) \qquad (10.22)$$

Regardless of the magnitude of V_{c} and V_{s}, the applied shear force must clearly never exceed V_{w}.

Design formulae recommended by EC2

Where shear reinforcement is required other than the minimum, EC2 offers two alternative methods of design. The first method, known as the **standard method**, assumes a model truss in which the concrete struts are inclined at a constant angle of 45°. The alternative method, known as the **variable strut inclination method**, allows the designer to select the inclination of the notional concrete struts. The Eurocode allows any values for θ within the range $22° < \theta < 68°$, provided certain conditions on curtailment are satisfied. As the principles of both methods are essentially the same, only the standard method is considered further in this text. A detailed description of the variable strut inclination method is given by McGregor (1992).

The shear capacity of a section with shear reinforcement is simply given as the sum of the contributions from the concrete, V_c, and that from the shear reinforcement, V_s. The contribution of the shear reinforcement is given by equation (10.18). However, for the standard method, the inclination of the struts, θ, is equal to 45° and the lever arm, z, can be approximated with $0.9d$. Therefore, equation (10.18) becomes:

$$V_s = \frac{0.9 A_{sw} f_y d}{\gamma_s s} \tag{10.23}$$

An expression for the area of shear reinforcement required at a given section can be found by rearranging the equations derived to date. Clearly the capacity to resist shear, V_{cap}, must equal or exceed the ultimate shear force due to the applied loads, V, that is:

$$V_{cap} \geq V$$

Separating the shear capacity into its components, V_c and V_s, and rearranging gives:

$$V_s \geq V - V_c$$

Substituting for V_s from equation (10.23) then gives:

$$\frac{0.9 A_{sw} f_y d}{\gamma_s s} \geq V - V_c$$

$$\Rightarrow \qquad \frac{A_{sw}}{s} \geq \frac{V - V_c}{0.9(f_y/\gamma_s)d} \tag{10.24}$$

where V_c is calculated from equation (10.8). It should be noted, when using these equations, that the spacing of the reinforcement, s, is restricted by detailing rules given in Section 5.4 of EC2.

Regardless of the quantity of shear reinforcement provided, EC2 requires that the member be checked for crushing of the concrete in the compression struts. Thus, the shear force must not exceed V_w, as given by equation (10.22). Letting $\theta = 45°$ and $z = 0.9d$ in this equation gives:

$$V_w = v \frac{f_{ck}}{\gamma_c} b_w (0.9d) \left(\frac{1}{\sqrt{2}}\right)\left(\frac{1}{\sqrt{2}}\right)$$

$$\Rightarrow \qquad V_w = 0.45 v \frac{f_{ck}}{\gamma_c} b_w d \tag{10.25}$$

The code further recommends that the effectiveness factor, v, be calculated as:

$$v = \text{greater of } (0.7 - f_{ck}/200) \text{ and } 0.5 \tag{10.26}$$

When checking that the applied shear force, V, is less than V_w, enhancement due to proximity to the support is not allowed.

Example 10.3 Reinforced concrete beam

Problem Use the standard method of EC2 to determine the shear capacity at sections A–A and B–B for the member illustrated in Fig. 10.35. The characteristic cylinder strength for the concrete is 35 N/mm². There is 1 per cent longitudinal reinforcement near the supports at the top and 0.75 per cent near mid-span at the bottom of the section. The effective depth for both hogging and sagging is $d = 650$ mm.

Fig. 10.35 Beam of Example 10.3: (a) elevation; (b) sections A–A and B–B

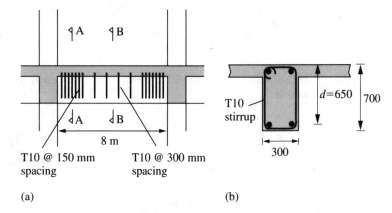

(a)

(b)

Solution For concrete with $f_{ck} = 35$ N/mm², the basic design shear strength is, from Table 10.1, 0.37 N/mm². As the effective depth exceeds 600 mm, there is no scale effect and the scale factor $k = 1$ (see equation (10.4)). Hence, the contribution of the concrete to the shear capacity at section A–A where there is 1 per cent reinforcement is, from equation (10.8):

$$V_c = [\tau_{rd}k(1.2 + 40\rho_l) + 0.15\sigma_{cp}]b_w d$$
$$= [0.37 \times 1(1.2 + 40 \times 0.01) + 0]300 \times 650$$
$$= 115\,440 \text{ N}$$
$$= 115 \text{ kN}$$

At section A–A, the shear reinforcement consists of 10 mm diameter high-yield stirrups at 150 mm centres. As there are two legs in each stirrup, the cross-sectional area of the two legs is:

$$A_{sw} = 2(\pi 10^2/4)$$
$$= 157 \text{ mm}^2$$

From equation (10.23), the contribution of the reinforcement to the shear capacity is:

$$V_s = \frac{0.9A_{sw}f_y d}{\gamma_s s}$$
$$= \frac{0.9 \times 157 \times 460 \times 650}{1.15 \times 150}$$
$$= 244\,920 \text{ N}$$
$$= 245 \text{ kN}$$

The sum of the contributions from the concrete and the reinforcement is:

$$V_{cap} = V_c + V_s$$
$$= 115 + 245$$
$$= 360 \text{ kN}$$

The possibility of crushing of the concrete in the web must also be checked. From equation (10.26), the effectiveness factor, v, is:

$$v = \text{greater of } (0.7 - f_{ck}/200) \text{ and } 0.5$$
$$= \text{greater of } (0.7 - 35/200) \text{ and } 0.5$$
$$= 0.525$$

Hence, from equation (10.25), the shear force required to cause crushing of the web is:

$$V_w = 0.45v \frac{f_{ck}}{\gamma_c} b_w d$$
$$= 0.45 \times 0.525 \left(\frac{35}{1.5}\right) 300 \times 650$$
$$= 1\,075\,000 \text{ N}$$
$$= 1075 \text{ kN}$$

Clearly this does not govern the design, and the shear capacity at A–A is 360 kN.

At B–B, the reinforcement ratio is 0.75 per cent and here the contribution of the concrete to the shear capacity is (equation (10.8)):

$$V_c = [\tau_{rd}k(1.2 + 40\rho_l) + 0.15\sigma_{cp}]b_w d$$
$$= [0.37 \times 1(1.2 + 40 \times 0.0075) + 0]300 \times 650$$
$$= 108\,225 \text{ N}$$
$$= 108 \text{ kN}$$

The contribution of the stirrups is:

$$V_s = \frac{0.9A_{sw}f_y d}{\gamma_s s}$$
$$= \frac{0.9 \times 157 \times 460 \times 650}{1.15 \times 300}$$
$$= 122\,460 \text{ N}$$
$$= 122 \text{ kN}$$

Hence the total shear capacity at B–B is:

$$V_{cap} = V_c + V_s$$
$$= 108 + 122$$
$$= 230 \text{ kN}$$

Example 10.4 **Beam with shear enhancement**

Problem For the beam of Example 10.1, determine the required shear reinforcement.

Solution In Example 10.1, the applied shear force at point C of $V = 200$ kN exceeds the capacity of the concrete, $V_c = 176$ kN. Hence shear reinforcement must be provided to resist the difference in accordance with inequality (10.24). Assuming mild steel stirrups ($f_y = 250$ N/mm^2):

$$\frac{A_{sw}}{s} \geq \frac{V - V_c}{0.9(f_y/\gamma_s)d}$$

$$\geq \frac{200 \times 10^3 - 176 \times 10^3}{0.9(250/1.15)500}$$

$$\geq 0.245 \text{ mm}^2/\text{mm}$$

The minimum required shear reinforcement ratio is, from Table 10.2, 0.0022. Thus (equation (10.14)), we get:

$$\rho_w = \frac{A_{sw}}{sb_w} \geq 0.0022$$

$$\Rightarrow \quad \frac{A_{sw}}{s} \geq 0.0022 b_w$$

$$\geq 0.0022 \times 300$$

$$\geq 0.66 \text{ mm}^2/\text{mm}$$

Thus, the minimum required reinforcement is sufficient to resist the applied shear. However, as mentioned in the previous section, EC2 suggests that, when the effect of shear enhancement is to reduce the reinforcement requirement to its minimum level, then the designer may wish to exercise caution and discount the beneficial effect of enhancement. Thus, in Example 10.1, τ_{rd} takes its basic unenhanced value of 0.37 N/mm^2 and V_c becomes 105 kN. Then inequality (10.24) becomes:

$$\frac{A_{sw}}{s} \geq \frac{V - V_c}{0.9(f_y/\gamma_s)d}$$

$$\geq \frac{200 \times 10^3 - 105 \times 10^3}{0.9(250/1.15)500}$$

$$\geq 0.971 \text{ mm}^2/\text{mm}$$

Stirrups of 10 mm diameter (two legs each) at a spacing centre to centre of 150 mm give a shear reinforcement area to spacing ratio of:

$$\frac{A_{sw}}{s} = 2\left(\frac{\pi 10^2}{4}\right)\frac{1}{150}$$

$$= 1.047 \text{ mm}^2/\text{mm}$$

which satisfies the reinforcement requirement for that part of the beam for which the shear force is 200 kN. Elsewhere in the beam, 10 mm diameter stirrups at 200 mm centres give an area to spacing ratio of:

$$\frac{A_{sw}}{s} = 2\left(\frac{\pi 10^2}{4}\right)\frac{1}{200}$$

$$= 0.785 \text{ mm}^2/\text{mm}$$

which satisfies the minimum reinforcement requirement. Hence, 10 mm diameter stirrups at 200 mm centres will be used elsewhere in this beam.

Interface shear

In flanged members subjected to bending, a shear stress develops in the flange and between the flange and the web. This is the process by which the compressive axial stress is transmitted from the web to the flange and, particularly in the case of thin flanges, its effect can be quite significant. The phenomenon is illustrated in Fig. 10.36 which shows a segment of beam where moment is increasing from zero at A to a high (sag) value at B. The interface shear stresses illustrated are necessary for equilibrium of the segment of flange abcd. The magnitude of the interface stresses can be calculated using equation (10.1) from which it can be seen that a large value for shear, V (rapidly changing moment), can result in a high interface shear stress. Reinforcement requirements are calculated on the basis of a truss analogy such as that illustrated in Fig. 10.36(b). In this case, the truss is used to transfer load in the horizontal plane between the flange and the web parts of the beam.

10.6 Design of slabs for punching shear

The punching shear failure mechanism illustrated in Fig. 10.37 can occur in any two-way spanning members which are supported directly by columns or which are acted on by heavy concentrated loads. The cross-sectional shape of the supporting column or the concentrated load, known as the **loaded area**, determines the shape of the failure surface in the concrete slab. In particular, circular loaded areas cause conical wedges to form and rectangular loaded areas cause pyramidal wedges to form. For design purposes, the inclined surfaces are represented by vertical ones. It is assumed that these vertical surfaces occur at a constant distance from the edges of the loaded area as illustrated in Fig. 10.38. The corresponding perimeter, illustrated in both figures, is known as the **critical perimeter**.

Location of the critical perimeter

There is disagreement among the codes of practice of various countries on the distance from the loaded area at which the critical perimeter is located. For

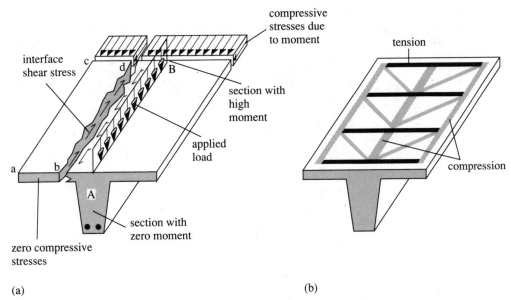

Fig. 10.36 Interface shear in flanged beam: (a) interface shear stresses; (b) truss model

instance, the code of the American Concrete Institute recommends that the critical perimeter be located at a distance of 0.5d from the edge of the loaded area whereas the British Standard, BS8110, recommends a distance of 1.5d, where d in both cases is the average effective depth of the longitudinal reinforcement spanning in two perpendicular directions. Consequently, the formula for shear resistance given by BS8110 gives a design shear strength significantly different from that given by the equivalent formula in the ACI code.

Fig. 10.37 Punching shear failure at column support

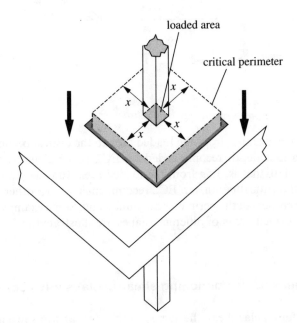

Fig. 10.38 Critical perimeters for typical loaded areas specified in EC2. Perimeters are located a constant distance, 1.5*d*, from edge of loaded area

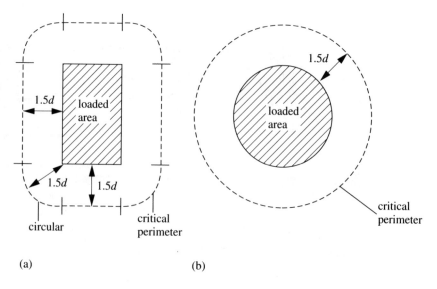

(a)   (b)

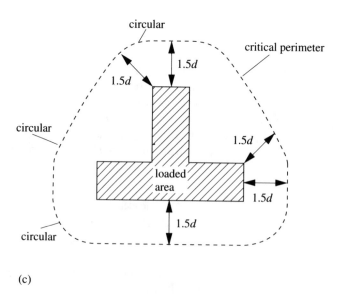

(c)

The distance from the loaded area to the critical perimeter for members with constant depth recommended by EC2 is the same as that recommended by BS8110, that is, 1.5*d* from the loaded area. In order to minimize the length, *u*, of the critical perimeter, EC2 recommends (unlike other codes of practice) that the corners of the perimeter be rounded off. Some examples of critical perimeters for loaded areas of different shapes are illustrated in Fig. 10.38.

Checking for punching shear in slabs with constant depth

As for regular shear, the average applied shear stress in a member with punching shear is given by equation (10.3), that is:

$$v = V/b_w d$$

However, in the case of members with punching shear, the breadth of the section in shear is equal to the length of the critical perimeter. Further, with punching shear, there are two effective depths, one for reinforcement running parallel to the X-axis and one for reinforcement running parallel to the Y-axis (the layers of reinforcement will, of necessity, be at different depths as one must rest on the other). Thus, for punching shear, the average shear stress becomes:

$$v = V/ud_{av} \tag{10.27}$$

where u is the length of the critical perimeter and d_{av} is the average of the two effective depths. The shear strength, v_c, for members subjected to punching, is calculated differently in EC2 to the strength for regular shear. The effect of prestress, in particular, is treated differently by the code. For regular shear, prestress is treated as an applied axial force and the factored design value, presumably with due allowance for losses, should be used in the calculation of the design axial stress, σ_{cp}. For punching shear, EC2 specifically stipulates that losses should not be allowed for. The axial stress for punching, thus calculated, is denoted σ_{cp0}. When a slab is stressed by different amounts in two directions, the average stress should be used. The influence of σ_{cp0} on the punching shear strength is effected through an adjustment in the reinforcement ratio. An effective longitudinal reinforcement ratio, ρ_l, is defined as:

$$\rho_l = \text{lesser of} \left(\sqrt{\rho_x \rho_y} + \frac{\sigma_{cp0}}{(f_y/\gamma_s)} \right) \text{ and } 0.015 \tag{10.28}$$

where ρ_x and ρ_y are the longitudinal tension reinforcement ratios in the X- and Y-directions respectively. Then, the punching shear strength for both pre-stressed and reinforced concrete members without shear reinforcement is:

$$v_c = \tau_{rd} k(1.2 + 40\rho_l) \tag{10.29}$$

where τ_{rd} = basic design shear strength. Values for τ_{rd} are given in Table 10.1. Unlike the case of regular shear, enhancement is not allowed for members subjected to punching.

k = scale effect parameter as defined in equation (10.4).

From equations (10.27) and (10.29), the capacity to resist punching shear in members without shear reinforcement is:

$$V_c = \tau_{rd} k(1.2 + 40\rho_l)ud_{av} \tag{10.30}$$

When the punching shear force due to applied loads is eccentric to the loaded area or the force is combined with a moment as would happen in an edge or corner column (see Fig. 10.39), the adverse effect of the applied load is significantly increased. This effect is allowed for in EC2 by the multiplication of the applied force, V, by a factor β. When no eccentricity of loading is possible, $\beta = 1$. In other cases, β can be taken from Fig. 10.39. (Note that even in an internal column, the variable portion of the load can be applied unsymmetrically which results in moment in the column.)

Fig. 10.39 Factor β to allow for eccentricity of loading

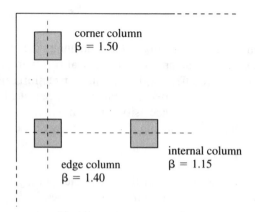

corner column
$\beta = 1.50$

edge column
$\beta = 1.40$

internal column
$\beta = 1.15$

Example 10.5   Punching shear in slab of constant depth

Problem   Check the slab of Fig. 10.40 for punching given that the ultimate factored reaction in the column due to permanent and variable loading is 300 kN. The characteristic cylinder strength of the concrete in the slab is 30 N/mm². The top reinforcement in both directions consists of 16 mm diameter high-yield bars at 150 mm centres.

Solution   As moment over the support will be hogging, the effective depths are measured from the bottom surface to the top layers of reinforcement. For reinforcement parallel to the *X*- and *Y*-axes, the respective effective depths are, from Fig. 10.40(b):

$$200 - 25 - \tfrac{16}{2} = 167 \text{ mm}$$

and:

$$200 - 25 - 16 - \tfrac{16}{2} = 151 \text{ mm}$$

Hence:

$$d_{av} = (167 + 151)/2 = 159 \text{ mm}$$

Fig. 10.40 Slab of Example 10.5: (a) plan view of column and portion of slab; (b) section A–A

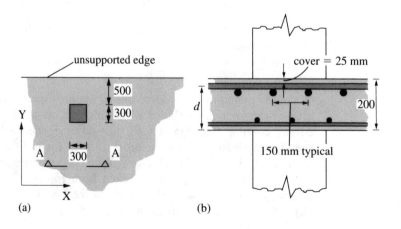

unsupported edge

500

300

Y

A   300   A

X

(a)

cover = 25 mm

d

200

150 mm typical

(b)

The area of top reinforcement provided in a 1 m length of slab is:

$$\frac{\pi 16^2}{4} \times \frac{1000}{150} = 1340 \text{ mm}^2$$

Hence, the reinforcement ratios are:

$$\rho_x = \frac{1340}{167 \times 1000} = 0.008\,02$$

and

$$\rho_y = \frac{1340}{151 \times 1000} = 0.008\,87$$

From equation (10.28), the effective reinforcement ratio is:

$$\rho_l = \text{lesser of} \left(\sqrt{\rho_x \rho_y} + \frac{\sigma_{cp0}}{(f_y/\gamma_s)} \right) \text{ and } 0.015$$

$$= \sqrt{0.008\,02 \times 0.008\,87} + 0$$

$$= 0.008\,44$$

From equation (10.4):

$$k = \text{greater of } (1.6 - d) \text{ and } 1$$

$$= 1.6 - 0.159 = 1.441$$

From Table 10.1, $\tau_{rd} = 0.34 \text{ N/mm}^2$. Two critical perimeters are considered as illustrated in Fig. 10.41. The perimeter illustrated in Fig. 10.41(a) consists of four straight segments, each of length 300 mm, and four quarter segments of circle of total length equal to the perimeter of a circle. Hence:

$$\text{length} = 4 \times 300 + 2\pi \times 238$$

$$= 2695 \text{ mm}$$

The perimeter length illustrated in Fig. 10.41(b) is:

$$\text{length} = 2 \times 500 + 3 \times 300 + \pi \times 238$$

$$= 2648 \text{ mm}$$

Fig. 10.41 Alternative critical perimeters for Example 10.5

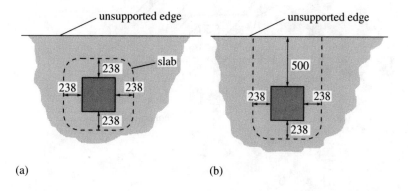

(a)   (b)

Taking the lesser of these two lengths ($u = 2648$ mm), the capacity to resist punching shear force is, from equation (10.30):

$$V_c = \tau_{rd}k(1.2 + 40\rho_l)ud_{av}$$
$$= 0.34 \times 1.441(1.2 + 40 \times 0.00844)2648 \times 159$$
$$= 317\,177 \text{ N}$$
$$= 317 \text{ kN}$$

The applied load must be factored in accordance with Fig. 10.39 to give an effective applied shear of:

$$\beta V = 1.4 \times 300$$
$$= 420 \text{ kN}$$

As $\beta V > V_c$, the slab is clearly not adequate without shear reinforcement.

Slabs with variable depth

For the member of Example 10.5, the factored applied shear force exceeded the punching shear capacity of the concrete. Therefore, punching shear failure will occur at the loaded area unless the capacity to resist it is increased. This can be achieved in a number of ways, including:

1. Increasing the depth of the member over its entire area.
2. Introducing shear reinforcement near the loaded areas.
3. Providing column heads or 'drops', as illustrated in Fig. 10.42, to thicken the member locally near the loaded area.
4. Introducing a gradual increase in depth which gives a deeper section near the loaded area.

Fig. 10.42 Column head

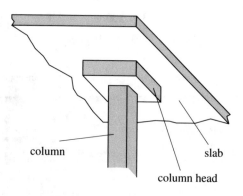

column   slab

column head

Where slabs subjected to punching are of continuously variable depth, important differences exist in the method of calculating shear capacity. Full details of the equations appropriate for such members are given in EC2. Another, more common, form of variable-depth slab is when the slab is thickened locally around the column. Where column heads (or drops) are provided to enhance the punching shear resistance, a check of the shear capacity should be made at two critical sections, one within the column head and one outside the column head, as illustrated in Fig. 10.43. The section within the column head should be taken on a perimeter a distance of $1.5d_h$ from the outer edge of the column, where d_h is the effective depth of the column head (see Fig. 10.43(a)). The section outside the column head should be located on a perimeter a distance $1.5d$ from the edge of the column head, where d is the effective depth of the slab. The adequacy of the two critical sections is then checked as described above. Of course, if the effective depth of the column head is such that the distance $1.5d_h$ extends beyond the column head, then only the outer critical perimeter needs to be checked.

Fig. 10.43 Critical perimeters for slab with column head: (a) elevation; (b) plan from below

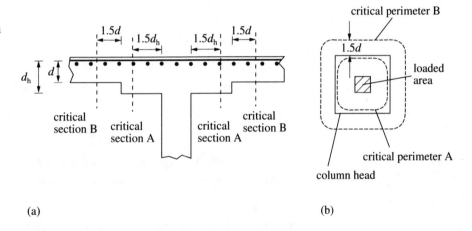

(a)   (b)

Example 10.6 Flat slab with column heads

Problem If the column of Example 10.5 is provided with a square head of side length 800 mm and depth (additional to that of the slab) of 100 mm, check the adequacy of the member against punching shear at the columns.

Solution The effective depth within the column head is increased by 100 mm over that in Example 10.5. Hence, the average value is:

$$d_h = 159 + 100 = 259 \text{ mm}$$

Hence the first perimeter to be checked is at a distance of $1.5 \times 259 = 388$ mm from the face of the column. However, as can be seen from Fig. 10.44, this is outside the column head and so does not need to be checked. The second perimeter to be considered is at a distance of $1.5d_{av} = 238$ mm outside the column head as illustrated in the figure. Because of the proximity of the

Fig. 10.44 Critical
perimeter of Example
10.6

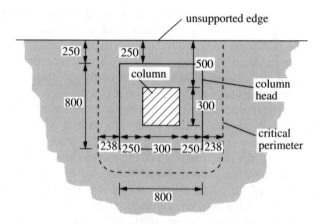

unsupported edge, the perimeter illustrated is clearly shorter than the perimeter which includes all four sides. The length of the perimeter illustrated is:

$$u = 800 \times 3 + 250 \times 2 + 238\pi$$

$$= 3648 \text{ mm}$$

Hence, referring to Example 10.5, the punching capacity is:

$$V_c = \tau_{rd} k (1.2 + 40\rho_l) u d_{av}$$

$$= 0.34 \times 1.441(1.2 + 40 \times 0.008\,44)3648 \times 159$$

$$= 436\,957 \text{ N}$$

$$= 437 \text{ kN}$$

This exceeds the effective applied shear force of 420 kN.

Members with punching shear reinforcement

For deeper slabs such as foundation pads and pile caps, punching shear reinforcement is commonly provided in the form of stirrups. These are fixed around longitudinal bars in two perpendicular directions, as illustrated in Fig. 10.45. For floor slabs, the relatively small depth makes it difficult to fix conventional stirrups. In recent years, proprietary systems which overcome these difficulties have increased in popularity. One such system consists of shear studs, as illustrated in Fig. 10.46. This system incorporates vertical steel rods capped with a circular plate welded to strips of flat steel plate. They have been found to be very effective and easy to fix on site. An alternative is the shearhoop system illustrated in Fig. 10.47 where hoops of stirrup reinforcement are placed around the column.

Whatever method is used, the shear reinforcement should extend from the loaded area at least as far as the critical perimeter. However, EC2 stipulates that only stirrups within a distance equal to the lesser of 1.5d and 800 mm from the loaded area are effective in resisting punching shear. Furthermore, only

Fig. 10.45 Punching shear reinforcement

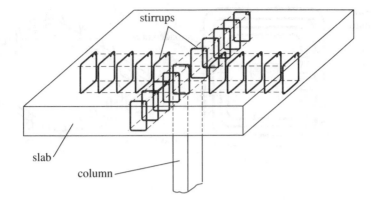

Fig. 10.46 *Decon* shear stud reinforcement: (a) shear studs attached to base plate; (b) shear studs in place around a column. Photographs courtesy of Decon, Canada

(a)

(b)

Design for Shear and Torsion

Fig. 10.47 Shearhoop reinforcement: (a) plan view; (b) basic shearhoop; (c) individual stirrup from prefabricated unit (from Chana and Clapson (1992))

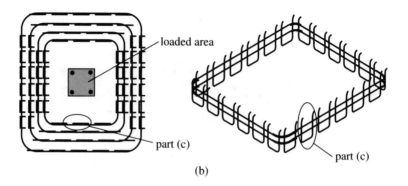

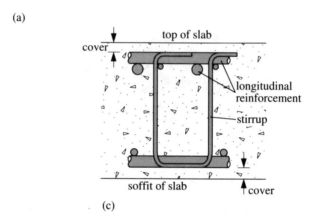

(a)

(b)

(c)

members having a depth of 200 mm or more can be provided with shear reinforcement.

As for members reinforced against regular shear, a minimum quantity of shear reinforcement must be provided for members subjected to punching shear. However (as for slabs with regular shear), this only applies when reinforcement is required. Thus, if the shear capacity of the concrete, V_c, exceeds the applied shear force, βV, then no reinforcement is necessary. If, on the other hand, βV exceeds V_c, the specified minimum quantity of reinforcement must be provided. The minimum shear reinforcement ratios which must be provided are 60 per cent of the values given in Table 10.2. For punching shear, the shear reinforcement ratio is defined as the ratio of the cross-sectional area of all stirrups to the area in plan of concrete. When tension reinforcement is not horizontal, as can be the case in slabs of variable depth, a variation on this definition applies (interested readers should refer directly to EC2). For slabs of constant depth:

$$\rho_w = \frac{\sum A_{sw}}{A_{crit} - A_{load}} \tag{10.31}$$

where A_{crit} is the plan area within the critical perimeter, A_{load} is the loaded area and the summation is over all reinforcement between the critical perimeter and the loaded area.

For a slab with shear reinforcement, the shear capacity is, as before:

$$V_{cap} = V_c + V_s \tag{10.32}$$

where V_s is the contribution of the shear reinforcement and V_c is given by equation (10.30). However, EC2 places an upper limit on the capacity of reinforced slabs of $1.6V_c$. The shear capacity provided by shear studs and other proprietary forms of reinforcement should be taken from manufacturers' specifications. For conventional shear reinforcement, the contribution of the reinforcement is given by:

$$V_s = \frac{\sum A_{sw} f_y}{\gamma_s} \tag{10.33}$$

where $\sum A_{sw}$ is the sum of the cross-sectional areas of all stirrups within a distance of $1.5d_{av}$ from the edge of the loaded area. In this calculation, the distance, $1.5d_{av}$, may not exceed 800 mm; that is, the distance is, in fact, the lesser of $1.5d_{av}$ and 800 mm.

For the punching shear capacity to equal or exceed the applied shear, V, we require:

$$V_{cap} \geq \beta V$$

$$\Rightarrow \qquad V_c + V_s \geq \beta V$$

$$\Rightarrow \qquad V_s \geq \beta V - V_c$$

Substitution from equation (10.33) then gives the required area of stirrup reinforcement:

$$\sum A_{sw} \geq \frac{\beta V - V_c}{f_y/\gamma_s} \tag{10.34}$$

If shear reinforcement is provided in the vicinity of the loaded area to prevent punching failure, it becomes necessary to check the shear capacity at a further critical perimeter a distance $1.5d_{av}$ from the outside of the shear reinforced area (as illustrated in Fig. 10.49). Similarly if this area is reinforced, a further critical perimeter, $1.5d_{av}$ outside this one, must be checked.

Example 10.7 Slab with punching shear reinforcement

Problem Determine the quantity of shear reinforcement required in the member of Fig. 10.48 to prevent punching failure due to the load from the circular column. The average effective depth of the slab in this region is 650 mm, $\tau_{rd} = 0.37$ N/mm^2, $k = 1$, $f_{ck} = 35$ N/mm^2 and the average effective reinforcement ratio is $\rho_l = 0.018$.

Solution The loaded area is of radius 200 mm. Hence the critical perimeter, $1.5d_{av}$ outside the loaded area, is of radius $200 + 1.5d_{av}$. The perimeter length is thus:

$$u = 2\pi(200 + 1.5d_{av})$$

$$= 2\pi(200 + 1.5 \times 650)$$

$$= 7383 \text{ mm}$$

Fig. 10.48 Geometry and loading for Example 10.7

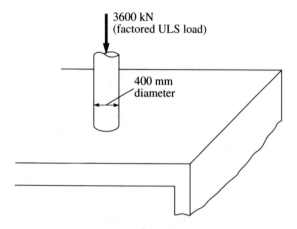

Hence the capacity of the concrete to resist punching shear is, from equation (10.30):

$$V_c = \tau_{rd}k(1.2 + 40\rho_l)ud_{av}$$

$$= 0.37 \times 1(1.2 + 40 \times 0.018)7383 \times 650$$

$$= 3\,409\,174 \text{ N}$$

$$= 3409 \text{ kN}$$

As the factored applied shear ($\beta V = 1.15 \times 3600$ kN) is less than $1.6V_c$, shear reinforcement can be provided to resist the difference between βV and V_c. High-yield stirrups will be used. From equation (10.34):

$$\sum A_{sw} \geq \frac{\beta V - V_c}{f_y/\gamma_s}$$

$$\geq \frac{1.15 \times 3600 \times 10^3 - 3409 \times 10^3}{460/1.15}$$

$$\geq 1828 \text{ mm}^2$$

As reinforcement is required, at least the minimum quantity must be provided. From Table 10.2, this is:

$$\rho_w \geq 0.6(0.0012)$$

$$\geq 0.000\,72$$

From equation (10.31):

$$\frac{\sum A_{sw}}{A_{crit} - A_{load}} \geq 0.000\,72$$

where:

$$A_{load} = \pi(400^2/4) = 125\,664 \text{ mm}^2$$

and:

$$A_{crit} = \pi(200 + 1.5d_{av})^2$$

$$= \pi(200 + 1.5 \times 650)^2$$

$$= 4\,337\,361 \text{ mm}^2$$

Hence:

$$\sum A_{sw} \geq 0.000\,72(A_{crit} - A_{load})$$

$$\geq 0.000\,72(4\,337\,361 - 125\,664)$$

$$\geq 3032 \text{ mm}^2$$

As this is greater than the area of reinforcement required to resist the applied load, it governs the design for punching on this perimeter. Detailing rules in EC2 require, for this situation, a maximum spacing of $0.6d_{av} = 390$ mm. A spacing of 325 mm is adopted. In the symmetrical arrangement of reinforcement illustrated in Fig. 10.49(a), 12 stirrups (with two legs each) are provided in all. A bar diameter of 16 mm will then result in a total area of:

$$\sum A_{sw} = 24\pi(16)^2/4$$

$$= 4825 \text{ mm}^2$$

The first critical perimeter, considered above, is circular as enhancement due to proximity to the column occurs at an equal distance in all directions from the (circular) column. As the reinforcement has been placed in the pattern of a cross, the next critical perimeter is not circular. A perimeter of shorter length is, in fact, illustrated in Fig. 10.49(b). It can be seen in this figure that a

Fig. 10.49 Shear reinforcement for Example 10.7: (a) plan view; (b) second critical perimeter

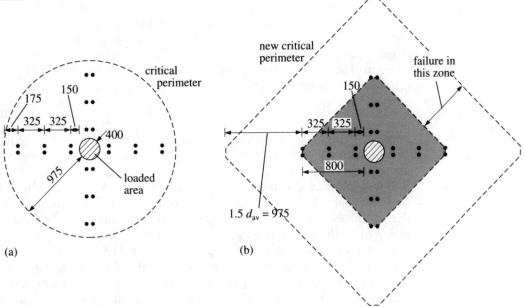

part-pyramid-shaped failure surface can be formed without crossing any of the reinforcement. The length of the diagonal of the (approximately) square area enclosed by the critical perimeter is $2(200 + 800 + 1.5d_{av})$, which is 3950 mm. Hence the length of the side of the square is $3950/\sqrt{2} = 2793$ mm and the perimeter length is approximately $4 \times 2793 = 11\,172$ mm. The capacity of the concrete for this perimeter is:

$$V_c = \tau_{rd}k(1.2 + 40\rho_l)ud_{av}$$

$$= 0.37 \times 1(1.2 + 40 \times 0.018)11\,172 \times 650$$

$$= 5\,158\,783\ N$$

$$= 5159\ \text{kN}$$

As this exceeds the effective applied shear, βV, no further shear reinforcement is required.

10.7 Behaviour of members with torsion

Torsional stresses in uncracked members

Members subjected to a torsional moment, commonly known as a torque, develop shear stresses. In general, these tend to increase in magnitude from the longitudinal axis of the member to its surface. If the shear stresses are sufficiently large, cracks will propagate through the member and, if torsion reinforcement is not provided, the member will collapse suddenly.

The elastic behaviour of uncracked concrete members with torsion, particularly non-circular members, is difficult to model precisely. In a circular member subjected to a torque T, such as the member in Fig. 10.50(a), the circumferential shear stress at a given cross-section varies linearly from the longitudinal axis of the member to a maximum value, τ_{max}, at the periphery of the section (Fig. 10.50(c)). The stress at any distance r from the longitudinal axis of a circular member is given by:

$$\tau_r = rT/I_p \tag{10.35}$$

where I_p is the polar second moment of area of the section and is equal to $\pi\phi^4/32$, where ϕ is the member diameter. The maximum shear stress, τ_{max}, is found by setting $r = \phi/2$ in equation (10.35).

For a non-circular member, the distribution of shear stress is not so straightforward. The rectangular member of Fig. 10.51(a), for instance, has the stress distribution illustrated in Fig. 10.51(b) at any given section when subjected to the torque illustrated. Unlike in the circular member, the stress distribution in a rectangular member is non-linear. The maximum shear stress occurs at the mid-point of the longer side and the shear stress at the corners of the section is zero indicating that the corners of the section are not distorted under torsion. Analytical studies have shown that the maximum shear stress,

Fig. 10.50 Member of circular section subjected to torsion: (a) geometry and loading; (b) section X–X

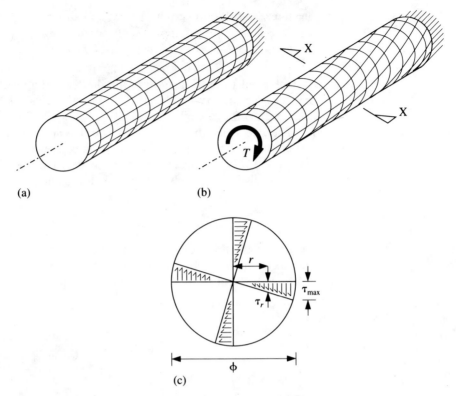

(a) (b)

(c)

Fig. 10.51 Member of rectangular section subjected to torsion: (a) geometry and loading; (b) section A–A ($y > x$)

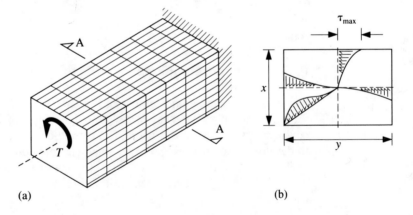

(a) (b)

rectangular section is given by:

$$\tau_{max} = \frac{T}{\alpha x^2 y} \qquad (10.36)$$

where x and y are the lengths of the shorter and longer sides, respectively. The value of the parameter α depends on the relative values of x and y. For a square section, $\alpha = 0.208$, while for a section with $x/y = 0.1$, $\alpha = 0.312$.

The stress distribution in thin-walled hollow members is much easier to determine than for solid non-circular members, even for a member with complex shape and varying thickness such as illustrated in Fig. 10.52(a). The shear stress in the walls is reasonably constant and is given by:

$$\tau = \frac{T}{2A_0 t} \qquad (10.37)$$

where t is the thickness of the wall of the member and A_0 is the area within a perimeter bounded by the centre line of the wall (Fig. 10.52(b)). On a given section, the shear stress is maximum where the thickness of the wall is minimum.

Fig. 10.52 Thin walled hollow section: (a) hollow bridge of box section; (b) definition of A_0

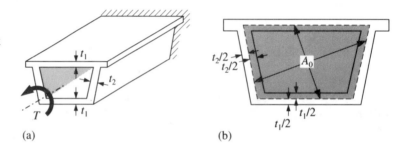

(a)   (b)

Failure of concrete members with torsion

Consider the rectangular member of Fig. 10.53 subjected to a torque T. Since there are no other external forces (and ignoring self-weight) the member is considered to be in pure torsion. The torque causes the member to twist and to develop shear stresses. Consider small elements on each face of the member, as illustrated in the figure. As for members with applied shear, shear stresses act on the sides of each element in the directions shown in Fig. 10.54(a). The

Fig. 10.53 Elements in member subjected to torsion

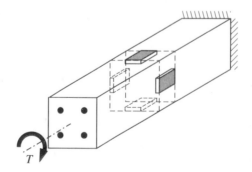

Fig. 10.54 Stresses and cracking due to torsion: (a) shear stresses; (b) principal stresses; (c) spiral cracking

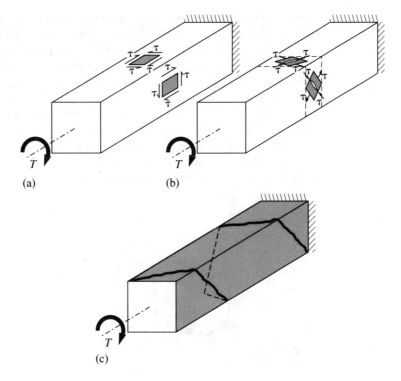

(a)   (b)

(c)

equivalent principal stresses, inclined at an angle of 45° to the horizontal, are illustrated in Fig. 10.54(b). In the same way as for shear, the principal tensile stresses cause the development of cracks inclined at a angle of 45°. However, in the case of torsion, they form a spiral all around the member, as illustrated in Fig. 10.54(c). Since the shear stresses in members with torsion are greatest at the surface, these cracks develop inwards from the surface of the member.

The member illustrated in Fig. 10.55 is subjected to a vertical force, V, in addition to the applied torque. This results in a combination of bending, shear and torsion and alters the orientation of the inclined cracks in a similar way to that described in section 10.2. As for members with shear, the introduction of prestress has the effect of delaying the onset of torsional cracking and altering the orientation of the inclined cracks.

For members with no form of reinforcement to prevent the opening of

Fig. 10.55 Combined shear, moment and torsion

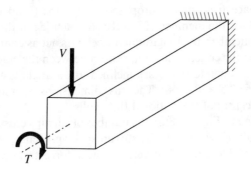

torsional cracks, failure of the member will occur almost as soon as the cracking begins. Therefore, torsional failure of a member without reinforcement is prevented only if the shear strength of the concrete exceeds the shear stress due to applied torsion. In practice, the shear strength is increased slightly through dowel action by the longitudinal reinforcing bars which cross the cracks.

The torsional strength of a concrete member can be significantly increased by providing suitable torsion reinforcement across the cracks. This is usually provided in the form of 'closed' four-sided stirrups, as illustrated in Fig. 10.56, in combination with longitudinal bars distributed around the periphery of the section. This reinforcement controls the propagation of cracks and ensures that when failure occurs due to yielding of the reinforcement, it is not sudden.

Fig. 10.56 Torsion reinforcement

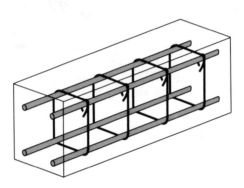

To quantify the behaviour of members with such torsional reinforcement, an equivalent space truss model, similar to the plane truss model for shear, can be used. This theory, developed by Lampert and Collins (1972), assumes that solid members can be designed as equivalent hollow members. Extensive tests indicate that this is a fair assumption since it has been found that the presence of the concrete at the centre of the member does not have a very significant effect on its torsional resistance. Thus, members are designed as equivalent thin-walled members. The thickness of the wall, t, is commonly taken as:

$$t = A_g/u \tag{10.38}$$

where A_g is the gross cross-sectional area of the member and u is the length of the perimeter.

The space truss model proposed by Lampert and Collins is illustrated in Fig. 10.57 for the member of Fig. 10.56. Each leg of the closed stirrups acts as a tension member, the longitudinal steel bars act as continuous top and bottom chords and the concrete in compression between the cracks acts as compression struts. The concrete struts are inclined at an angle θ which, like shear, varies in the range $22° < \theta < 68°$. The truss dimensions, x_0 and y_0, are measured from centre to centre of the notional thin walls.

Referring to Fig. 10.58(a), the number of stirrups crossing a side face of height y_0 is $y_0(\cot \theta)/s$ where s is the stirrup spacing. Hence, assuming that reinforcement has yielded, the total vertical shear force transmitted across the cracks

Fig. 10.57 Space truss model

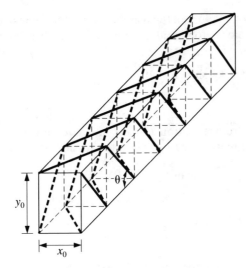

Fig. 10.58 Details of truss model: (a) stirrups transferring forces across crack; (b) equivalent thin walled member

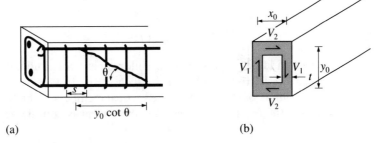

(a)   (b)

by the stirrups on one side face is:

$$V_1 = \left(\frac{y_0(\cot \theta)}{s}\right) A_{\text{leg}}\left(\frac{f_y}{\gamma_s}\right) \tag{10.39}$$

where A_{leg} is the area of one leg of the stirrup. Similarly the shear force transferred across the top or bottom face is:

$$V_2 = \left(\frac{x_0(\cot \theta)}{s}\right) A_{\text{leg}}\left(\frac{f_y}{\gamma_s}\right) \tag{10.40}$$

The shear force on the side wall of a thin-walled member is the product of the average stress due to applied load and the surface area. Hence the shear on a side wall is (see Fig. 10.58(b)):

$$V_1 = \tau t y_0$$

where t is the wall thickness. Substituting for τ from equation (10.37) gives:

$$V_1 = \left(\frac{T}{2A_0 t}\right) t y_0$$

$$\Rightarrow \quad V_1 = \frac{T}{2x_0} \tag{10.41}$$

463

Hence:

$$T = 2x_0 V_1 \tag{10.42}$$

Similarly it can be shown that:

$$T = 2y_0 V_2 \tag{10.43}$$

Substituting from equation (10.39) into equation (10.42) (or from equation (10.40) into equation (10.43)) gives the torsion at which the stirrups yield:

$$T = 2x_0 \left[\left(\frac{y_0 \cot \theta}{s} \right) A_{\text{leg}} \left(\frac{f_y}{\gamma_s} \right) \right]$$

$$\Rightarrow \quad T = \left(\frac{2x_0 y_0 \cot \theta}{s} \right) A_{\text{leg}} \left(\frac{f_y}{\gamma_s} \right) \tag{10.44}$$

Thus the stirrup reinforcement required to resist a torsion of T is:

$$A_{\text{leg}} = \frac{Ts}{2x_0 y_0 \cot \theta (f_y/\gamma_s)} \tag{10.45}$$

In addition to the stirrup reinforcement, longitudinal reinforcement is required to resist torsion. As can be seen in the truss model of Fig. 10.59(a), diagonal compression struts join the vertical members of the truss. Equilibrium at a joint of the truss where these members meet is considered in Figs 10.59(b) and (c).

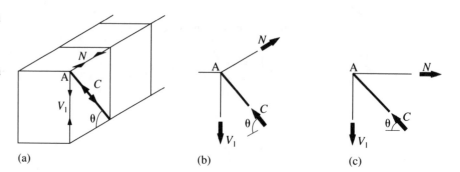

Fig. 10.59 Equilibrium at joint in truss model: (a) part of truss model; (b) free body diagram for joint A; (c) elevation of free body diagram at A

(a)   (b)   (c)

The compressive force in the diagonal members, C, must equal $V_1/\sin \theta$. There must also be a force in the longitudinal members of $N = C \cos \theta = V_1 \cot \theta$. Substitution from equation (10.39) gives:

$$N = \left(\frac{y_0}{s} \right) A_{\text{leg}} \left(\frac{f_y}{\gamma_s} \right) \cot^2 \theta \tag{10.46}$$

The total force in the longitudinal members from all four joints at a given cross-section is:

$$2V_1 \cot \theta + 2V_2 \cot \theta = 2(V_1 + V_2) \cot \theta$$

If the total area of longitudinal reinforcement is A_{long}, then:

$$A_{\text{long}}\frac{f_y}{\gamma_s} = 2(V_1 + V_2)\cot\theta$$

Substituting from equations (10.42) and (10.43) gives:

$$A_{\text{long}}\frac{f_y}{\gamma_s} = 2\left(\frac{T}{2x_0} + \frac{T}{2y_0}\right)\cot\theta$$

$$= T\frac{(x_0 + y_0)}{x_0 y_0}\cot\theta$$

$$\Rightarrow \qquad A_{\text{long}} = T\frac{(x_0 + y_0)\cot\theta}{x_0 y_0(f_y/\gamma_s)} \qquad (10.47)$$

This is the total area (all bars) of longitudinal reinforcement required to resist an applied torsion of T. It is additional to whatever longitudinal reinforcement is required to resist bending moment. Alternatively, equation (10.47) can be rearranged to give the maximum torsion possible without leading to yielding of the longitudinal reinforcement:

$$T = A_{\text{long}}\frac{x_0 y_0(f_y/\gamma_s)}{(x_0 + y_0)\cot\theta} \qquad (10.48)$$

Regardless of how much stirrup and longitudinal reinforcement is provided, the torsion must not be of such magnitude as to cause crushing of the concrete in the diagonal struts. As mentioned above, equilibrium at the joints of the truss requires a compressive force in the struts of $C = V_1/\sin\theta$. This force is resisted by stresses in the concrete between the diagonal cracks. The surface area of concrete to which this is applied is, from Fig. 10.60, $y_0(\cos\theta)t$, where t is the thickness of the notional wall. Hence the stress in the struts is:

$$\frac{V_1/\sin\theta}{y_0(\cos\theta)t} = \frac{V_1}{y_0 t(\sin\theta)(\cos\theta)}$$

Fig. 10.60 Breadth of compression struts

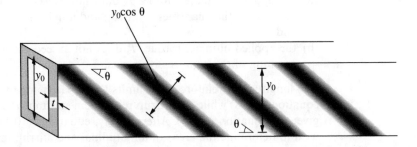

This must not exceed the compressive strength of the concrete, vf_{ck}/γ_c, where v is the effectiveness factor for torsion, that is:

$$\frac{V_1}{y_0 t(\sin\theta)(\cos\theta)} \leq \frac{vf_{ck}}{\gamma_c} \qquad (10.49)$$

Substitution for V_1 from equation (10.41) gives:

$$\frac{T/2x_0}{y_0 t(\sin\theta)(\cos\theta)} \leq \frac{vf_{ck}}{\gamma_c}$$

from which the torsion which would cause crushing of the concrete struts is T_w, where:

$$T_w = \frac{2x_0 y_0 t(\sin\theta)(\cos\theta)vf_{ck}}{\gamma_c} \tag{10.50}$$

10.8 Design of members for torsion in accordance with EC2

Where the static equilibrium of a structure relies on the torsional resistance of individual members, that is, in the case of equilibrium torsion, EC2 stipulates that a full design for torsion is necessary. The torsional resistance of members is calculated on the basis of an equivalent thin-walled section (i.e. the truss analogy) as described above. As for shear, the strut inclination angle, θ, can have any value in the range $22° \leq \theta \leq 68°$. The thickness of the equivalent wall is given by equation (10.38) but must not be less than twice the cover to the longitudinal steel. In the case of hollow members, the equivalent wall thickness should not exceed the actual wall thickness.

For sections of complex (solid) shape, such as T-sections, the torsional resistance can be calculated by dividing the section into individual elements of simple (say, rectangular) shape. The torsional resistance of the section is equal to the sum of the capacities of the individual elements, each modelled as an equivalent thin-walled section.

Members with pure torsion

For members with pure equilibrium torsion, EC2 requires that:

(a) the applied ultimate torque, T, does not exceed the torsional capacity, as dictated by the quantities of stirrup and longitudinal reinforcement present; and

(b) the applied ultimate torque, T, does not exceed the level that would cause crushing of the compressive struts, T_w.

The longitudinal reinforcement limits the capacity for torsion to that given by equation (10.48) while the stirrup reinforcement limits the capacity to the value given by equation (10.44). Alternatively, equations (10.47) and (10.45) can be used to determine the areas of longitudinal and stirrup reinforcement required to resist a torque T. The torque that would cause crushing of the compression struts, T_w, is calculated from equation (10.50). However, for torsion, the effectiveness factor, v, is restricted to 70 per cent of the level allowed for shear, that is:

$$v = \text{greater of } 0.7(0.7 - f_{ck}/200) \text{ and } 0.35 \tag{10.51}$$

Example 10.8   Member with pure torsion

Problem   Determine the maximum torque which can be applied to the member of Fig. 10.61 given that $f_{ck} = 30$ N/mm², the yield strength for the longitudinal reinforcement is $f_y = 460$ N/mm² and the yield strength for the stirrup reinforcement is $f_y = 250$ N/mm².

Fig. 10.61 Beam of Example 10.8

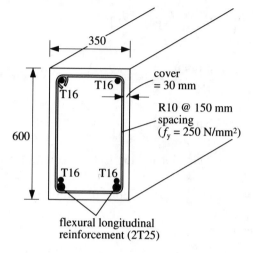

cover = 30 mm

R10 @ 150 mm spacing ($f_y = 250$ N/mm²)

flexural longitudinal reinforcement (2T25)

Solution   The total area of longitudinal reinforcement available to resist torsion is:

$$A_{long} = 4(\pi 16^2/4) = 804 \text{ mm}^2$$

From Fig. 10.61, the dimensions of the analogous thin walled section are:

$$t = \frac{A_g}{u} = \frac{350 \times 600}{2(350 + 600)} = 110 \text{ mm}$$

$$x_0 = 350 - t = 240 \text{ mm}$$

$$y_0 = 600 - t = 490 \text{ mm}$$

Hence assuming a compression strut angle of 45°, equation (10.48) gives:

$$T = \frac{A_{long} x_0 y_0 (f_y/\gamma_s)}{(x_0 + y_0) \cot \theta}$$

$$= \frac{804(240)(490)(460/1.15)}{(240 + 490)(1)} = 51\,808\,000 \text{ Nmm}$$

$$= 52 \text{ kNm}$$

Similarly, equation (10.44) gives the torsional capacity as dictated by the area of stirrup reinforcement. The label 'R10' indicates a 10 mm diameter mild steel stirrup with a characteristic yield strength of $f_y = 250$ N/mm^2. The area of one leg is:

$$A_{leg} = \pi 10^2/4 = 78.5 \text{ mm}^2$$

Hence:

$$T = \left(\frac{2x_0 y_0 \cot \theta}{s}\right) A_{leg} \left(\frac{f_y}{\gamma_s}\right)$$

$$= \frac{2(240)(490)(1)}{150}(78.5)\left(\frac{250}{1.15}\right) = 26\,758\,000 \text{ Nmm}$$

$$= 27 \text{ kNm}$$

The effectiveness factor for torsion is, from equation (10.51):

$$v = \text{greater of } 0.7(0.7 - 30/200) \text{ and } 0.35$$

$$= 0.385$$

Hence the torque that would cause crushing of the compression struts is, from equation (10.50):

$$T_w = \frac{2x_0 y_0 t(\sin \theta)(\cos \theta) v f_{ck}}{\gamma_c}$$

$$= \frac{2(240)(490)(110)(1/\sqrt{2})(1/\sqrt{2})(0.385)(30)}{1.5}$$

$$= 99\,607\,200 \text{ Nmm}$$

$$= 99 \text{ kNm}$$

Thus, with a compression strut inclined at an angle of 45°, the torsion capacity is governed by the area of stirrup reinforcement. However, the strut inclination angle can have any value in the range $22° \le \theta \le 68°$. Hence, $\cot \theta$ can vary in the range $2.5 \ge \theta \ge 0.4$. By trial and error (or by equating the two equations for T), an optimum value for $\cot \theta$ can be found. Taking $\cot \theta = 1.4$ ($\theta = 36°$), the longitudinal reinforcement dictates a torsional capacity of 37 kNm and the stirrup reinforcement dictates a capacity of 38 kNm. The corresponding value for T_w is 94 kNm. It can therefore be concluded that this beam has the capacity to resist a torsion of 37 kNm.

Members with combined actions

For members subjected to combined moment and torsion, EC2 recommends that the requirements for each action be determined separately and that the following rules are then applied:

(a) In the flexural tension zone, the longitudinal reinforcement required for torsion should be provided in addition to the amount required for moment.
(b) In the flexural compression zone, if the tensile stress in the concrete due to torsion is less than the compressive stress due to moment, then no longitudinal torsion reinforcement need be provided.

For members with combined torsion and shear, the ultimate torque, T, and the ultimate shear force, V, should satisfy the condition:

$$\left(\frac{T}{T_w}\right)^2 + \left(\frac{V}{V_w}\right)^2 \le 1 \tag{10.52}$$

where T_w and V_w are the torque and shear force respectively that would, acting alone, cause crushing of the concrete struts. The calculations for the design of stirrups may be made separately for torsion and shear. However, the angle, θ, for the concrete struts must be the same in both cases. The requirements for shear and torsion are, of course, additive.

Problems

Section 10.4

10.1 The ordinary reinforced section illustrated in Fig. 10.62 is subjected to a sagging moment. Determine whether stirrup reinforcement other than the minimum is required to resist an applied factored shear force of 100 kN, given that the concrete cylinder strength is $f_{ck} = 35$ N/mm^2.

Fig. 10.62 Section of Problem 10.1

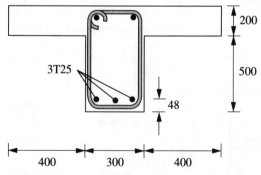

10.2 For the beam illustrated in Fig. 10.63, determine if shear reinforcement is required at B, given that the factored ULS shear is 300 kN and the factored ULS moment is 500 kNm. The concrete cylinder strength is $f_{ck} = 40$ N/mm^2.

10.3 The beam in Fig. 10.64 is of rectangular section and is prestressed with ten horizontal strands of area 150 mm^2 each, at an average distance of 150 mm above the soffit. The

Fig. 10.63 Beam of
Problem 10.2

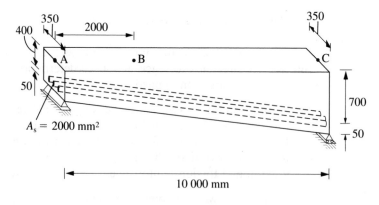

Fig. 10.64 Beam of
Problem 10.3

$f_{ck} = 45$ N/mm²

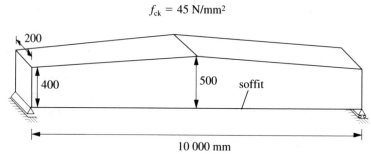

strands are tensioned to a stress of 970 N/mm² after all losses. If the factored ULS moment 2.5 m from the support is 200 kNm, determine the shear capacity of the concrete at this point.

Section 10.5

10.4 Calculate the shear reinforcement required for the beam and loading of Problem 10.2.

10.5 For the section illustrated in Fig. 10.65, calculate the shear reinforcement required. The total factored ULS shear force is 600 kN, of which 76 kN is resisted by the concrete. Check for crushing of concrete in the compression zone of the section.

Fig. 10.65 Section
of Problem 10.5

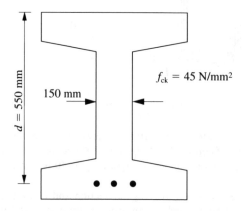

Section 10.6

10.6 Determine the capacity of the 200 mm slab illustrated in Fig. 10.66 to resist punching of the corner column given that it is reinforced with T12 bars at 150 mm centres in one direction (inner layer) and T16 at 200 mm centres in the other (outer layer). Assume $f_{ck} = 35$ N/mm^2 and that the cover is 30 mm.

Fig. 10.66 Plan view of portion of slab of Problem 10.6

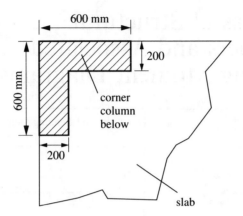

Appendix A

Stiffness of Structural Members and Associated Bending Moment Diagrams

No.	Force/moment per unit displacement	Bending Moment Diagram (positive sag)
1	$M = \dfrac{4EI}{l}$ 1 l	$\dfrac{2EI}{l}$ $\oplus$ $\ominus$ $\dfrac{4EI}{l}$
2	$M = \dfrac{3EI}{l}$ 1 l	$\ominus$ $\dfrac{3EI}{l}$
3	$M = \dfrac{EI}{l}$ 1 l	$\ominus$ $\dfrac{EI}{l}$
4	$F = \dfrac{3EI}{l^3}$ 1 l	$\dfrac{3EI}{l^2}$ $\ominus$
5	$F = \dfrac{12EI}{l^3}$ 1 l	$\ominus$ $\oplus$ $\dfrac{6EI}{l^2}$ $\dfrac{6EI}{l^2}$

6	$F = \dfrac{48EI}{l^3}$ $\dfrac{Fl^2}{16EI} = \dfrac{3}{l}$ $\dfrac{l}{2}$ $\dfrac{l}{2}$ 1	$\dfrac{12EI}{l^2}$
7	$F = \dfrac{AE}{l}$ 1 l	No bending moment

Appendix B

Reactions and Bending Moment Diagrams due to Applied Load

No.	Loads and Reactions	Bending Moment Diagram (positive sag)
1	ω $\frac{\omega l}{2}$ l $\frac{\omega l}{2}$	$\oplus$ $\frac{\omega l^2}{8}$
2	P $\frac{P}{2}$ l $\frac{P}{2}$	$\oplus$ $\frac{Pl}{4}$
3	a P b $\frac{Pb}{l}$ l $\frac{Pa}{l}$	a b $\oplus$ $\frac{Pab}{l}$
4	$\frac{\omega l^2}{2}$ ω ωl l	$\frac{\omega l^2}{2}$ $\ominus$
5	Pl P P l	Pl $\ominus$

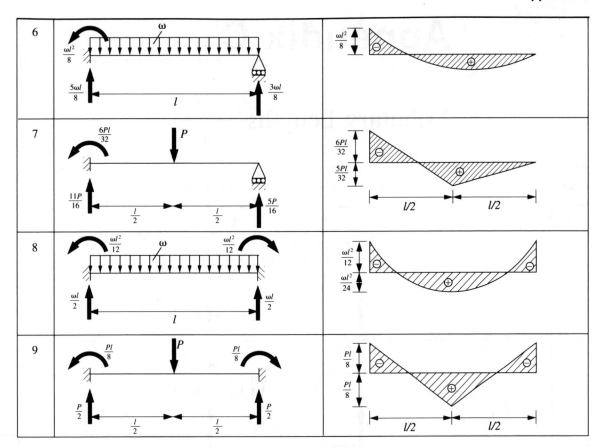

Appendix C

Tributary Lengths

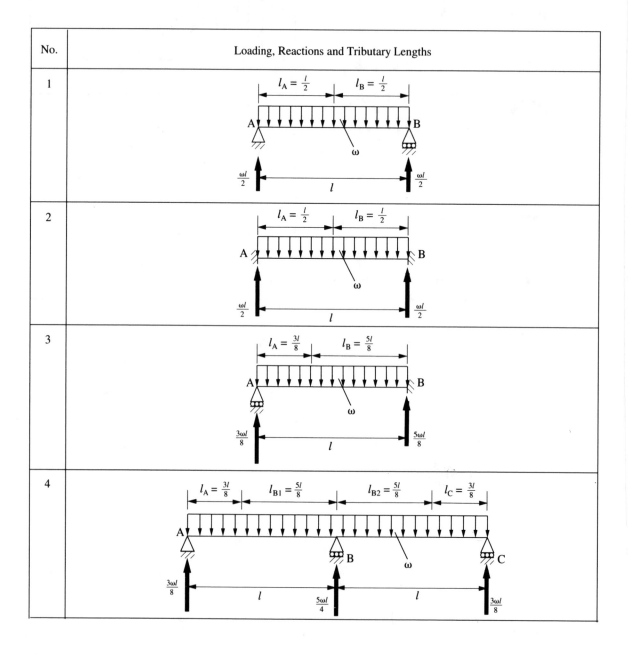

No.	Loading, Reactions and Tributary Lengths
1	
2	
3	
4	

Appendix D

Formulae for Analysis of Continuous Beams (from Reynolds and Steedman 1988)

Internal moment at support $= \alpha Q U F$ (base span length) where

$\alpha = 1$ for uniform loading or 1.5 for central point loading
$Q =$ Support moment coefficient, given below for each support
$U =$ Moment multiplier given below ($=$ unity if all spans are equal)
$F =$ Total applied load
Base span length $=$ length of 'base span', as identified below

Note: $K_i =$ ratio of span i to base span length.

For load arrangements other than those shown, superimpose the moments due to individual loadings. For moments between supports, superimpose the bending moment diagrams for simply supported spans given in Appendix B (Nos 1 and 2).

Appendix D

	Loading	Support Moment Coefficients			Moment Multipliers	
		Q_A	Q_B	Q_C		
two-span	$\triangle$A $\triangle$B $\triangle$C; base span l, K_1l (load on base span)	—	−0.063	—	$U_B=\dfrac{2}{1+K_1}$	
	$\triangle$A $\triangle$B $\triangle$C; base span l, K_1l (load on K_1l span)	—	−0.063	—	$U_B=\dfrac{2K_1^{\,2}}{1+K_1}$	
	$\triangle$A $\triangle$B $\triangle$C; base span l (full load)	—	−0.125*	—		
three-span	$\triangle$A $\triangle$B $\triangle$C $\triangle$D; K_1l, base span l, K_2l (load on K_1l)	—	−0.067	+0.017	$U_B=0.5yU_C$ $U_C=3K_1^{2}H$	$x=K_1+1$ $y=K_2+1$ $H=\dfrac{5}{4xy-1}$
	$\triangle$A $\triangle$B $\triangle$C $\triangle$D; K_1l, base span l, K_2l (load on base span)	—	−0.050	−0.050	$U_B=H(y+K_2)$ $U_C=H(x+K_1)$	
	$\triangle$A $\triangle$B $\triangle$C $\triangle$D; K_1l, base span l, K_2l (load on K_2l)	—	+0.017	−0.067	$U_B=3K_2^{2}H$ $U_C=0.5xU_B$	
	$\triangle$A $\triangle$B $\triangle$C $\triangle$D; l, base span l, l (full load)	—	−0.100*	−0.100*		

Appendix E

Slab Design Moment Equations

Formulae have been developed by Wood and Armer for design moments in slabs (see, for example, the paper by Wood (1968)). When reinforcement is parallel to the coordinate axes, the required moment capacities per unit breadth are:

$$m_x^* = m_x + |\tan \alpha||m_{xy}| \tag{E.1}$$

$$m_y^* = m_y + \frac{|m_{xy}|}{|\tan \alpha|} \tag{E.2}$$

where m_x, m_y and m_{xy} are the applied moments per unit breadth and α is the angle between the failure plane and the Y-axis, as illustrated in Fig. E.1. Selection of the moment capacities dictates the orientation of the failure plane. Hence, any value for α may be selected provided equations E.1 and E.2 are satisfied. Selecting $\alpha = 45°$ results in the following simplified form of the equations:

$$m_y^* = m_x + |m_{xy}| \tag{E.3}$$

$$m_y^* = m_y + |m_{xy}| \tag{E.4}$$

Fig. E1 Plan view of portion of slab

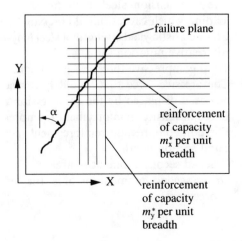

failure plane

Y

α

reinforcement of capacity m_x^* per unit breadth

X

reinforcement of capacity m_y^* per unit breadth

Appendix F

General Notation for Chapters 7–10

General Reinforced Concrete

A_g	Gross cross-sectional area
A_s	Area of longitudinal tension reinforcement
A'_s	Area of longitudinal compression reinforcement
b	Breadth of section
b_w	Breadth of section or, in the case of flanged sections, minimum web breadth
d	Effective depth equal to distance from extreme fibre in compression to the centre of tension reinforcement
E_c	Elastic modulus (Young's modulus) for concrete
$E_{c,\,eff}$	Effective elastic modulus for concrete taking account of the effect of creep
E_s	Elastic modulus (Young's modulus) for steel
f_c	Stress in concrete
f_{ck}	Characteristic cylinder compressive strength (capacity to resist compressive stress)
f_{ct}	Tensile strength of concrete (capacity to resist tensile stress)
f_s	Stress in tension reinforcement (compression positive)
f'_s	Stress in compression reinforcement (compression positive)
f_{sr}	Stress in tension steel assuming a cracked section, due to loading which causes initial cracking
f_y	Characteristic yield strength of steel (capacity to resist stress)
f_{yk}	Alternative notation for f_y
F_c	Total compressive force in (gross) concrete section
F_{disp}	Compressive force that would have been in concrete had it not been displaced by the compression reinforcement
F_s	Force in tension reinforcement (compression positive)
F'_s	Force in compression reinforcement (compression positive)
h	Total section depth
I_c	Cracked second moment of area
I_g	Gross second moment of area (neglecting presence of reinforcement)

I_u	Uncracked second moment of area (allowing for presence of reinforcement)
$l_{b, net}$	Design anchorage length
m	Modular ratio equal to E_s/E_c
M	Applied (factored ULS) moment
M_{cr}	Cracking moment equal to moment at which section first cracks
M_{ult}	Ultimate moment capacity
x	Distance from extreme fibre in compression to neutral axis
y	Distance from neutral axis measured in direction of extreme fibre in compression
γ_c	Material factor of safety for concrete
γ_s	Material factor of safety for steel
δ_c	Deflection assuming a cracked section
δ_u	Deflection assuming an uncracked section
ε_0	Strain in extreme fibre in compression
ε_{ult}	Strain at which concrete crushes equal to 0.0035
ε_c	Strain in concrete (compression positive)
ϕ	Bar diameter
ρ_l	Reinforcement ratio equal to $A_s/(bd)$
φ	Creep coefficient (Table 7.1)
ζ	Distribution factor (Eq. 7.7)

Additional Notation for Chapter 8

A	Cross-sectional area
A_p	Area of prestressing steel
e	Eccentricity of prestress
E_p	Elastic modulus of prestressing steel
$f_{ck, 0}$	Concrete characteristic cylinder strength at transfer
M_0	Moment due to applied loading at transfer
M_p	Moment due to prestress
M_s	Moment due to applied loading at service (SLS)
p_{0min}, p_{0max}	Minimum and maximum permissible total stress in concrete at transfer
p_{Smin}, p_{Smax}	Minimum and maximum permissible total stress in concrete at service (SLS)
P	Prestress force at transfer
P_i	Prestress force at transfer at section i
P_{jack}	Prestress force applied at jack
x	Distance along length of beam
y	Distance from centroid of beam measured in direction of top face for sag moment
Z_b, Z_t	Elastic section moduli for bottom and top fibres (Z_b always negative)
θ	Angle of inclination of prestressing tendon

ρ	Ratio of prestress force at service to prestress force at transfer
σ_b, σ_t	Stress due to prestress at transfer at bottom and top fibres
$\sigma_{bmin}, \sigma_{bmax}$	lower and upper allowable limits on σ_b
$\sigma_{tmin}, \sigma_{tmax}$	lower and upper allowable limits on σ_t

Additional Notation for Chapter 9

$A_{s, min}, A_{s, max}$	Minimum and maximum allowable total areas of reinforcement
$A_{s, tot}$	Total area or reinforcement ($A_s + A_s'$)
e	Eccentricity of applied force
e_a	Additional eccentricity due to imperfections
l_e	Effective length
M	Applied (factored ULS) moment
M_{max}	Ultimate moment capacity in absence of force
M_{ult}	Ultimate moment capacity (in presence of N_{ult})
M_1	First order applied moment (neglecting effect of deflection)
M_2	Second order applied moment (due to deflection)
N	Applied (factored ULS) force
N_{bal}	Axial force for which design is balanced
N_{max}	Ultimate force capacity in absence of moment
$N_{t, ult}$	Ultimate tensile force capacity
N_{ult}	Ultimate force capacity in presence of M_{ult}
r	Radius of gyration equal to $\sqrt{\dfrac{I}{A}}$
R_{min}	Minimum radius of curvature ($1/R_{min}$ = maximum curvature)
κ	Curvature equal to M/EI
λ	Slenderness ratio (Eq. 9.1)

Additional Notation for Chapter 10

A_{leg}	Area of one leg of stirrup
A_{long}	Total area of longitudinal reinforcement (all bars) provided for resistance of torsion
A_{sw}	Area of stirrup reinforcement
A_0	Product $x_0 y_0$
a_v	Shear span equal to distance between points of zero and maximum moment
d_{av}	Average effective depth (of two layers of reinforcement in coordinate directions)
k	Scale effect factor (Eq. 10.4)
T	Applied (factored ULS) torque
T_w	Torque that will cause crushing of compression strut
t	Wall thickness in (equivalent) hollow section
u	Perimeter length

v	Average shear stress equal to $V/(b_w d)$
v_c	Average shear strength of concrete equal to capacity to resist v
V	Applied (factored ULS) shear force
V_c	Shear force that can be resisted by concrete section without shear reinforcement
V_w	Shear force required to cause crushing of compression strut
x_0	Shorter cross-sectional dimension measured between centres of notional thin walls
y_0	Longer cross-sectional dimension
β	Shear enhancement factor/Factor to account for difference between effective applied shear in the presence of moment and simple shear
v	Effectiveness factor equal to ratio of compressive strength of concrete in web to compressive strength of cylinder (there are different definitions in EC2 for shear and torsion)
θ	Angle of inclination of compression strut
ρ_w	Reinforcement ratio for stirrups equal to area of stirrup crossing crack divided by $(b_w d)$
σ_{cp}	Axial stress as defined in text following Eq. 10.5
σ_{cpo}	Axial stress for punching shear as defined in text preceding Eq. 10.28
τ_{rd}	Basic design shear strength given in Table 10.1

References and Bibliography

Abeles, P. W. and Bardhan-Roy, B. K. (1981), *Prestressed Concrete Designer's Handbook*, 3rd Edition, Viewpoint Publications, Slough

Allen, A. H. (1988), *Reinforced Concrete Design to BS8110*, E. & F. N. Spon, London

Beedle, L. S. (1983) (Editor in Chief), *Developments in Tall Buildings 1983*, Van Nostrand Reinhold Company, New York

Bhatt, P. and Nelson, H. M. (1990), *Marshall and Nelson's Structures*, 3rd Edition, Longman Scientific and Technical, Harlow

BS8110 (1985), *British Standard, Structural use of Concrete, Part 1*, British Standards Institution, London

Chana, P. and Clapson, J. (1992), Innovative shearhoop system for flat slab construction, *Concrete*, The Concrete Society, Slough, Vol. 26, No. 1, Jan/Feb 1992, pp 21–24

CP110 (1974), *Code of Practice for the Structural use of Concrete, Part 1*, British Standards Institution, London

Cranston, W. B. (1972), *Analysis and Design of Reinforced Concrete Columns* (Research Report No. 20), Cement and Concrete Association, Slough

EC1 (1993), *Eurocode 1: Basis of Design and Actions on Structures*, Drafts: Part 1 March 1993, Part 2.1 June 1993 and Part 2.3 August 1993, European Committee for Standardisation TC250, Brussels

EC2 (1991), *Eurocode 2: Design of Concrete Structures – Part 1: General Rules and Rules for Buildings*, European Prestandard ENV 1992-1-1:1991, European Committee for Standardisation TC 250, Brussels

Ellingwood, B. and Leyendecker, E. V. (1978), Design against progressive collapse, *J. Struct. Div.*, ASCE, Vol. 104, No. 3, March 1978, pp. 413–423

Gilbert, R. I. and Mickleborough, N. C. (1990), *Design of Prestressed Concrete*, Unwin Hyman, London

Gross, J. K. and McGuire, W. (1983), Progressive collapse resistant design, *J. Struct. Engrg.*, ASCE, Vol. 109, No. 2, Feb. 1983, pp. 1–15

Hambly, E. C. (1991), *Bridge Deck Behaviour*, 2nd Edition, E. & F. N. Spon, London

Heyman, J. (1971), *Plastic Design of Frames, Vol. 2: Applications*, Cambridge University Press

ISE Manual (1985), I. Struct. E./ICE Joint Committee, *Manual for the Design of Reinforced Concrete Building Structures*, Institution of Structural Engineers, London

Kong, F. K. and Evans, R. H. (1987), *Reinforced and Prestressed Concrete*, 3rd Edition, Chapman and Hall, London

Kotsovos, M. D. (1988), Design of reinforced concrete deep beams, *The Structural Engineer*, Vol. 66, No. 2, Jan. 1988, pp. 28–32

Lampert, P. and Collins, M. P. (1972), Torsion bending and confusion – an attempt to establish the facts, *ACI Journal, Proceedings*, Vol. 69, No. 8, Aug. 1972, pp. 500–504

Lin, T. Y. and Burns, N. H. (1982), *Design of Prestressed Concrete Structures*, 3rd Edition, John Wiley & Sons, Chichester

McGinley, T. J. and Choo, B. S. (1990), *Reinforced Concrete, Design Theory and Examples*, E. & F. N. Spon, London

McGregor, J. G. (1992), *Reinforced Concrete Mechanics and Design*, 2nd Edition, Prentice-Hall

Park, R. and Paulay, T. (1975), *Reinforced Concrete Structures*, John Wiley & Sons, Chichester

Reynolds, C. E. and Steedman, J. C. (1988), *Reinforced Concrete Designer's Handbook*, 10th Edition, E. & F. N. Spon, London.

Timoshenko, S. P. and Woinowsky-Krieger, S. (1970), *Theory of Plates and Shells*, 2nd Edition, McGraw-Hill, New York

Wood, R. H. (1968), The reinforcement of slabs in accordance with a predetermined field of moments, *Concrete*, Feb. 1968, pp. 69–76

Index